AF821664

Kakadu Region

Geobotany 23

Series Editor

M.J.A. WERGER

The titles published in this series are listed at the end of this volume.

Landscape and Vegetation Ecology of the Kakadu Region, Northern Australia

edited by

C. MAX FINLAYSON and ISABELL VON OERTZEN
Environmental Research Institute of the Supervising Scientist,
Jabiru, Northern Territory, Australia

Kluwer Academic Publishers
Dordrecht / Boston / London

Library of Congress Cataloging-in-Publication Data

Landscape and vegetation ecology of the Kakadu Region, Northern Australia / edited by C. Max Finlayson & Isabell von Oertzen.
p. cm. -- (Geobotany ; v. 23)
ISBN-13:978-94-010-6547-4 e-ISBN-13:978-94-009-0133-9
DOI:10.1007/978-94-009-0133-9

1. Botany--Australia--Kakadu National Park Region (N.T.)--Ecology.
2. Phytogeography--Australia--Kakadu National Park Region (N.T.)
3. Physical geography--Australia--Kakadu National Park Region (N.T.)
I. Oertzen, Isabell von. II. Series: Geobotany ; 23.
QK451.L37 1995
581.5'099429'5--dc20 95-39765

ISBN-13:978-94-010-6547-4

Published by Kluwer Academic Publishers,
P.O. Box 17, 3300 AA Dordrecht, The Netherlands.

Kluwer Academic Publishers incorporates
the publishing programmes of
D. Reidel. Martinus Nijhoff, Dr W. Junk and MTP Press.

Sold and distributed in the U.S.A. and Canada
by Kluwer Academic Publishers,
101 Philip Drive, Norwell, MA 02061, U.S.A.

In all other countries, sold and distributed
by Kluwer Academic Publishers Group,
P.O. Box 322, 3300 AH Dordrecht, The Netherlands.

Printed on acid-free paper

Table of Contents

Preface

The Kakadu region of northern Australia is steeped in cultural history and natural grandeur. Over the past few decades the rich cultural and natural heritage of this fascinating region has become increasingly known to more and more people. At the same time as the natural heritage of the region was being recognised by conservationists and tourists alike the mineral wealth was being recognised by mining enterprises. Almost inevitably, the mix of conservation and mining interests led to conflict that is still not completely resolved. However, much has happened over the years and we now have a major national park that is largely leased from the Aboriginal traditional owners under a management agreement. The park is recognised internationally for both its cultural and natural heritage and is now well known to tourists from around the world. We have also witnessed the development of two uranium mines and support services alongside an expanding tourist infrastructure.

When the conservationists and miners came into the region in increasing numbers some 20 years ago they entered an amazing natural realm that was almost scientifically unknown. Spurned by the rapidly developing interest in the natural wealth of the region a determined scientific effort to describe and record details of the main features started. It was not long before a seemingly inordinate number of scientists and technicians was swarming over the landscape with their meters and notebooks and a vast store of information was gathered. This book is a summary of the immense amount of information collected on the geobotanic features of the region. The cultural heritage of the traditional Aboriginal inhabitants of the region and the diverse and populous fauna were also investigated, but both these subjects warrant their own separate volumes and are not treated here.

Throughout this period of intense scientific interest the very nature of the region has changed. Besides changes in human habitation the physical and biological environment has come under challenge and even threat. We now have more weed species. We no longer have the large numbers of feral buffalo that once roamed freely. We still have angry debates over fire management! We know the region is changing ecologically and we still do not know the extent of these changes. We are managing more and more and we still need more scientific information. Further research is needed, but at the same time we need to make better use of that information which is already available. The scientific information collected in the past is of little value if it is not made available to the individuals and agencies responsible for managing the natural heritage of the region. We need to manage on the basis of current knowledge as we collect further information to ensure this management is effective.

C.M. Finlayson and I. von Oertzen (eds), Landscape and Vegetation Ecology of the Kakadu Region, Northern Australia, vii.

Acknowledgements

The idea for this book came about a decade ago from Steve Morton before he left Kakadu for the drier climes of Central Australia. It took a long time for the idea to become a reality, but thanks for the prompt all those years ago Steve. The information contained in this book represents the research efforts of many individuals and many agencies over the last 1–2 decades. Thank you to all of them. Thank you also to the authors who have become more and more dispersed across Australia as this book has progressed.

C.M. Finlayson and I. von Oertzen (eds), Landscape and Vegetation Ecology of the Kakadu Region, Northern Australia, viii.

CHAPTER 1

The Kakadu region

C. MAX FINLAYSON and ISABELL VON OERTZEN

Abstract. The Kakadu region in northern Australia has become well known for its spectacular wilderness and nature conservation values. These values are enshrined in the management regime for the 20 000 km^2 Kakadu National Park which comprises a large proportion of the region. The remainder of the region takes in part of the vast area of Aboriginal land known as Arnhem Land. The landscape is spectacular and is visibly dominated by a sandstone escarpment and plateau, extensive lowlands and seasonally flooding plains along the major rivers. The climate is highly seasonal with a hot, humid wet season and a mild to warm dry season. The soils are generally low in nutrients and organic material but they do support a highly diverse vegetation comprised of large tracts of savanna woodlands and forests, restricted monsoonal forests and extensive wetlands. The vegetation has been greatly influenced by the frequency and intensity of fires, and the occurrence of alien weeds and feral animals, especially the Asian water buffalo. Whilst the numbers of buffaloes have been severely reduced in recent years weed species are still a major problem. The vegetation patterns are also greatly influenced by interactions with native animals such as termites and ants.

The development of pastoralism, uranium mining and tourism in recent decades has greatly altered the pattern of human settlement throughout the region. The current management structures need to take into account Aboriginal land ownership, conservation, tourism and mining interests. Two uranium mines have been developed and tourist facilities expanded throughout the Park. An administrative centre now caters for both residents and visitors and provides services for smaller settlements throughout the region. The traditional Aboriginal occupants still have ownership over much of the land under various arrangements including leasing the Park to a government agency. Mining is conducted under arrangements with both the Aboriginals and the government whilst all pastoral leases have been resumed and incorporated into the National Park.

1. Introduction

Some 150 km to the east of the city of Darwin in the Northern Territory of Australia (Fig. 1) is a vast wilderness area. Isolated and relatively undisturbed the area is of immeasurable value for the world's natural and cultural heritage and has attracted great interest over the past 20–30 years for its conservation values (Christian & Aldrick 1977, Fox *et al.* 1977, Ovington 1986, ANPWS 1986, 1991a). It also contains immensely rich mineral deposits (Christian & Aldrick 1977, Fox *et al.* 1977, Needham 1988). With this combination of conservation values and mineral wealth it is not surprising that the area has been embroiled in controversy that continues to this day (Fox *et al.* 1977, Lea & Zehner 1986, Braithwaite & Woinarski 1990, Galligan & Lynch 1992).

The area, awe inspiring by its vastness and natural beauty, is known as the Alligator Rivers Region (Christian & Aldrick 1977, ANPWS 1986). The name of the region is derived from the main rivers that form the hydrological network (Fig. 1); the large South Alligator and East Alligator Rivers and the smaller West Alligator River. The Region encompasses the 20 000 km^2 Kakadu National Park and a further 8 000 km^2 of land within the western portion of Arnhem Land (a large tract of Aboriginal owned land that stretches to the Gulf of Carpentaria). It also includes four uranium mining leases (Fig. 1). Outside of local and official circles the area is more often referred to as Kakadu; a name that does not reflect the boundaries of the region, but which is well known by the general public. In recognition of this fact and given that the vast majority of scientific endeavour within the region has centred on Kakadu National Park this treatise on the geobotany of the Alligator Rivers Region makes use of the popular and better known descriptor "Kakadu region" whilst recognising that the boundaries of the region extend beyond those of the National Park and into western Arnhem Land.

C.M. Finlayson and I. von Oertzen (eds), Landscape and Vegetation Ecology of the Kakadu Region, Northern Australia, 1-15.

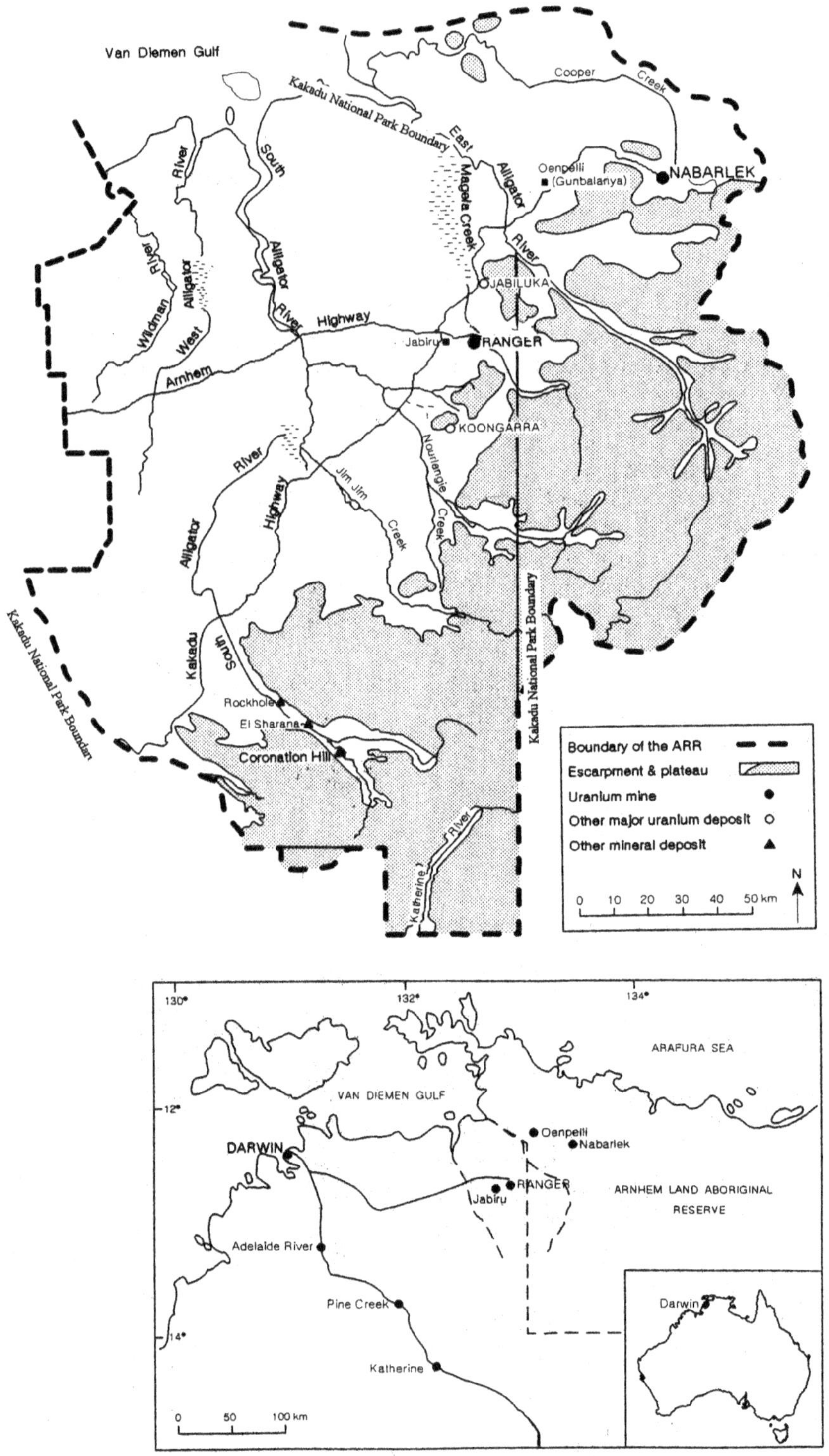

Fig. 1. The Kakadu region. Inset shows location in northern Australia.

As a prelude to specific descriptions of the geobotanic features of the Kakadu region a brief overview is presented. This is intended to place the management of the physical (climate and hydrology, landform evolution and stability) and botanical features of the region into the framework provided by recent human history. This framework has largely been moulded by the conservation significance recently given to the region. The papers that follow present in depth coverage of the vast amount of geobotanic knowledge that has been largely collected over the past two decades. There is a degree of overlap betweeen the papers in recognition of the connections that exist in the natural environment, for example the links between climate and the growth of plants and the occurrence and frequency of fire.

Aspects of the fauna are considered in some chapters, but a comprehensive description of the diverse animal life of the region is not within the scope of this treatise. Interactions between the fauna and the vegetation and soils, especially those related to the former presence of large numbers of feral water buffalo, are considered in two papers. Similarly, the rich cultural heritage of the Aboriginal occupants of the region while a significant feature of the region (Ovington 1986, ANPWS 1991a) and well illustrated in the numerous rock painting sites (Fig. 2) is also beyond the scope of this book.

2. Natural environment

The Kakadu region is widely acclaimed for its natural heritage (Christian & Aldrick 1977, Brennan 1986, Ovington 1986, Miles 1986). The natural heritage (along with the cultural heritage) is recognised nationally with the establishment of the 20 000 km^2 Kakadu National Park. This recognition extends to the international arena with listings under the UNESCO World Heritage Convention and the Ramsar Convention on Wetlands of International Importance. The former is accompanied by the exclaimer "There is simply no other protected area on Earth like Kakadu" (see Dugan 1993).

2.1. Climate

The climate of the region is classified as "summer rainfall – tropical" (Bureau of Meteorology 1989) with two broad seasons – the wet season from November–March and the dry season from

Fig. 2. Aboriginal painting on rock surfaces (photographs C.M. Finlayson).

Fig. 3. The rocky plateau (a), escarpment (b,c) and waterfall (d) (photographs C.M. Finlayson).

April–October. The former is characterised by heavy periodic rains and generally hot and humid conditions and the latter by dry and mild to warm conditions. Most of the 1300–1500 mm of annual rainfall comes from monsoonal depressions and intense downpours during the wet season. It is recognised that this is a very simplistic description of the climate which is far more complex in terms of weather patterns that include rain bearing monsoonal fronts and cyclones. The Aboriginal people have a more refined perception of the climate that recognises six major seasons. This is based on the relationship between changes in the weather and the availability of food (bush tucker) items (see description in Ovington 1986).

2.2. *Landscape*

The Kakadu region extends southwards from the coast of van Diemen Gulf (Fig. 1), through flood plains and lowland hills to the sandstone escarpment and dissected Arnhem Land plateau (Fig. 3). The main physiographic units are the dissected sandstone plateau and escarpment, a lowland plain, flood plains and flood basins along the rivers, estuarine flood plains and a narrow coastal plain. The geological foundations of the region extend back some 2000 million years (Needham 1988) with subsequent geological and geomorphic events shaping the present landscape (Williams 1969). In recent times, the landscape has been modified by pastoral, mining and tourism activities, the introduction of exotic flora and fauna and changes in the vegetation burning practices (Haynes *et al.* 1991).

2.3. *Soils*

The soils of the lowlands are commonly red and yellow earths (sandy and silty loams) with large fractions of quartz and ferricrete nodules and gravel (Hooper 1969). They are acidic, highly leached and deficient in organic matter and nutrients (Chartres *et al.* 1991). The surface is often covered by a lag of quartz gravel and ferricrete nodules. The flood plain soils are clayey, silty and sandy alluvium, and include silty brown earths, yellow earths, gradational red earths, texture contrast (duplex) soils and cracking clays. The flood basins and coastal plains have fine-textured cracking clay soils (Hooper 1969). Duplex soils also occur along the footslopes of the sandstone hills and ridges whilst in the south-east of the region hills and strike ridges have clayey or loamy textured skeletal soils with a high gravel content. The escarpment soils are predominantly siliceous sands along with skeletal soils and bare rock (Aldrick 1976).

2.4. Hydrology

The hydrological zones in the region include the escarpment, the lowlands and the flood plains (a term that generally but inaccurately combines the flood plains and flood basins; T.J. East personal communication). During the wet season runoff from the escarpment and lowlands is collected in the creeks and thence in the large tidal rivers (Fig. 4). Once the creeks and rivers are full freshwater spills across the extensive flood plains and covers them to a depth of several metres. The creeks have a base flow of less than 5 $m^3 s^{-1}$ with peak flows late in the wet season reaching 1000 $m^3 s^{-1}$. Freshwater flow in the creeks and rivers ceases within a few months of the end of the rains and the creeks and flood plains dry out except for a few permanent swamps and billabongs (Finlayson 1991). The groundwater level in the lowlands is recharged by the wet season rains, but can fall 2–4 m during the dry season (Chapman 1988). The spring tidal range in van Diemen Gulf is 5–6 m and estuarine water can extend more than 100 km upstream in the large rivers (Woodroffe *et al.* 1989).

2.5. Terrestrial vegetation

The Kakadu region is considered to be the most floristically diverse area of monsoonal Australia (Lazarides *et al.* 1988) with more than 1800 species being recognised (Brennan 1992). This diversity reflects the variety of landforms and associated plant habitats and localised vegetation patterning. The terrestrial vegetation across 55 % of the region is eucalypt-dominated open-forest and woodland with a 1–2 m Andropogonoid grassy understorey (Story 1969, 1976). These communities are generally known as savannas, or more specifically, tropical tallgrass savannas (Mott *et al.* 1985). A further 30% of the region is covered by heaths and open-woodlands with sparse Andropogonoid-Eragrostoid grass understoreys. Closed-canopy monsoon rain forests are of particular note although they are not extensive, being confined to lowland springs, rock outcrops, flood plain and beach levees and sandstone areas (Russell-Smith 1991).

Whilst the vegetation formations within the region are not unique they are considered to be outstanding examples of the most important formations found in northern Australia (Bowman & Wilson 1987). Furthermore, many rare species are

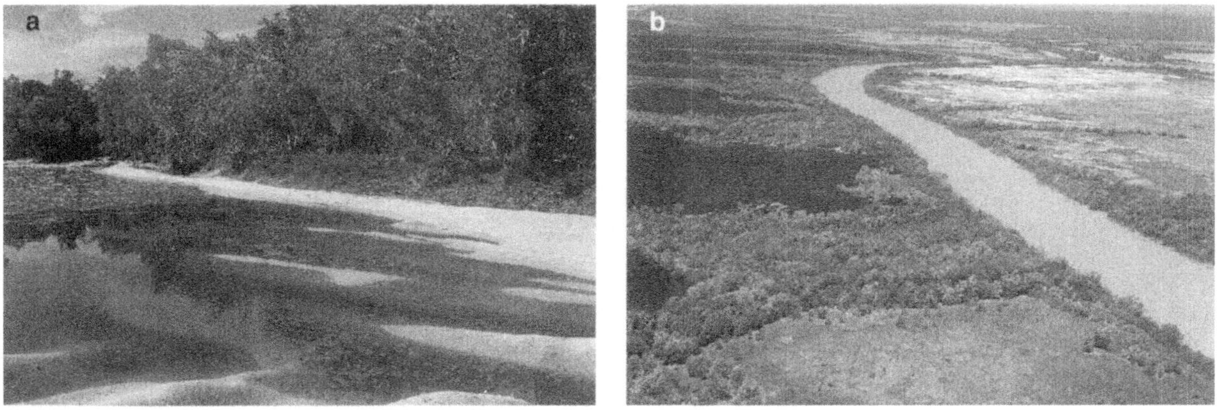

Fig. 4. Sandy creeks (a) and large tidal rivers (b) that form the drainage network (photographs C.M. Finlayson).

Fig. 5. Woodland fire (photographs C.M. Finlayson).

found in the region, particularly in the dissected sandstone environments (Wilson *et al.* 1990). The monsoon rain forests in the sandstone environments contain a larger proportion of pantropical genera compared with the eucalypt-dominated communities in the region and also more endemic and Australasian taxa than the lowland monsoon rain forests (Russell-Smith 1986).

The terrestrial vegetation is also greatly influenced by the occurrence of fire (Fig. 5) (Andersen & Braithwaite 1992) and the presence of some 90 alien species (Cowie *et al.* 1988, Cowie & Werner 1987). Interactions with animals are also important factors affecting the vegetation. In the past, feral buffalo were a major detrimental influence, particularly in riparian areas alongside the streams and fringing the flood plains (Letts *et al.* 1977, Braithwaite *et al.* 1984). Plant-animal interactions go far beyond those shown by the buffalo with termites and ants for instance, playing vital roles in nutrient cycling and seed dispersal (Andersen & Lonsdale 1990, Andersen 1991).

2.6. *Wetland vegetation*

Wetlands include mangroves along the coast (Fig. 6a) and tidal streams, flood plains (Fig. 4b, 6b) and billabongs along the freshwater reaches of the

Fig. 6. Mangroves lining a coastal creek (a) and a freshwater flood plain (b) (photographs C.M. Finlayson).

rivers and creeks, and small lakes (Finlayson *et al.* 1989, 1990, Story 1969, 1976). Mangroves are now confined to a narrow strip along the coast and the tidal reaches of the major streams; however, some 5000–7000 years ago they were far more extensive and may have extended over 80 000 ha (Woodroffe *et al.* 1985). The mangroves occur in zoned communities, especially those located along the coast, and are extremely productive (Woodroffe *et al.* 1988). On the landward side of the mangroves, especially along the coast, there are broad unvegetated saline flats that are separated from the freshwater wetlands by low levees.

The freshwater wetlands comprise seasonally inundated flood plains and billabongs, and small permanent lakes (Finlayson 1988, Finlayson *et al.* 1988). Flood plains cover about 195 000 ha along the major rivers and creeks. The extremes of the seasonal climate have resulted in a diverse flora with characteristic wet and dry season plant communities (Finlayson *et al.* 1989, 1990). In the wet season the flowering plants are diverse and profuse, but in the dry season the vegetation is sparse and far less diverse. Overall, the plant communities are highly dynamic and highly productive (Finlayson 1991, Finlayson *et al.* 1993).

The wetlands were greatly affected by the large numbers of feral buffalo (Letts *et al.* 1977), although there is evidence of ecological recovery following the removal of most of these animals (Finlayson *et al.* 1995). Another threat to wetlands comes from alien plants. The floating fern *Salvinia molesta* is well established (Fig. 7) (Finlayson *et al.* 1988, 1994) and great efforts are being made to prevent the prickly shrub *Mimosa pigra* from establishing and spreading across the flood plains (Skeat *et al.* 1987).

Fig. 7. Salvinia molesta covering the surface of a billabong (photograph C.M. Finlayson).

3. Human settlement

Much of the general information on human occupancy and settlement in the Kakadu region has been summarised in Fox *et al.* (1977). The region has possibly been occupied by Aboriginal people for 60 000 years; a date not only considered to be the oldest recorded for Aboriginal occupation but also most probably the time of initial human arrival on the Australian continent (Roberts *et al.* 1990). The Aboriginal people have a traditional bond with the land and individuals have specific responsibilities and rights as owners. These rights are nowadays generally referred to under the broad descriptor of "traditional ownership" and are recognised within Kakadu National Park under a lease with the Director of the Australian National Parks and Wildlife Service that provides for lease payments, a share in revenue generated from the Park and protection of Aboriginal interests (ANPWS 1991b).

The Aboriginals basically lived a semi-nomadic lifestyle with the rivers and wetlands, the open forests and woodlands of the lowlands, and the sandstone gorges and plateux providing a great variety of habitats and food resources. The wetlands were an important source of food for the traditional Aboriginal hunter-gatherers with large numbers of people sited seasonally along the wetland margins, moving to higher ground in the wet season (Jones 1985, Meehan 1988). Aboriginal people still utilise the food resources commonly known as "bush tucker", although the extent of utilization of these resources has changed greatly since Europeans settled in the region (Meehan 1988). When feral buffalo were present in large numbers they were regularly hunted. Hunting for feral pigs and various native animals still occurs.

Aboriginal contact with people from outside the region commenced several centuries ago with the Macassan fishermen who came in search of the trepang sea slug that they processed and traded with the Chinese (Fox *et al.* 1977). Contact with Europeans started in the first half of the nineteenth century when the British made two abortive attempts to establish settlements at Raffles Bay (1827) and Port Essington (1838) to the north of the region. However, it was not until after the establishment of Palmerston (later renamed Darwin) in 1869 that Europeans ventured into western Arnhem Land with the advent of buffalo hunting and drawn by the prospect of finding gold.

In response to the deleterious consequences to Aboriginals from increased contact with Europeans the Western Arnhem Land Aboriginal Reserve was declared in 1920 and in 1931 incorporated into the 96 089 km^2 Arnhem Land Reserve. The 505 km^2 Woolwonga Aboriginal Reserve was declared in 1936 and the boundaries revised in 1968. Entry to these reserves was normally restricted to Aboriginal people.

Oenpelli, in the east of the region (Fig. 1) has been occupied intermittently by Europeans since 1906 when a pastoral lease was taken up and held until 1916 (O'Ferrall 1969). In 1925 a church mission was established at Oenpelli and soon became the focus for Aboriginal groups from within the region and also from further east. The church mission continued to administer Oenpelli until 1975 when a local government structure was established. Oenpelli is now known by the Aboriginal name Gunbalanya and is administered by a local community council. Currently, an estimated 800–900 people live at the townsite with smaller numbers living in decentralised communities known as "outstations".

Pastoral leases and buffalo hunting attracted small numbers of Europeans to the region until these activities ceased with the closure of the Mudginberri abattoir in 1988. By this time the uranium mines at Nabarlek and Ranger (Fig. 8) were being operated and a townsite existed at Jabiru near the Ranger mining lease (Fig. 1) since 1982. The townsite was established exclusively for people associated with the mining activities at Ranger and possibly, in the future, those required for the Jabiluka and Koongarra mines (Fox *et al.* 1977). Initially, it was not intended to utilise Jabiru as a centre for tourism, although this has changed and tourist accommodation and services now exist in the town to supplement those specifically for tourists elsewhere in the park. The town currently has a population of around 1200 and provides services for surrounding small settlements established specifically for national park staff and for groups of Aboriginals.

Fig. 8. Ranger uranium mine near Jabiru (photograph C.M. Finlayson).

4. Land use

Before 1965, non-Aboriginal land use and settlement in the area was restricted to buffalo hunting, the cattle station and Aboriginal mission at Oenpelli, and several small uranium mines in the upper reaches of the South Alligator River. Pastoral activities and mineral exploration expanded in the 1970s; much of this was superceded by conservation developments and associated tourism in the 1980s.

4.1. Pastoralism and buffalo hunting

As with much of northern Australia cattle were introduced into the Kakadu region in the 1880s and 1890s. However, these activities were short-lived, mainly due to the combined problems of isolation and the tick-borne redwater disease that severely affected the cattle. Pastoralism was revived with the granting of leases for the 2996 km^2 Gimbat and 3701 km^2 Goodparla stations in the south of the region in 1962. Further leases were given for the 1106 km^2 Mudginberri and 1010 km^2 Munmarlary stations in 1969 which had been held since 1965 as Occupation (Development) Licences and before that as Grazing Licences (Christian & Aldrick 1977, Fox *et al.* 1977). These leases were for 50 years and did not appear to give any cognizance to concurrent moves to establish a large conservation reserve in the region. The leases were resumed by the government in 1978 as part of the process that led to the establishment of Kakadu National Park and in February 1984 were fully incorporated into the Park. However, mustering of buffaloes on the former lease areas continued until 1988.

Prior to World War II the basis of the buffalo industry in the region was hunting for hides which were exported throughout the 1920–30s to India (Fox *et al.* 1977, Fisher 1988). As prices for hides fell after the war attention was given to producing meat, first for pet food and later for human consumption. Mobile abattoirs were used initially and later an export standard abattoir was constructed at Mudginberri for the processing of buffaloes mainly from Mudginberri and Munmarlary stations. The abattoir continued operating until the end of 1988. An abattoir at Gunbalanya was used to process meat for the domestic market.

At various times plans were drawn up to alter the basis of the grazing industry from one that relied primarily on feral buffaloes to one based on either domesticated buffaloes or introduced strains of cattle, especially Brahman-crosses, more suited to the climate (Fox *et al.* 1977). Such plans were generally limited by costs and the vagaries of the meat markets and also by the availability of suitable pastures. Schemes to provide better nutrition through improved pastures were short-lived and introduced pasture species are now regarded as weeds in Kakadu National Park (Cowie *et al.* 1988, Cowie & Werner 1987).

The future of a buffalo industry in the region was also threatened by the prevalence of bovine tuberculosis in feral animals (Hein & Tomasovic 1981). An eradication program was commenced in the 1980s and despite being successful it attracted a howl of protest which has continued to this day. Within Kakadu National Park buffalo numbers have been reduced to very low numbers as a result of a systematic eradication program involving buffalo-catching to supply the abattoir at Mudginberri coupled with shooting from helicopters. This effort built on even earlier efforts in the 1970s to reduce buffalo numbers in the Woolwonga Wildlife Reserve. Buffalo are now rarely encountered within the park, although stray animals do wander in from Arnhem Land where the control program has not been as intensive. Except for a herd of around 600 domesticated buffalo kept under disease free and controlled conditions on the flood plain of the South Alligator the buffalo industry in the region came to an end in 1987 when the Federal Government resumed Goodparla and Gimbat stations and incorporated them into Stage III of Kakadu National Park. The herd on the South Alligator is managed by the local Gagudju Aboriginal association to provide meat for the local Aboriginal population.

4.2. Mining

A brief history of mining exploration and development in the South Alligator Valley has been provided by Fisher (1988) and other information is available in Christian & Aldrick (1977), Fox *et al.* (1977) and Lea & Zehner (1986). Records of mining exploration in the region go back to 1898, and in the 1920s gold bearing reefs and shallow alluvial deposits were worked, but only for short periods as they were not overly profitable. A series of galena and silver deposits were worked in the late 1940s and early 1950s and a number of mining tenements were secured over small copper lodes.

The first uranium deposit was discovered at Coronation Hill alongside the South Alligator River on Gimbat station (Fig. 1) in 1953 and a few years later the Sliesbeck deposit on the headwaters of the Katherine River a little further south was also discovered. More deposits were then found further to the north. Over the following decade the uranium deposits in the South Alligator valley were consolidated and mining continued. However, by 1964, due to adverse market factors, all ore production ceased and the mines closed.

Interest in the East Alligator uranium province was rekindled in the 1960s and by 1973 mining exploration leases, which give their holders exclusive rights to explore but not to recover minerals, covered a large part of the region (Fox *et al.* 1977). By this stage four large uranium deposits had been discovered (Fig. 1). The deposit at Nabarlek in Arnhem Land was the first in May 1970, followed by those at Koongarra and Ranger later in the same year and the Jabiluka deposit in November 1971. Following an investigation into uranium mining in the region (Fox *et al.* 1977) the Nabarlek and Ranger deposits were developed (Fig. 8). Before mining could commence the lease holders had to negotiate agreements with the traditional Aboriginal owners of the land. Agreements were reached with the owners of all four leases, but mining has not occurred at Jabiluka and Koongarra as a consequence of Federal Government policy limiting the extent of uranium mining in Australia.

The rich but very small orebody at Nabarlek was mined by an open pit method in 1979 and the ore stockpiled. Processing of the ore commenced in 1980 and was all but complete in 1988 and the site is now being rehabilitated (Supervising Scientist for the Alligator Rivers Region 1988). Mining and commercial production of uranium concentrate has been underway at Ranger since 1981 and steps are being taken to develop a further ore body on the same lease.

Gold and platinate minerals have been discovered at Coronation Hill, but these have not been developed as a consequence of Federal Government policy which supported the beliefs and attitudes of the Aboriginal traditonal owners who were opposed to any disturbance of this area (Galligan & Lynch 1992). A number of exploration leases in this area were resumed and incorporated into Kakadu National Park in November 1989. The leases at Coronation Hill were incorporated into the Park in June 1991.

4.3. Conservation

The conservation value of the Kakadu region has been recognised for some years (see ANPWS 1986) and this is now shown in the 20 000 km^2 World Heritage listed Kakadu National Park. However, it took some years for the conservation value of the region to be given the protected status that it now has.

The first step in the long process that resulted in the consolidation of the area that is now known as Kakadu National Park occurred in 1964 with the establishment of the Woolwonga Wildlife Reserve (Christian & Aldrick 1977). In 1965 the Northern Territoty Reserves Board made a proposal for a 6410 km^2 National Park in the area between the South Alligator River and Arnhem Land; the proposal was amended in 1967 and 1968 before approval to secure 2590 km^2 of land for conservation purposes was gained in 1969. Two further areas in the vicinity of Deaf Adder Gorge (233 km^2) and Jim Jim Falls (100 km^2) were excluded from mining in 1971 and much of the area proposed for the National Park was declared as the 3290 km^2 Alligator Rivers Wildlife Sanctuary by 1972 (Christian & Aldrick 1977, Fox *et al.* 1977). This area corresponded to the 1975 Kakadu National Park proposal except that the Park also included small parts of Mudginberri and Gimbat stations.

Further small reserves had also been established in the south of the region; Waterfall Creek Falls in the south of the region was declared as a 235 ha Recreation Reserve in 1972 (Fisher 1988) and the 4.5 ha Christmas Creek Cave Paintings Reserve was proclaimed in 1974 as a reserve to protect a place of historic interest (Fox *et al.* 1977).

The decision to establish a park under the name of Kakadu was taken in May 1975, but declaration was delayed by the competing claims for mining leases in the same area (Fox *et al.* 1977). The Park was eventually declared in three stages: Stage I (6144 km^2) in April 1979; Stage II (6929 km^2) in February 1984; and Stage III in June 1987 (which together with later additions in November 1989 and June 1991 added a total of 6930 km^2).

The natural and cultural value (see Figs 2,3,6) of Kakadu National Park is also recognised as being of international importance (ANPWS 1991b). In 1981 Stage I of the park was the first Australian site to be included under the Convention Concerning the Protection of the World Cultural and Natural Heritage. Stage II was included in 1987 and Stage III in 1991. Inclusion under the so-called World Heritage Convention not only recognised the high value of the Park but carried an obligation to both conserve and preserve the natural and cultural heritage of the Park. Further international recognition is given under the Convention on Wetlands of International Importance especially as Waterfowl Habitat. The wetlands of Stage I of the park were placed on the Ramsar Convention List of Internationally Important Wetlands in 1980, followed by those in Stage II in 1987 (Ramsar Convention Bureau 1990, ANPWS 1991b).

4.4. *Tourism*

Tourist and recreational use of the region soon followed in the footsteps of the miners with a number of rock pools becoming popular destinations (e.g. Waterfall Creek Falls). Several attempts were made in the 1960s to develop safari camps, but these were not viable (Fisher 1988). With the completion of the Arnhem Highway from Darwin to Jabiru in 1974 the region became increasingly popular for tourists and the number of visitor days had increased from 19 000 in 1972 to 60 000 in 1976 and were predicted to increase further to 2 100 000 over the next 20 years (Fox *et al.* 1977). A lack of tourist accommodation was considered as a limiting factor for further development of tourism. Fox *et al.* (1977) considered the potential for developing the proposed mining townsite of Jabiru as a centre for tourism, but recommended instead that tourist accommodation be located outside the park.

With the development of Kakadu National Park in the 1980s tourism based on the cultural and natural heritage of the region boomed and visitation increased considerably, with an estimated 46 000 visitors in 1982 swelling to 100 000 in 1985 (ANPWS 1986) and to 222 000 in 1993 (H. Moorcroft personal communication). Tourist facilities are now provided in Jabiru with a hotel and camping ground to supplement those at the South Alligator and Cooinda and the camping areas throughout the park. Visitors also utilise facilities located outside the park. Tourism is centred on a number of cultural sites (Fig. 2), the rugged and spectacular escarpment (Fig. 3), and the wetlands (Fig. 6) with boat tours (Fig. 9) being particularly popular.

Fig. 9. Tourist boat (photograph C.M. Finlayson).

5. Administrative structure

5.1. *Kakadu national park*

Kakadu National Park is administered by the Australian National Parks and Wildlife Service (ANPWS), now known as the Australian Nature Conservation Agency (ANCA). Much of the Park is owned by Aboriginal people who have leased their land to the Director of ANCA for 99 years. Whilst ANCA subscribes to the principles underlying the definition and management of parks proposed by IUCN (The World Conservation Union) they also take into account the rights and expectations of the Aboriginal people. Since July 1989 the management arrangements for the Park have centred on the 14 member Kakadu National Park Board of Management; 10 of these members are Aborigines nominated by the traditional land owners. The Board works closely with ANCA staff and liaises with appropriate Aboriginal people and organisations.

The strategy adopted for managing the Park is laid out in the Kakadu National Park Plan of Management (ANPWS 1991b) which specifies the operations to be carried out over the next five years. The current Plan of Management came into effect on 1 January 1992 and will remain in effect until 31 December 1996. ANCA is also responsible for ensuring that the park is managed in accordance with its status under World Heritage and Internationally Important Wetland Listings.

The Park has been divided into five management zones: intensive management, intermediate management, scientific research, minimum management and wilderness. These represent a progression from a high level of intervention and provision of facilities to a low level of intervention and no provision of facilities. The scientific research zone is known as the Kapalga Research Station and covers 670 km^2 of savanna woodland and flood plain. A few residential and commercial leases have been issued within the park boundaries in recognition of specific activities or claims that existed before the park was declared.

5.2. Aboriginal communities

Much of the land within Kakadu National Park is Aboriginal land held under title by the Gagudju, Djabulukagu and Jawoyn Associations and leased to the Director of ANCA. At present about 300 Aboriginal people reside in the Park, comprised of traditional owners and Aboriginals with recognised attachments to the area (ANPWS 1991b). These people reside mainly at a number of small settlements under arrangements between individuals and ANCA. Underpinning many of these arrangements is a wish by residents to pursue elements of their traditional life sytle in relative privacy. A living area for Aboriginal people has also been established in Jabiru and is administered by the Jabiru Town Council.

The adjoining Aboriginal lands in Arnhem Land are administered as land trusts by the Northern Land Council (NLC). The NLC is a statutory authority established by the Federal Government to represent Aboriginal interests associated with legislative and social changes in relation to land ownership and management in the Northern Territory. Settlements in Arnhem Land are locally administered by councils, such as that as Gunbalanya, or associations that represent the traditional owners.

5.3. Jabiru townsite

The Jabiru townsite is located on a 13 km^2 area (Fig. 1) leased from the Director of ANCA. The initial townplan was drawn by consultants on behalf of ANCA; however, from early in 1979 all development was undertaken by the Jabiru Town Development Authority (JTDA). The JTDA is a statutory corporation established by the Northern Territory Government which had only recently taken over many administrative functions from the Australian Federal Government. The JTDA comprises three government representatives, three from Energy Resources of Australia who run the nearby Ranger uranium mine and one from the Jabiru Town Council. Originally, the companies that owned the uranium mining leases at Jabiluka and Kongarra were to also be represented on the JTDA, but as they did not establish a significant presence in the town this did not occur. Planning and financing of the town development was carried out through the JTDA after much discussion over the cost sharing arrangements for the facilities between the government and mining companies (Lea & Zehner 1986).

The JTDA governed Jabiru until May 1984 when elections for the Jabiru Town Council were first held for five elected members who joined three appointed members. The number of appointed members was reduced to one in 1985. Despite the establishment of an elected council the JTDA still retained power over local finances, collected from taxes levied on property owners, and the lease arrangements with ANCA, which had previously leased the land from the Aboriginal traditional owners (Lea & Zehner 1986). Under the lease arrangements with ANCA the Jabiru townsite is affected by a number of regulations and laws applicable to the park, such as those governing the introduction of exotic plants and animals. The nature of local government within Jabiru and its relationship with other small communities spread throughout the Park is currently being reassessed.

5.4. Mining development

Mining is not permitted in Kakadu National Park (ANPWS 1991b). Mining leases at Ranger, Jabiluka and Koongarra are not included within the Park. Mineral exploration leases that existed prior to the current boundaries of the Park being declared have been resumed by the government. The mineral deposits at Coronation Hill had been mined previously and then the leases abandoned. An application to recommence mining was not permitted after a protracted national debate (Calligan & Lynch 1992). Mining is only permitted in Arnhem Land under specific arrangements negotiated with the traditional owners through the Northern Land Council. The Nabarlek uranium mine was developed under conditions which governed both the operation of the mine and mill and its rehabilitation once the operation came to an end.

Regulation of uranium mining at Ranger and Nabarlek has been carried out under Northern Territory legislation. The Federal Government does not directly regulate these activities, although it has indirectly exerted a strong influence under its power to control exports. As Australia does not have a domestic market for uranium this has proved to be an effective method of exerting a large amount of influence over the uranium mining industry. To ensure the adequacy of the regulatory arrangements implemented by the Northern Territory the Federal Government appointed a Supervising Scientist for the Alligator Rivers Region who was required to report on environmental aspects of uranium mining. The Supervising Scientist was a statutory officer who reported to the Federal Minister for the Environment on the effects of uranium mining on the environment of the region, supervised the implementation of Federal environmental requirements for mining developments and coordinated research relating to uranium mining through a research institute based in Jabiru. The Supervising Scientist is now a statutory officer within the Federal Environmental Protection Agency.

The mining enterprises in the region have been subject to a large amount of scrutiny and regulation that has at times reached acrimonious levels. The high level of regulation is directly related to the high cultural and natural heritage values of the region (Fox *et al.* 1977).

7. References

Aldrick, J.M. 1976. Soils of the Alligator Rivers area. In: Story, R., Galloway, R.W., McAlpine, J.R. Aldrick, J.M. & Williams, M.A.J. (eds) Lands of the Alligator Rivers Region, Northern Territory. pp. 71–88. Land Research Series No. 38, CSIRO, Melbourne.

Andersen, A.N. 1991. Seed harvesting by ants in Australia. In: Huxley, C.R. & Cutler, D.F. (eds) Ant-plant Interactions. pp. 493–503. Surrey Beattie & Sons, Sydney.

Andersen, A.N. & Braithwaite, R.W. 1992. Burning for conservation of the Top End's savannas. In: Moffat, I. & Webb, A. (eds) Conservation and Development Issues in Northern Australia. pp. 117–122. North Australian Research Unit, Darwin.

Andersen, A.N. & Lonsdale, W.M. 1990. Herbivory by insects in Autralian tropical savannas: a review. J. Biogeog. 17: 433–444.

ANPWS 1986. Kakadu National Park Plan of Management. Canberra Publishing and Printing Co., Canberra. 109 pp.

ANPWS 1991a. Nomination of Kakadu National Park by the Government of Australia for inscription in the World Heritage List. Unpublished report, Australian National Parks and Wildlife Service, Department of the Arts, Sport, the Environment, Tourism and Territories, Canberra. 104 pp plus bibliography.

ANPWS 1991b. Kakadu National Park Plan of Management. Commonwealth of Australia, Canberra. 171 pp.

Bowman, D.J.M.S. & Wilson, B.A. 1987. Fuel characteristics of coastal monsoon forests, Northern Territory, Australia. J. Biogeog. 15: 807–817.

Braithwaite, R.W. & Woinarski, J.C.Z. 1990. Coronation Hill Kakadu Stage III — assessing the conservation value. Aust. Biol. 3: 3–13.

Braithwaite, R.W., Dudzinski, M.L., Ridpath, M.G. & Parker, B.S. 1984. The impact of water buffalo on the monsoon forest ecosystems in Kakadu National Park. Aust. J. Ecol. 9: 309–322.

Brennan, K. 1986. Wildflowers of Kakadu. Kym Brennan, Jabiru, Australia. 127 pp.

Brennan, K. 1992. Checklist of vascular plants of the Alligator Rivers Region, Northern Territory. Supervising Scientist for the Alligator Rivers Region, Second Edition. Open File Record 62. 44 pp.

Bureau of Meteorology 1989. Climate of Australia. AGPS, Canberra. 49 pp.

Chapman, T.G. 1988. Groundwater recharge and discharge in the Magela Creek catchment. Northern Territory Department of Mines and Energy, Internal Progress Report, Darwin. 15 pp.

Chartres, C.J., Walker, P.H., Willett, I.R., East, T.J., Cull, R.F., Talsma, T. & Bond, W.J. 1991. Soils and hydrology of Ranger Uranium Mine sites in relation to the application of retention pond water. Supervising Scientist for the Alligator Rivers Region Technical Memorandum 34. 61 pp.

Christian, C.S. & Aldrick, J.M. 1977. Aligator Rivers Study: A review report of the Alligator Rivers Region environmental fact-finding study. Australian Government Publishing Service, Canberra. 174 pp.

Cowie, I.D. & Werner, P.A. 1987. Weeds in Kakadu National Park — a survey of alien flora. Unpublished report to Australian National Parks and Wildlife Service, Canberra. 42 pp. plus appendices.

Cowie, I.D., Finlayson, C.M. & Bailey, B.J. 1988. Alien plants in the Alligator Rivers Region, Northern Territory, Australia. Supervising Scientist for the Alligator Rivers Region Technical Memorandum 23. 26 pp.

Dugan, P. (ed.) 1993. Wetlands in Danger. Reed International Books Limited, London. 187 pp.

Finlayson, C.M. 1988. Productivity and nutrient dynamics of seasonally inundated flood plains in the Northern Territory. In: Wade-Marshall, D. & Loveday, P. (eds) North Australia: Progress and Prospects, volume 2 Flood plains Research. pp. 58–83. North Australia Research Unit, Australian National University, Darwin.

Finlayson, C.M. 1991. Primary production and major nutrients in three grass species on a tropical flood plain in northern Australia. Aquat. Bot. 41: 263–280.

Finlayson, C.M., Bailey, B.J. & Cowie, I.D. 1989. Macrophyte vegetation of the Magela Creek flood plain, Alligator Rivers Region, Northern Territory. Supervising Scientist for the Alligator Rivers Region Research Report 5. 38 pp.

Finlayson, C.M., Cowie, I.D. & Bailey, B.J. 1990. Characteristics of a seasonally flooded freshwater system in monsoonal Australia. In: Whigham, D.F., Good, R.E. & Kvet, J. (eds) Wetland Ecology and Management: Case Studies. pp. 141–162. Kluwer Academic Publsihers, Dordrecht, The Netherlands.

Finlayson, C.M., Cowie, I.D. & Bailey, B.J. 1993. Litterfall in a Melaleuca forest on a seasonally inundated flood plain in tropical northern Australia. Wetland Ecol. & Manage. 2: 177–188.

Finlayson, C.M., Bailey, B.J., Freeland, W.J. & Fleming, M. 1988. Wetlands of the Northern Territory. In: McComb, A.J. & Lake, P.S. (eds) The Conservation of Australian Wetlands. pp. 103–126. Surrey Beatty & Sons, Sydney.

Finlayson, C.M. Thompson, K., von Oertzen, I. & Cowie, I.D. 1995. Vegetation of five Magela Creek billabongs, Alligator Rivers Region, northern Australia. Supervising Scientist for the Alligator Rivers Region Technical Memorandum (in press).

Finlayson, C.M., Julien, M.H., Russell-Smith, J. & Storrs, M.J. 1994. Summary of a workshop on *Salvinia molesta* in Kakadu National Park, Northern Territory, Australia. Plant Protect. Quart. 9: 114–116.

Fisher, W.J. 1988. A brief histroy of mining, exploration and development in the South Alligator River Valley. Unpublished report to Coronation Hill Joint Venture, W.J. Fisher & E.E. Fisher Pty Ltd, Darwin. 26 pp.

Fox, R.W., Kelleher, G.G. & Kerr, C.B. 1977. Ranger Uranium Environmental Inquiry. Australian Government Publishing Service, Canberra. 415 pp.

Galligan, B. & Lynch, G. 1992. Integrating conservation and development: Australia's Resource Assessment Commission and the testing case of Coronation Hill. In: Moffat, I. & Webb, A. (eds) Conservation and Development in Northern Australia. pp. 239–249. North Australia Research Unit, Australian National University, Darwin.

Haynes, C.D., Ridpath, M.G. & Williams, M.A.J. 1991. Monsoonal Australia: Landcape, Ecology and Man in the Northern Lowlands. A.A. Balkema, Rotterdam. 231 pp.

Hein, W.R. & Tomasovic, A.A. 1981. An abattoir survey of tuberculosis in feral buffalo. Aust. Vet. J. 57: 543–547.

Hooper, A.D.L. 1969. Soils of the Adelaide-Alligator area. In: Story, R., Williams, M.A.J., Hooper, A.D.L., O'Ferrall, R.E. & McAlpine, J.R. (eds) Lands of the Adelaide-Alligator Area, Northern Territory. pp. 95–113. Land Research Series 25, CSIRO, Melbourne.

Jones, R. 1985. Archealogical conclusions. In: Jones, R. (ed.) Archaelogical Research in Kakadu National Park. pp. 17–29. Australian National Parks and Wildlife Service Special Publication No. 13. AGPS, Canberra.

Lazarides, M., Craven, L.A., Dunlop, C.R., Adams, L.G. & Byrnes, N. 1988. A checklist of the flora of Kakadu National Park and environs, Northern Territory, Australia. Occasional Paper No. 15. Australian National Parks and Wildlife Service, Canberra. 42 pp.

Lea, J.P. & Zehner, R.B. 1986. Yellowcake and Crocodiles. Allen & Unwin Australia Pty Ltd, Sydney. 200 pp.

Letts, G.A., Bassingthwaigthe, A. & de Vos, W.E. 1977. Feral Animals in the Northern Territory. Report of the Board of Inquiry to the Northern Territory Legislative Assembly. Northern Territory Government Printer, Darwin. 234 pp.

Meechan, B. 1988. Changes in Aboriginal exploitation of wetlands in northern Australia. In: Wade-Marshall, D. & Loveday, P. (eds). North Australia: Progress and Prospects, volume 2 Flood plains Research, appendix 2, 23 pp. North Australia Research Unit, Australian National University, Darwin.

Miles, G. 1986. Wildlife of Kakadu. Barker Souvenirs, Alice Sprngs, Australia. 96 pp.

Mott, J.J., Williams, J., Andrew, M.H. & Gillison, A.W. 1985. Australian savanna ecosystems. In: Tothill, B.J. & Walker, B.H. (eds) Ecology of Tropical Savannas, pp. 80–100. Springer-Verlag, Berlin.

Needham, R.S. 1988. Geology of the Alligator Rivers uranium field, Northern Territory. Bureau of Mineral Resurces Bulletin 224. AGPS, Canberra. 96 pp.

O'Ferrall, R.E. 1969. Introduction. In: Story, R., Williams, M.A.J., Hooper, A.D.L., O'Ferrall, R.E. & McAlpine, J.R. (eds) Lands of the Adelaide-Alligator Area, Northern Territory. pp. 7–15. Land Research Series 25, CSIRO, Melbourne.

Ovington, D. 1986. Kakadu: A World Heritage of Unsurpassed Beauty. Commonwealth of Australia, Canberra. 116 pp.

Ramsar Convention Bureau 1990. Directory of Wetlands of International Importance. Ramsar Convention Bureau, Gland, Switzerland. 782 pp.

Roberts, R.G., Jones, R. & Smith, M.A. 1990. Thermoluminescence dating of a 50,000-year-old human occupation site in northern Australia. Nature 345: 153–156.

Russell-Smith, J. 1986. The Forest in Motion: Exploratory Studies in Western Arnhem Land, Northern Territory. Unpublished PhD thesis, ANU, Canberra. 601 pp.

Russell-Smith, J. 1991. Classification, species richness, and environmental relations of monsoon forests in northern Australia. J. Veg. Sci. 2: 259–278.

Skeat, A.J., von Oertzen, I. & Maddison, J. 1987. Control of *Mimosa pigra* in Kakadu National Park. In: Lemerle, D. & Leys, A.B. (eds) Proceedings of the eighth Australian Weeds Conference. pp. 464–466. Weed Society of New South Wales, Sydney.

Story, R. 1969. Vegetation of the Adelaide-Alligator area. In: Story, R., Williams, M.A.J., Hooper, A.D.L., O'Ferrall, R.E. & McAlpine, J.R. (eds) Lands of the Adelaide-Alligator Area, Northern Territory. pp. 114–130. Land Research Series 25, CSIRO, Melbourne.

Story, R. 1976. Vegetation of the Alligator Rivers area, Northern Territory. In: Story, R., Galloway, R.W., McAlpine, J.R. Aldrick, J.M. & Williams, M.A.J. (eds) Lands of the Alligator Rivers Region, Northern Territory. pp. 89–111. Land Research Series No. 38, CSIRO, Melbourne.

Supervising Scientist for the Alligator Rivers Region 1988. Annual Report 1987–88. Australian Government Publishing Service, Canberra. 134 pp.

Williams, M.A.J. 1969. Geomorphology of the Adelaide-Alligator area. In: Story, R., Williams, M.A.J., Hooper, A.D.L., O'Ferrall, R.E. & McAlpine, J.R. (eds) Lands of the Adelaide-Alligator Area, Northern Territory. pp. 71–94. Land Research Series 25, CSIRO, Melbourne.

Wilson, B.A., Brocklehurst, P.S., Clark, M.J. & Dickinson, K.J.M. 1990. Vegetation survey of the Northern Territory. Technical Report 49, Conservation Commission of the Northern Territory, Darwin. 222 pp.

Woodroffe, C.D., Thom, B.G. & Chappell, J. 1985. Development of widespread mangrove swamps in mid-Holocene times in northern Australia. Nature 317: 711–713.

Woodroffe, C.D., Bardsley, K.N., Ward, P.J. & Hanley, J.R. 1988. Production of mangrove litter in a macrotidal embayment, Darwin Harbour, Northern Territory. Estuar. Coastal Shelf Sci. 26: 581–598.

Woodroffe, C.D., Chappell, J., Thom, B.G. & Wallensky, E. 1989. Depositional model of a macrotidal estuary and flood plain, South Alligator River, northern Australia. Sedimentology 36: 737–756.

CHAPTER 2

Climate and hydrology

CHRIS V. MCQUADE, JIM T. ARTHUR and IAN J. BUTTERWORTH

Abstract. The climate of the Kakadu region is classified as "summer rainfall tropical" with heavy periodic rains and generally hot and humid conditions from November–March and almost rainless and mild to warm conditions from April–October. The major climatic controllers for the region are the sub-tropical high pressure cells and a broad low pressure area. Seasonal variations are brought about by the north-south movements of Hadley Cells which are greatly affected by the 2–7 year cycle of the east-west movement of the El Niño Southern Oscillation. Temperatures are generally warm with mean monthly maxima and minima ranging from about 31–38 °C and 13–25 °C, respectively. Relative humidity is also high with 09.00 h values ranging from about 60–80% and 15.00 h values from 30–70%. In addition, the region is greatly influenced by the effects of tropical depressions, although relatively few have passed directly over the region. Most of the 1300–1500 mm rainfall comes in intense downpours or from monsoon depressions. Evaporation can reach 6–7 mm d^{-1} in the generally warmer months of October–November. The climatic conditions greatly affect the frequency and behaviour of fires in the region with the greatest danger from fires occurring near the end of the dry season and before the onset of the wet season rains.

The region contains three distinct hydrologic zones: the escarpment, the lowlands and the flood plains. The soils and geology of these zones determine the availability of water, which in turn affects the type and extent of vegetation in each zone. Billabongs are characteristic features of the drainage system and form isolated sources of water during the dry season. Variation in stream discharge is large with peak flows of more than 200 $m^3 s^{-1}$ occurring through the wet season. The groundwater in the lowlands can fall 2–4 m during the dry season and is recharged early in the wet season. Intense rainfall results in rapid runoff from the escarpment as water storage is limited to the faults and fractures in the rocks. Runoff from the escarpment and lowlands is collected in the creeks that have a base flow of 1–5 $m^3 s^{-1}$ for most of the wet season. Once the creeks are full, water spills across the extensive flood plains. Modelling of the hydrological cycle in the region has been attempted, but the models are not yet comprehensive and more attention to the interactions that affect water movement is needed. Similarly, the rapid response of the vegetation to the annual flooding cycle is well recognised, but poorly understood.

1. Introduction

Climate and hydrology are dynamic inter-related components of the suite of physical factors influencing the evolution of the landscape and biota of the Kakadu region. The region is a tectonically stable landscape and the climatic and associated hydrological variability represent the major mechanisms for environmental change over recent and future times (Nanson *et al.* 1990).

Climate is an abstract concept; it represents the summation of all interacting atmospheric processes affecting a locality. Since not all meteorological elements are omnipresent, climatic data are usually expressed in terms of individual calendar months or seasons, and are determined over a period long enough to ensure representative values for the month or season are obtained. The main factors governing the climate of a location are latitude, position relative to continents and oceans, position relative to large-scale atmospheric circulation patterns, altitude and local geographic features (Meteorological Office 1991). In the Kakadu region these factors combine to give a very distinct climatic pattern with two broad seasons; one mild to warm and dry, the other humid, hot and wet. The distinct seasonality of temperature and rainfall are reflected in the seasonal nature of the hydrological cycle in the region.

The hydrological cycle illustrates the complex interactions between the atmosphere, the surface of the earth and the lithosphere. The cycle has no beginning nor end; water is continuously cycled between the atmosphere and the surface of the earth by the physical processes of evaporation and precipitation. The terrestrial water components are controlled largely by the geomorphologic

C.M. Finlayson and I. von Oertzen (eds), Landscape and Vegetation Ecology of the Kakadu Region, Northern Australia, 17-35.

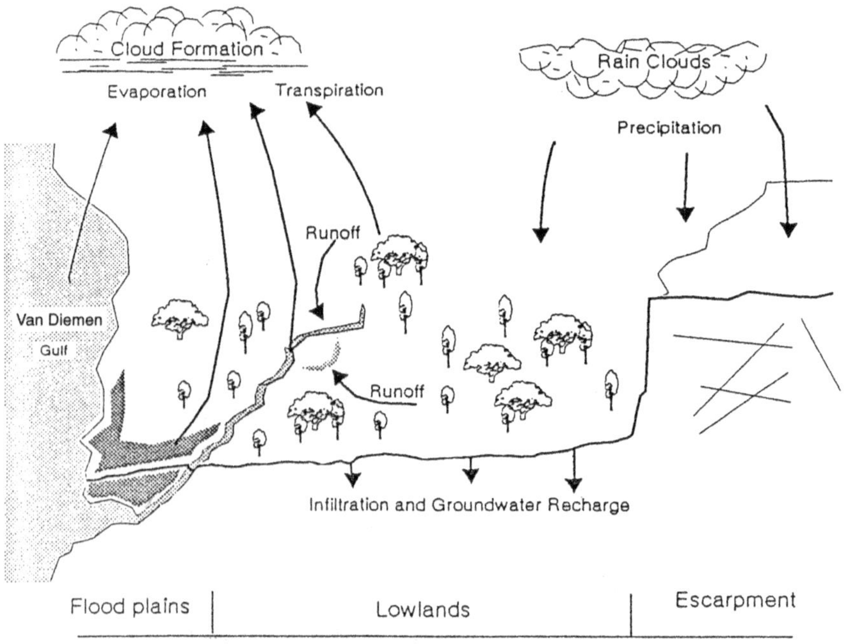

Fig. 1. Hydrological cycle.

characteristics of the land and include the slope of the land surface, the water drainage network, the permeability of the soils, and the hydraulic properties of the underlying rock material.

Understanding the hydrology of streams and wetlands is a fundamental requirement for understanding the biological and chemical processes that characterise stream and wetland ecosystems (Winter & Llamas 1993). A descriptive representation of the hydrological cycle for the Kakadu region is shown in Fig. 1; however, the complexities of this cycle are not understood, especially in relation to groundwater and interactions with the aquatic and wetland ecosystems. Effective management practices for such aquatic ecosystems are often limited by an inadequate understanding of the underlying hydrological processes (Winter & Llamas 1993). Although the Kakadu wetlands have undergone major ecological change over the past few decades (see Finlayson 1991a, Finlayson *et al.* 1988) and controversy still surrounds plans by mining companies in the region to release excess runoff water to the aquatic ecosystem (Johnston 1991) this has not provided sufficient impetus to thoroughly investigate the complex hydrology of the region.

2. Climate

2.1. Classification

The existence of different climatic environments over the surface of the earth has been noted since early historic times (Oliver 1972). Deriving suitable classification schemes to describe these environments is difficult; classification generally involves selecting simple criteria to represent complex systems. If different criteria are used the classification of a region could well change. Climatic classifications are often based on rainfall, temperature, crop ecology, humidity, vegetative and geographic criteria or a combination of criteria. Gaffney (1971) noted that climatic classifications developed in other parts of the world have limited application in Australia because they do not take account of the seasonal atmospheric circulation peculiar to the Australian region. The preferred classification for Australia is based on rainfall (Bureau of Meteorology 1988) and contains seven climatic zones (Fig. 2).

The Kakadu region is classified as "summer rainfall-tropical" and is characterised by heavy periodic rains, generally hot and humid conditions from November–March (corresponding to the

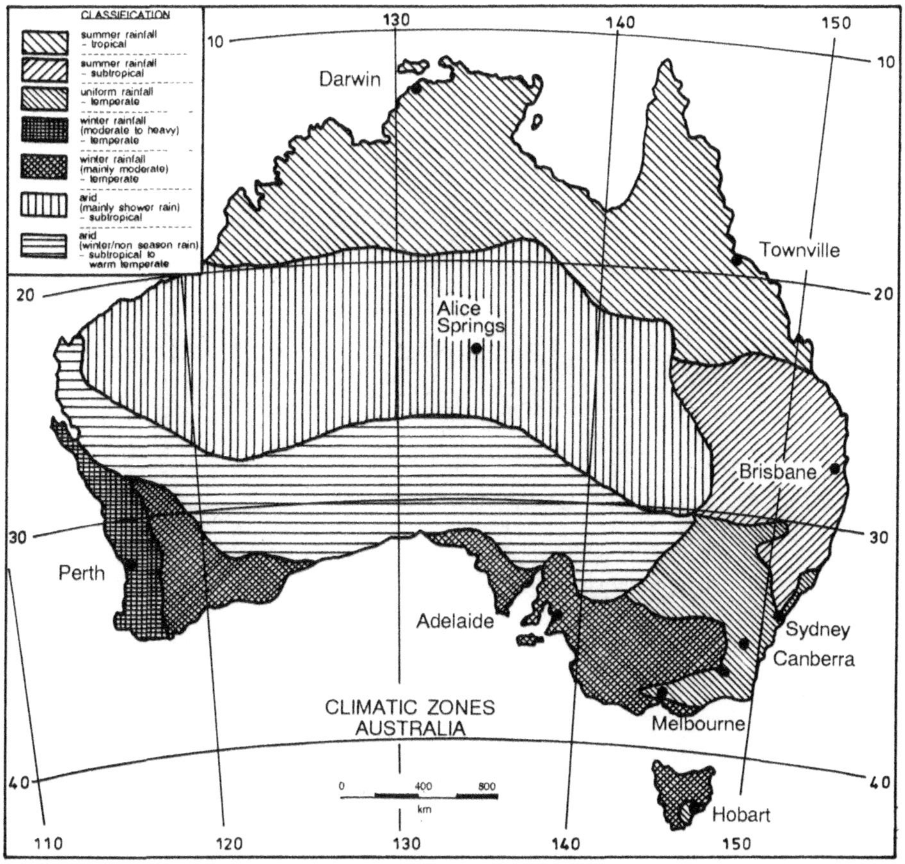

Fig. 2. Australian climatic zones (after Bureau of Meteorology 1989).

summer months of southern Australia, but known as the wet season in northern Australia) and mostly rainless, mild to warm conditions from April–October (the winter months in southern Australia, but known as the dry season in northern Australia). As in all climate classifications, this is a simplistic description, but it does present the basis for comparing the climate to that in both nearby and distant regions. Important features of this tropical, summer rainfall climate are described in detail in the following sections and include: rainfall variability, intensity and reliability; temperature and humidity; the incidence of cyclonic and tropical depressions; the manner in which climatic elements combine to cause high evapotranspiration rates and influence vulnerability to wildfires.

2.2. *Major climate controllers*

Looking at the climate in terms of classical surface or "weather map" features, it is controlled largely by two major systems: the subtropical high pressure cells (anti-cyclones), and a broad low pressure area in the tropical region, called the monsoon trough.

The subtropical high pressure cells traverse in a west-east direction across the southern Australian continent during winter; in summer, they are normally found to the south of the continent. These high pressure cells are the driving mechanism behind the south-east trade wind regime that significantly affects the Kakadu region in the dry season. The monsoon trough is normally situated over the northern part of the Australian continent during the wet season (southern Australian summer), and effectively separates a north-westerly airstream (north-west monsoon)

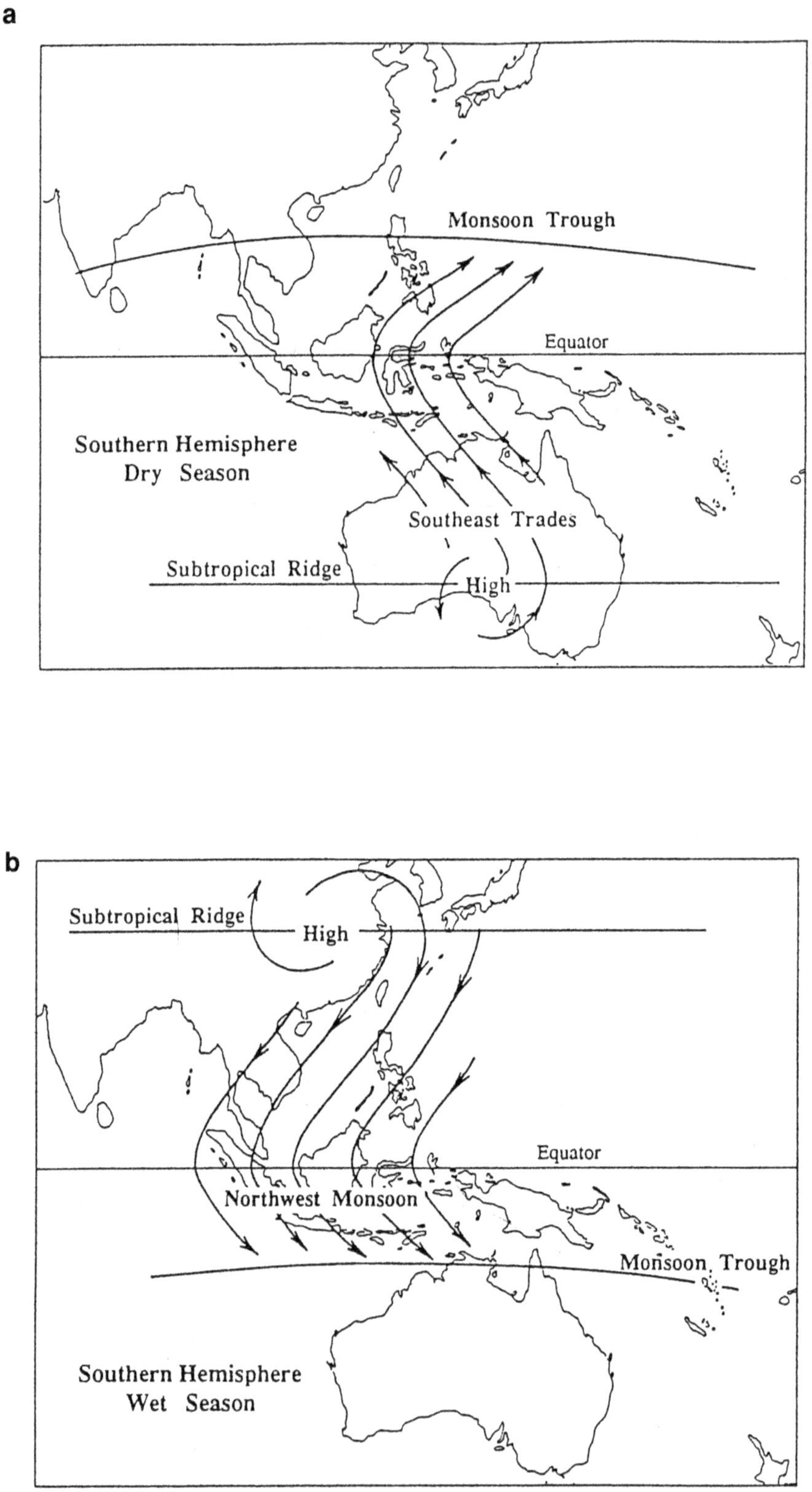

Fig. 3. Schematic presentation of an idealised mid-season low-level wind regime over the Kakadu region for a) the dry season and b) the wet season.

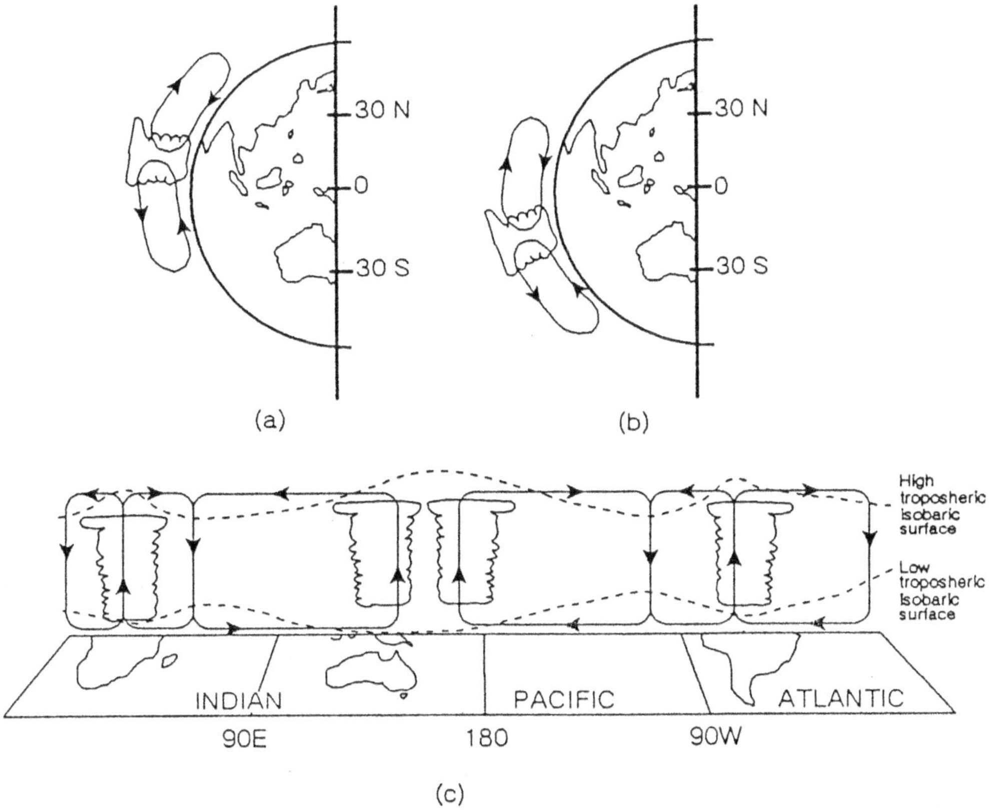

Fig. 4. Schematic representation of a) the Hadley cell circulation for the northern hemisphere summer, b) the Hadley cell circulation for the southern hemisphere summer and c) the east-west circulation.

on its northern side from a south-easterly airstream on its southern flank. Major disturbances (e.g. tropical cyclones) are spawned in the trough. Fig. 3 shows a schematic representation of respective mid-dry and mid-wet season, low-level wind regimes over the Kakadu region.

The above described broad-scale view does not lend any insight into the extent of either year to year or intra-seasonal climatic variability. Such variability is an important feature of the climate of the region. To understand the causes of this variability it is necessary to consider the interactions between the broad-scale, three-dimensional atmospheric circulation systems that greatly influence the climate.

The three-dimensional general atmospheric circulation system in the tropics is composed of relatively stable (from year to year) north-south (inter-hemispheric) circulations known as Hadley Cells, with more variable (from year to year) east-west circulations superimposed (Fig. 4). It is where the rising segments of these circulations occur that most rainfall will occur while the subsiding segments suppress active rain making weather events. The strong seasonal contrasts of the climate of the Kakadu region are caused by movements of the Hadley Cells as they follow the apparent north-south movements of the sun. In the dry season, the region is under the subsiding segment of the Hadley Cells, while during the wet season it is under the rising segment. The strength of these vary during the seasons; however, they can be considered stable over the course of a year, varying only over periods of a decade or longer. The seasonal variations in the position of the two segments of the Hadley Cells (Fig. 4) are reflected on the surface in the sub-tropical high pressure systems, the monsoon trough and their linking south-east trade wind and a north-west monsoon as in Fig. 3..

The rising (rain producing) segments of the Hadley Cells are suppressed or enhanced by the intensity and preferred locations of the rising and subsiding segments of the east-west circulation cells. These cells are subject to a 2–7 year periodic variation known as the El Niño-Southern Oscillation which results in variations from year to year in intensity and locations of the areas favoured for tropical convection, including enhanced rainfall and the formation of tropical depressions The onset of wet season rainfall is particularly affected by these inter-annual variations.

A further complication superimposed on the general circulation model is a quasi-periodic variation of 40–50 days (Madden & Julian 1971). This variation may be considered as a wave which moves across the tropics from west to east, alternatively enhancing and suppressing tropical rainstorm activity. It is considered that this particular oscillation is strongly linked to the well established intermittent nature of the Australian monsoon (Holland 1986). The intermittent nature of the monsoon (referred to as the "active-break" concept) is an important feature of the climate of the Kakadu region and brings with it the notion of extended drought periods during the wet season, with most rain coming in a small number (one, two or three) of active downbursts.

2.3. *Temperature and humidity*

Temperatures peak in October/November when the sun is almost directly overhead and the monsoon trough is still to the north. During those months, mean daily maximum temperatures are in the mid to high thirties. It is rare for temperatures to exceed 40 °C. The mid-dry season months of June and July are the coldest (Table I). Mean daily maximum temperatures are commonly in the low thirties and minimum temperatures normally range between 13–20 °C. Minimum temperatures below 10 °C are rare and confined to the southern parts of the region. Table I lists mean daily maximum and minimum temperatures together with extreme values for each month of the year at Jabiru. The values are considered indicative of the Kakadu region, although some moderation in extreme temperatures would be expected around the coastal fringe, while slightly greater extremes may be expected in the more southerly areas of the region.

Humidity is a measure of the water vapour content in the atmosphere. The more commonly used meteorological element is relative humidity which, at a given temperature, is the ratio (expressed as a percentage) of the actual vapour pressure to the saturated vapour pressure (Wiesner 1970, Meteorological Office 1991). Essentially, relative humidity is a measure of the percentage moistness of the air, making it a convenient figure for comparative purposes. Relative humidity has a marked diurnal variation opposite in phase to that of temperature; that is, relative humidity has a daily maximum around dawn, the normal time of temperature minimum. Temperature and relative humidity are similarly out of phase in mid-afternoon where relative humidity would normally be at a minimum. Indicative of the mean monthly variation of relative humidity over the Kakadu region are values at Gunbalunya (Fig. 5); highest values at both 09.00 h and 15.00 h, occur during

Table I. Mean daily maximum and minimum temperatures (°C) together with extreme values for each month at Jabiru

Month	Mean Maximum	Maximum	Mean Minimum	Minimum
January	33.5	37.7	24.5	20.5
February	33.1	37.7	24.2	20.6
March	33.1	37.0	24.1	20.4
April	34.4	37.1	23.3	16.0
May	33.4	36.5	21.7	13.9
June	31.3	35.5	18.9	9.9
July	31.5	35.9	13.3	8.8
August	33.6	37.4	19.3	12.3
September	35.7	39.8	21.4	12.0
October	37.5	41.6	23.7	13.7
November	36.5	42.0	24.8	19.0
December	34.9	39.3	24.7	21.1

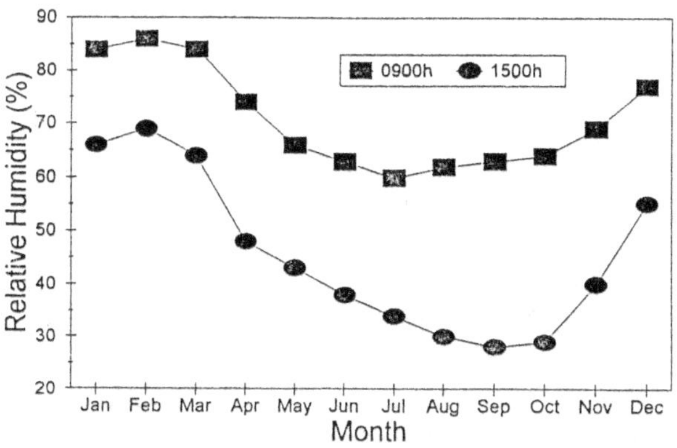

Fig. 5. Mean monthly relative humidity at Gunbalunya for 09.00 h and 15.00 h.

the height of the wet season when moist (maritime) northerly winds predominate. As would be expected, minimum values occur in the dry season months when the region is influenced by dry, (continental) east to south-easterly trade winds. Irrespective of the comparatively large annual variation relative humidity values are high when compared with inland arid Australia.

2.4. Wind

The seasonality of wind over the Kakadu region is illustrated in Fig. 6 which shows scaled directional components of 09.00 h surface wind at Jabiru for July and January, respectively. The contrast between the two is clearly identifiable: 85% of all winds in July at 09.00 h are from the east or south-east (continental); in January, only 17% of winds are east or south-easterly, while those with a westerly or northerly (maritime) component predominate. Significant afternoon northerly wind components, associated with the sea breeze, would be experienced throughout the year along the coastal fringes of the region.

2.5. Tropical cyclones

Lourensz (1981) showed that from July 1959, when dependable detection of tropical cyclones commenced, to June 1980 the annual average incidence of cyclones in the Australian region was 10.5. During that same period, an average of 1.6 tropical cyclones per year made landfall on the central northern region of the Australian coast, compared with 4.8 for Australia (multiple crossings not included). The most notable landfall was made by cyclone Tracy which devastated Darwin, the capital of the Northern Territory, on Christmas Day 1974.

Coastal crossings by tropical cyclones in the immediate vicinity of the Kakadu region (southern coast of van Diemen Gulf, west of the East Alligator River) have been rare, with available records revealing only eight crossings from 1827 to the end of 1992 (Lourensz 1981, Murphy 1984,

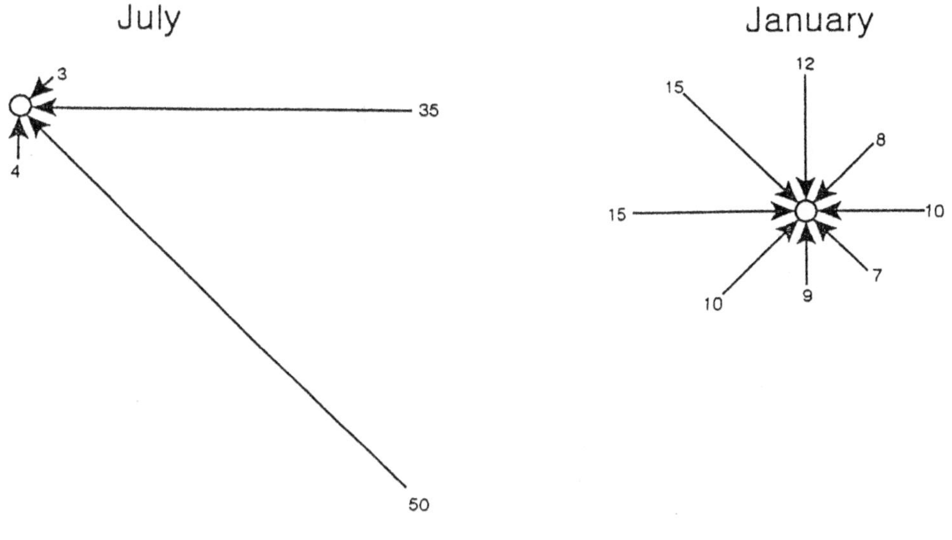

Fig. 6. Scaled surface wind directional components for 09.00 h at Jabiru for January and July.

Bureau of Meteorology, personal communication). No direct observational evidence relating to the few tropical cyclones that have crossed the Kakadu coast is readily available; however, those that have affected nearby coastal regions are considered comparable.

According to unpublished records held by the Bureau of Meteorology, estimated maximum wind gusts associated with tropical cyclone Tracy as it passed over Darwin were 240 km h^{-1}; the highest recorded gust was 217 km h^{-1} before the anemometer failed. A 12-hour rainfall total of 250 mm was recorded at Darwin airport on the same day. Cyclone Max crossed the Cobourg Peninsula into van Dieman Gulf and skirted the coast of the Kakadu region in March 1991. The maximum wind gust recorded at Darwin airport was 107 km h^{-1} with later estimates reaching 170 km h^{-1}. The heaviest 24-hour rainfall associated with this cyclone was 426 mm at Maningrida (130 km east of Kakadu). In April 1985 Cyclone Gretel followed a similar path to Max and dropped 270 mm of rainfall in a 24-hour period at Point Stuart (130 km west of Kakadu).

Despite this infrequency of coastal crossings, the region has felt the impact of many systems through intense rainfall and high winds as they traverse the Arafura and Timor Seas or cross other coastal segments and pass through or adjacent to the region as cyclonic depressions. An example of a tropical depression over the region is shown in Fig. 7, where extensive cloud, associated with the depression, covers almost the entire region.

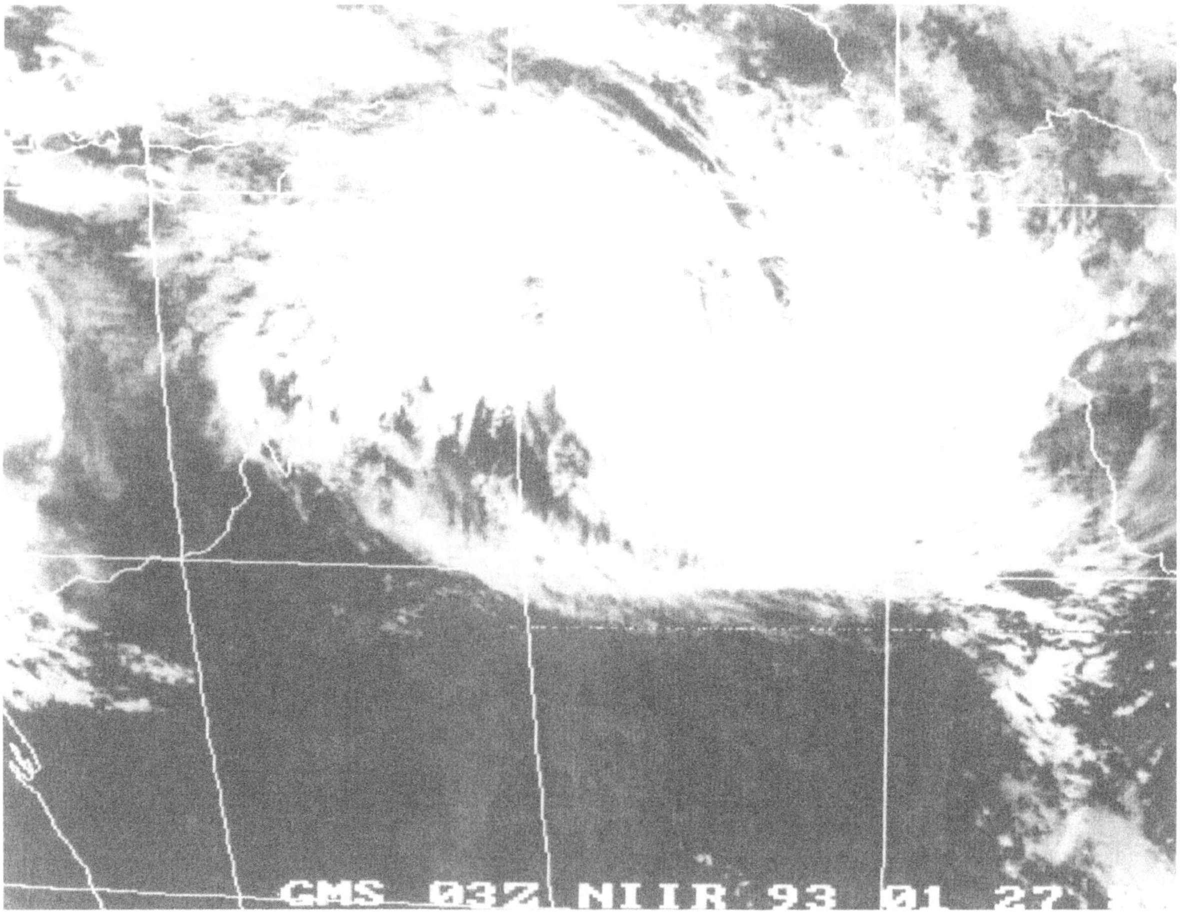

Fig. 7. A High Resolution Infrared satellite picture from the Japanese Geostationary Satellite GMS-4 on 27 January 1993 showing a tropical depression situated over the Kakadu region.

2.6. *Rainfall*

2.6.1. *Variability*

Rainfall, unlike many other meteorological elements is non-continuous in time and space, and as a result, its statistical description can be complex. The best known and most commonly used statistic is the arithmetic mean (often called the 'average' or 'normal'); however, this statistic is inappropriate for periods of less than a year. The median or 50 percentile value, which is the value that is exceeded by half the occurrences and not exceeded by the other half, is more appropriate for shorter periods, with the spread or variability of rainfall described by the 10 percentile and 90 percentile values. An index for assessing the variability of annual rainfall (VI) is given by the ratio: VI =(90 percentile – 10 percentile)/50 percentile (Bureau of Meteorology 1989). Rainfall at Jabiru has a VI value of 0.72. This is consistent with other tropical regions in the Northern Territory and is placed in the low to moderate range of rainfall variability. By contrast, Variability Indexes in excess of 1.50 (very high to extreme) are common throughout the desert areas of central Australia (Bureau of Meteorology 1989).

Stability in annual rainfall of the Kakadu region is implied by the Variability Index. However, single season totals may not always be representative of the climate as shown by Butterworth & Arthur (1993) in a study of rainfall trends over the north-west Northern Territory, including the Kakadu region; oscillations of wet and dry periods approaching a decade were identified. Taylor & Tulloch (1985) while emphasising the reliability of rainfall in the north Australian tropics, noted that variability in the occurrence of extreme rainfall events is also a feature of the climate.

Table II lists average and median monthly rainfall values at Jabiru for the period 1971–92. From these data it can be deduced that 92% of the annual average total (about 1460 mm) is recorded in the wet season months of November-March. It is also evident that rain has been recorded in each month of the year at some time during the period of record, but dry season rainfall is the exception rather than the rule. McQuade (1993) shows annual average totals over the region to range from 1300–1500 mm, and the clear seasonal rainfall pattern at Jabiru can be considered representative of the region. Further, McQuade (1993) also showed that over the long-term, regional annual rainfall ranges from 500 mm to 2700 mm.

Table II. Monthly mean and median rainfall (mm) at Jabiru for the period 1971-92.

Month	Mean Rainfall	Median Rainfall
January	341	327
February	332	267
March	298	302
April	66	49
May	12	7
June	1	0
July	3	0
August	4	0
September	9	0
October	28	26
November	158	139
December	211	169

2.6.2. *Rain producing systems*

Six discrete atmospheric systems that produce significant rainfall over northern Australia have been identified (J. Gill personal communication); each is an extension of the more general convective or cyclonic rain types and each applies in varying degrees of frequency and spatial and temporal scales over the Kakadu region. Fig. 8 illustrates these systems in terms of their typical spatial and temporal scales. At the lower end of the time-space scale, air-mass convection systems (single-cell thunderstorms) are around 10 km^2 in size, and rain duration ranges from several minutes to around two hours. At the other end of the scale, monsoon bursts typically affect areas of 100 000 km^2 (five-times the size of Kakadu National Park), and can last from a week to periods in excess of a month.

Air-mass convective systems are by far the most frequently occurring rain producing events. Their behaviour is dominated by microscale and mesoscale effects, and it is not possible to resolve actual numbers with currently available meteorological observation techniques. However, it is probable that many hundreds of individual air-mass convective events occur in the Kakadu region during the course of a wet season. Convergence lines and squall lines are convective by nature, but their occurrence relies partly on external meteorological

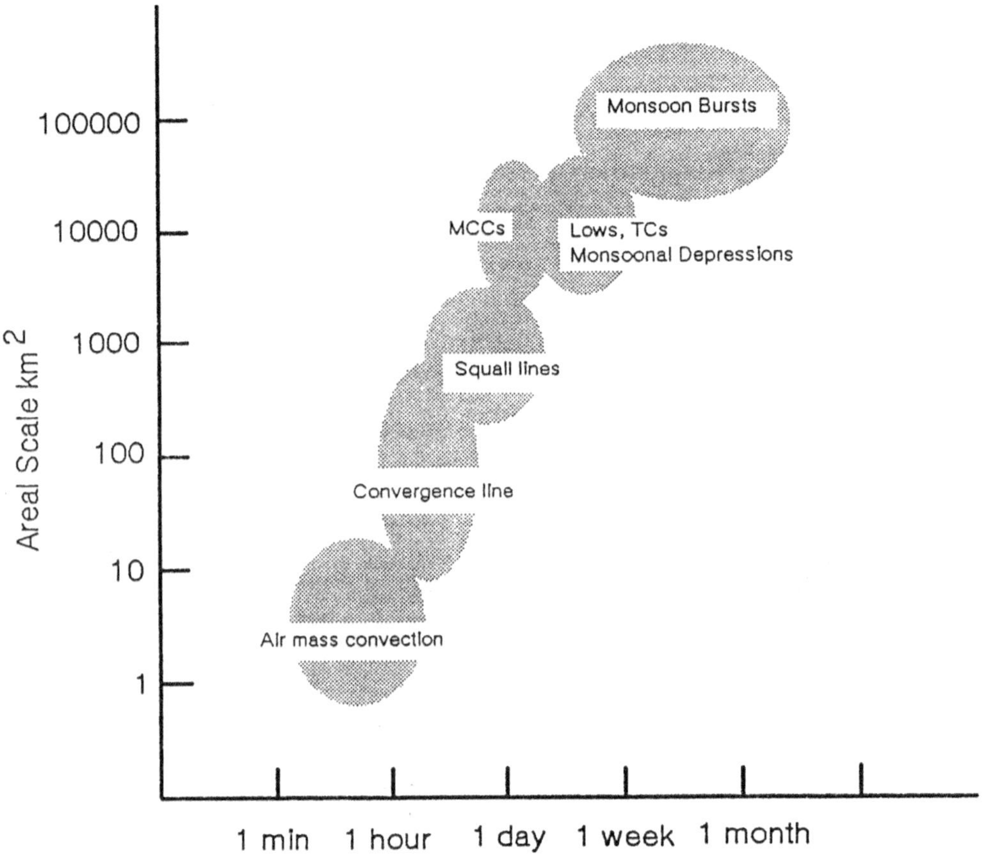

Fig. 8. Space-time scales of significant rain producing phenomena over northern Australia (J. Gill personal communication).

influences. The estimated frequency of convergence lines and squall lines during a wet season is 25 and 12, respectively. Lightning is also a significant feature associated with these tropical convective systems and 30 000 lightning flash registrations, either cloud-to-cloud or cloud-to-ground, could be expected in a season (Bureau of Meteorology unpublished data). By comparison with the convective systems, meso-scale convective complexes, tropical lows and monsoon bursts are rare events with respective seasonal frequencies ranging from zero to three.

2.6.3. *Intensity*

Rainfall intensity is a measure of rainfall amount over a period of time. By its very nature, rainfall associated with convection is usually the most intense, and each of the systems noted above has the potential to generate intense rainfall. Taylor & Tulloch (1985) found that high intensity falls in the Australian tropics have been common and have important ecological consequences (e.g. local flooding, erosion, improved water relations and redistribution of seeds and nutrients).

Rainfall from monsoon bursts is largely from stratiform cloud and is of moderate intensity; embedded convective elements provide locally enhanced rainfall. High rainfall registrations from these systems are due mainly to their prolonged existence. Monsoon bursts are often responsible for large scale (sometimes seasonal) flooding of the rivers and tidal flats, flood plains, lowlands and basins in the region. Fig. 7 is an example of a tropical depression over land; this particular depression caused rainfall extremes over northern Australia, including the highest recorded 24 hour total at Jabiru of 164 mm on 26 January 1993.

It is well known that extreme low rainfall (drought) events have significant consequences for the biota of a region; similarly, extreme high rainfall events can have important ecological implications. When extreme rainfall events coincide with the key establishment and early growth and reproductive phases of a plant species, they can have a marked effect on the recruitment, distribution and survival of a species (Taylor & Tulloch 1985).

2.7. *Potential evapotranspiration*

Evapotranspiration is the actual transfer of water to atmospheric vapour by both evaporation from the soil and water surfaces plus plant transpiration processes. On the other hand, potential evapotranspiration is the complete "drying power" of the climate system, assuming there is no limit to the supply of water to vaporise and it may be considered an index of the heat stress applied to an ecosystem. The very high year round potential evapotranspiration rate is a fundamental feature of the climate of the Kakadu region.

Evapotranspiration cannot be directly measured, only estimated by indirect techniques. McQuade (1993) applied a number of techniques – evaporation pan, water budget (Morton 1983), Webb correction for pan estimates (Webb 1960), and the Priestly & Taylor correction (Priestly-Taylor 1972) – to the estimation of evapora-transpiration from a 15 ha billabong near Jabiru. These estimates are included in Table III, together with estimates by the Penman (Oliver & Fairbridge 1987) and Vardavas (1987) techniques. The water budget technique is generally considered to provide the best estimates of evapotranspiration (Morton 1983) and is followed closely by the Penman and Vardavas techniques. However, a complete water budget was not calculated for all months of the year (Table III) and the superiority of this technique can not be shown in this instance. The comparison between techniques is further limited as the Penman and Vardavas calculations are from 1987 and the water budget is from 1983.

It is important to note that the pan evaporation technique significantly overestimates evaporation; however, these estimates are improved by the Webb correction. Potential evaporation is high year round (3.9–6.8 and 5.3–6.5 mm d^{-1} estimated by the Penman and Vardavas techniques, respectively) and but reaches a peak during October and November Table III). These months have the ideal combination of longer hours of sunshine and more direct solar radiation, which is a significant element affecting evapotranspiration as can be demonstrated by a simple analysis of the relative contributions of the various elements of the Penman equation for temperate and tropical latitudes.

2.8. *Fire weather*

The influence of fires on the ecosystems of tropical Australia is a matter of continuing controversy and debate (Lonsdale & Braithwaite 1991, 1992, Bowman 1992). That fire is an integral part of the Kakadu environment is no surprise considering

Table III. Daily mean potential evapotranspiration rate (mm $^{-1}$) for the Kakadu region.

Month	Water budget	Pan evaporation	Webb correction	Priestly-Taylor	Penman	Vardavas
January	Na	6.3	5.9	6.1	5.7	5.8
February	Na	5.8	5.2	6.1	5.5	5.1
March	Na	5.7	4.6	6.4	5.1	5.4
April	Na	6.9	5.5	7.0	5.3	5.3
May	3.4	7.2	5.2	5.8	4.8	5.0
June	5.3	7.1	6.1	Na	3.9	5.0
July	6.0	7.3	6.7	5.0	4.2	4.8
August	5.9	8.3	7.0	5.4	5.1	5.3
September	6.0	9.3	8.4	5.8	6.1	6.1
October	7.7	9.8	8.5	6.3	6.8	6.5
November	Na	8.2	7.5	6.2	6.7	6.2
December	Na	7.1	6.8	6.3	6.1	6.0

Na = insufficient data for the calculation.

the meteorological elements (wind speed, temperature and humidity) conducive to the spread of fires. The effects of these elements are generally combined with the quantity of fuel and degree of curing of the fuel into a single fire danger index: the McArthur Fire Danger Rating. These ratings are scaled from 1 to 100 and the index number is directly related to the rate of spread, ignition probability and suppression difficulty (Luke & McArthur 1978). Although derived for use in temperate parts of Australia, the same ratings are used by the Bureau of Meteorology and the Bush Fires Council of the Northern Territory in fire related operations, and may be used as an indication of fire danger climatology. An index number as low as 8 is classified as a High Fire Danger because suppression is difficult and a significant resource is needed to control fires.

Fig. 9 shows that the Fire Danger in 1991 (considered a typical year) peaked during the months August to October when 50–70% of all days were classified with a Fire Danger Rating of High or above. By August, the high biomass of grass species produced during the wet season is totally cured, while humidities remain low, temperatures rise and the wind in the afternoon, at the time of maximum temperature, maintains its early dry season strength.

The incidence of dry lightning strikes towards the start of the wet season provides a natural ignition factor which increases the overall Fire Danger although the effect is not included in the calculated Fire Danger Ratings.

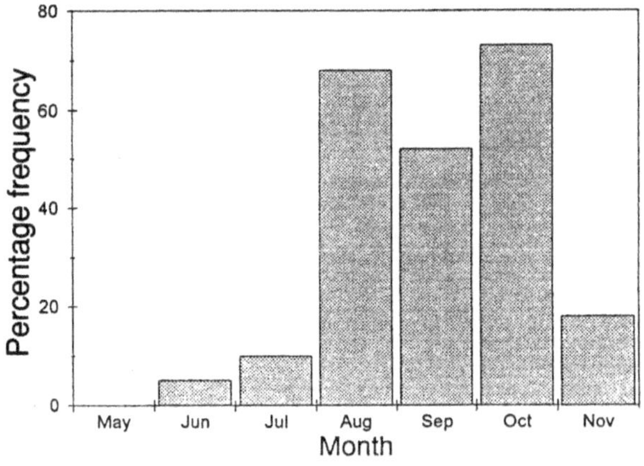

Fig. 9. Percentage frequency (days per month) on which a Fire Danger Rating of HIGH or greater was observed at Jabiru during 1992.

Despite the favourable meteorological regime of high fire vulnerability, experience suggests that the Kakadu region does not receive days where the rating could approach 100, unlike southern states of Australia, notably South Australia, Victoria and Tasmania. An analysis of the fire conditions in the Kakadu region clearly indicates that fires could start and spread on a large number of days, but days of potential catastrophe, as in some other parts of Australia, are not known to occur.

2.9. Climate variability and change

Climate is not a static feature, it is a dynamic regime subject to natural variations on all timescales, from minutes to millenia and beyond. Climate variability over periods of decades was discussed earlier in terms of rainfall and climate controls that affect the Kakadu region. Larger scale climatic variations, with periods of around 100 000 years, over the past two or three million years have been postulated (Oliver & Fairbridge 1987). These variations are related to changes in the orbital parameters of the earth.

In contrast, climate change is the term commonly used to describe long-term trends (probably irreversible) which are expected to result, or have resulted, from human interference. An example of climate change in today's world is the phenomenon commonly known as the "greenhouse effect".

Global climate models form the basis for prediction of greenhouse climate change and, despite their many weaknesses (including inadequate links to the ocean circulation, poor representation of many physical components of the climate system, and coarse vertical and horizontal resolution), the models do tend to show similar gross features (Suppiah *et al.* 1992). Common features include a predictive increase in mean temperatures of 0–2 °C in the tropics in the next forty years and an increase in the rainfall by as much as 20% in the same time. Direct application of these trends to the Kakadu region cannot be assumed, not only because of the acknowledged weaknesses in the models, but also because significant regional variations in the effect can be expected. However, the issue of potential sea-level rise under the enhanced greenhouse effect is one with dramatic ramifications for the region. Current estimates indicate global sea-level rises, above the 1990 level, of 5–35 cm by the year 2030 and 10–80 cm by 2070 (Wigley & Raper 1992).

Some recent studies have suggested that the greenhouse scenario has already caused climate change (Henderson-Sellers & Blong 1989). However, no significant increase in annual rainfall totals in the Kakadu region has occurred in the past 50 years (Butterworth & Arthur 1993). In any event, it is unlikely that changes of the order suggested by global climate models will be confirmed for decades, because of the masking effect of natural climate variability.

3. Hydrology

3.1. General features

The major part of the Kakadu region is drained by the South Alligator and East Alligator Rivers with the smaller West Alligator and Wildman Rivers draining the north-western portion of the region. (The Mary and Katherine Rivers drain a minor portion of the south-westerly part of the region and they are not considered further.) The rivers are fed by a network of ephemeral creeks and drain into van Diemen Gulf, in the north (Fig. 10). The combined catchment area of the four major rivers is approximately 28 000 km^2 (about

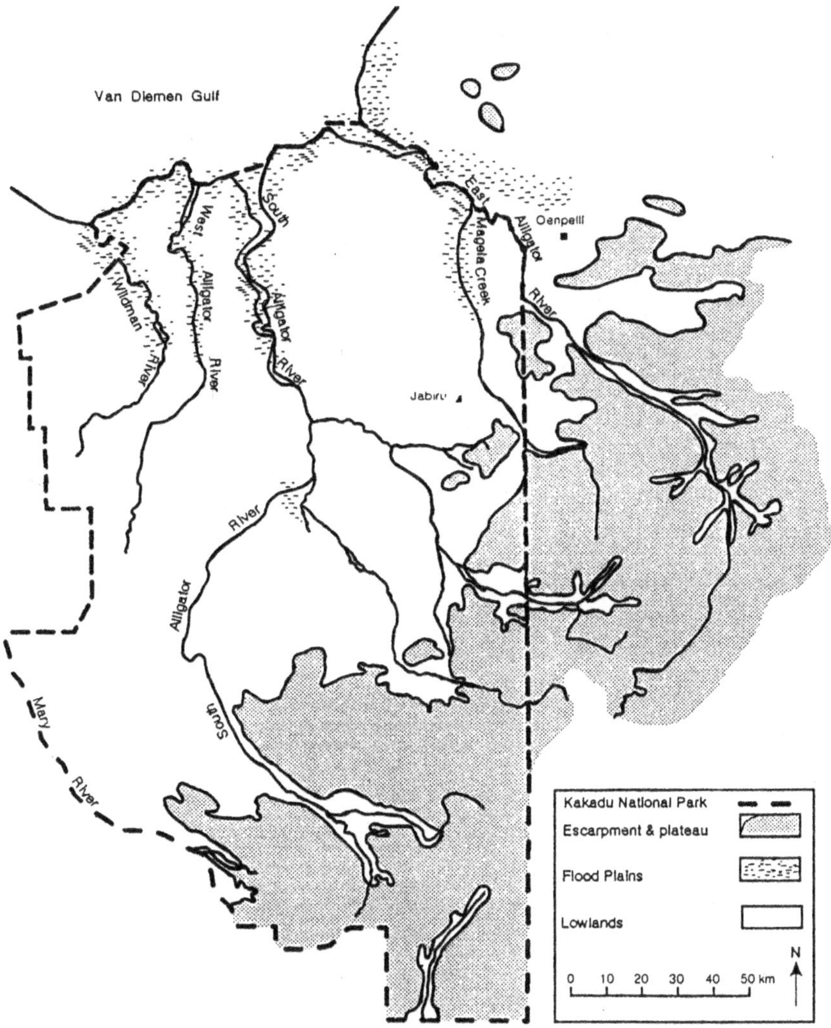

Fig. 10. Drainage network of the Kakadu region (modified from Supervising Scientist for the Alligator Rivers Region 19

8000 km^2 greater than the size of Kakadu National Park). In describing the hydrology of the region reference is made to three of the major physiographic land surface units: i) the escarpment; ii) the lowlands; and iii) the flood plains. Much of the information on the hydrology of the region comes from Chartres *et al.* (1991), Kingston (1981), Nanson *et al.* (1990) and Roberts (1991).

The drainage network of the region is not considered dense, reflecting the absence of impermeable soils (Chapman 1988); in general, a dense drainage network has a preponderance of impermeable soils. Kingston (1981) estimated the lag between rainfall and runoff for the region as around 2 months, which supports the conclusions of Chapman (1988) that the drainage network is not dense. The water storage areas in the generally permeable soils are filled by the initial falls of rain. Once these storage areas are filled runoff to the small streams following further rainfall is rapid. Storage also occurs in the deep sand beds of the creek channels.

The spring tidal range in van Diemen Gulf is 5–6 m, affecting the river flow up to 105 km inland (Woodroffe *et al.* 1989). During the wet season, water in the estuaries is predominantly fresh, becoming more saline in the dry season when rainfall runoff from the catchments declines and eventually ceases. Discharges from the rivers have not been measured because the tidal influence makes hydraulic rating extremely difficult; however, average flows between 400–700 $m^3\ s^{-1}$ for the South Alligator River were estimated from a comparison with the nearby Daly River (Woodroffe *et al.* 1986).

An indication of the variation in stream discharges is given in Fig. 11 for Magela Creek which has a catchment of 605 km^2. The variations in discharge are a direct response to the seasonal variations in climate with discharge peaks occurring late in the wet season when the flood plain and billabongs are full with water. Peak discharges of more than 200 $m^3\ s^{-1}$ are not uncommon, although most are less than 100 $m^3\ s^{-1}$ (Fig. 11). The relationship between rainfall and creek flow in Magela Creek was also shown in a comparison of flow rate with water depth in the creek and rainfall in Jabiru, carried out by Finlayson (1991b) for the period October 1893 to February 1985. Whilst an empirical relationship was evident in this comparison, it has not been quantified.

The water quality of the rivers and creeks typically follows a cyclic pattern in direct response to the seasonal cycle, with the ionic concentrations gradually increasing during the dry season (Hart & McGregor 1980, Brown *et al.* 1985, Walker &

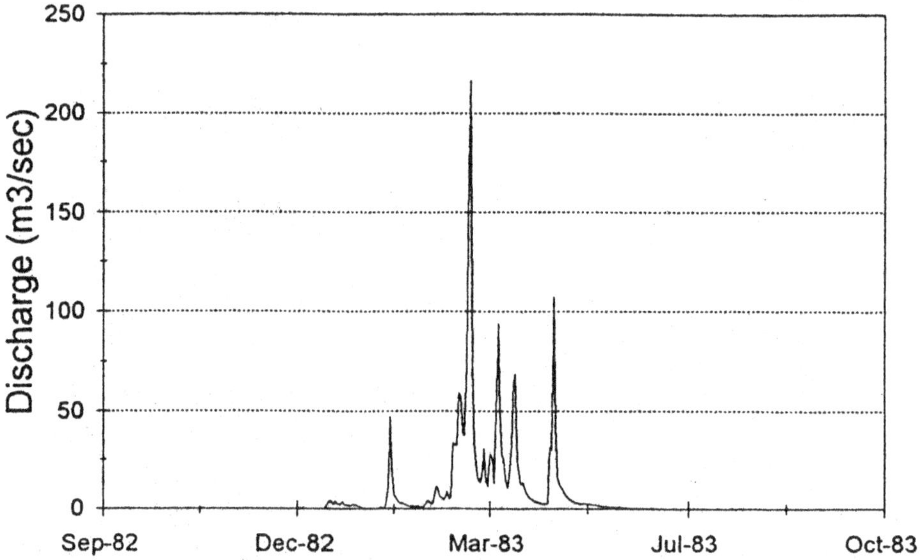

Fig. 11. Discharge hydrograph for Magela Creek.

Tyler 1985). This cyclic effect is more pronounced for water bodies with no groundwater inflow. Given the seasonal variability in water quality the detection of a change in water quality due to discharges of excess water stored on minesites could be difficult. In order to detect any such changes a statistically rigorous sampling program that characterised the extent of the natural variability in water quality in relation to the hydrological variability would be required. This baseline monitoring would also need to take into account other factors such as the eradication of feral buffaloes which has presumably had a dramatic indirect effect on water quality of billabongs. However, as with the assumed biological changes following the eradication of buffalo (Finalyson *et al.* 1994a), this has not been thoroughly investigated.

3.2. Escarpment area

The Arnhem Land escarpment comprises dissected plateau remnants in the east of the region. The plateau is mainly resistant Carpentarian sandstone of the Kombolgie formation, generally between 250–430 m above sea level. In the central and northern sections of the region the Cretaceous rocks have been eroded away and creeks dissect the sandstone block along joints and faults. About one third of the plateau is bare rock. Elsewhere, soils are less than a metre thick and are sandy and skeletal. Consequently, runoff response to rainfall is rapid, resulting in spectacular waterfalls that occur along the escarpment draining the plateau to the lowland area (Fig. 12).

The majority of the plateau is massive rock and the storage of water is restricted to the faults and fracture systems (C.V. McQuade, personal observation). Water fills the faults and fractures during the wet season, but these are typically drained within two months after the end of the wet season. The groundwater resource of the rock material is considered poor to negligible.

The footslopes from the escarpment to the lowlands vary in slope from about 45° to 20° above the lowland floor (C.V. McQuade, personal observation). Often, these rock screes start a third to half way down the escarpment face, comprising layers of sand mixed with large boulders from the face of the escarpment.

3.3. Lowland area

Erosion and weathering of the escarpment cliffs have exposed Archaean and Proterozoic rocks, and has produced sheets of sand of a different character from the generally much thicker and ferruginous Tertiary sands of the Koolpinyah Surface (Needham 1988). The lowland area has slopes as little as 1 in 1500 (Needham 1988), which is conducive to the formation of braided streams and billabongs.

Fig. 12. Jim Jim Creek waterfall (photo C.M. Finlayson).

The soils of the lowland regions are gravelly red and yellow massive earths, typically having more than 30% of coarse fractions (Hooper 1969). Their vertical permeability is in excess of 100 mm h^{-1} and is 3–12 times greater than the horizontal permeability (Chartres *et al.* 1991). These soils overlie a discontinuous and extremely weathered layer approximately 500 mm below the surface, which is characterised with permeabilities of less than 10 mm h^{-1}. This extremely weathered layer acts as an aquitard that hydrologically divides the surface soils from the deeper fractured rock. Water yield from the fracture systems vary from about 1 l s^{-1} to more than 8 l s^{-1}, which is sufficient to supply water to small township communities and mining operations in the region (C.V. McQuade unpublished data).

At the beginning of the wet season, rainfall rapidly infiltrates the soil and begins to recharge the groundwater table. The groundwater table can fall 2–4 m over a year which, considering soil porosity, is equivalent to 150–300 mm of water (Chapman 1988). As the wet season progresses, a perched water table develops over the weathered soil layer and leads to surface runoff during intense rainfall events. These intense rainfall events are accompanied by rapid increases in creek flows which are superimposed on a wet season base flow of 1–5 m^3 s^{-1} that ceases within a few months of the end of the wet season (C.V. McQuade personal observation).

In the dry season, the surface soils dry well below wilting point for plants (C.V. McQuade personal observation), assumedly because of evaporation and transpiration from the vegetation. Annual grasses, such as *Sorghum intrans* flower and set seed at the end of the wet season and cure before becoming a fuel source for fires that can burn across vast areas of the region. Within Kakadu National Park fire is used to maintain and enhance the biological diversity of the region (ANPWS 1991), an aim that could be greatly confounded by the extent of water retained in the lowland soils during the dry season.

The streams that cross the lowlands are braided and contain washed, coarse, quartzose sands that vary in depth from less than a metre to over 5 m deep along the length of the streams (Roberts 1991). At the commencement of the wet season, the channels begin recharging with most of the water coming from the escarpment. As the wet season progresses the creeks continue to fill with water from the escarpment and a distinct wave-front moves downstream. This wave-front carries fish and organic material out of the permanent water holes located along the creek lines. The water quality of the first flush in Magela Creek differs greatly from that later in the wet season and is characterised by a pH of 3–4 and an electrical conductivity > 1000 μS cm^{-1} compared with mid-season pH values of 5–6 and electrical conductivity less than 30 μS cm^{-1} (Brown *et al.* 1985, Walker & Tyler 1985).

Vardavas (1988) developed a water balance model for the Magela Creek catchment and provided an estimate of daily creek discharge given daily rainfall and monthly evaporation. This model represented an improvement on the previously used Boughton (1966) water balance model that segregated the hydrological process into water storage cells, such as interception and groundwater stores, and provided a set of rules for the movement of water from one storage cell to another. Vardavas (1988) identified three distinct phases to describe the main elements of the hydrological cycle. The initial wetting phase is rapid in response to rainfall, followed by a slow soil infiltration phase and then an even slower drying phase that is governed by the rate of evaporation. Modelling discharge from the creeks onto the flood plains is an important element in understanding the complex interactions between soils, hydrology and climate, and how they interact and influence the biota on the flood plains. Overall, however, these interactions are poorly understood (Finlayson *et al.* 1989).

Billabongs are an important component of the wetland ecosystems found on the lowlands of the region (Finlayson *et al.* 1988, 1994a). In the context of the Kakadu region a billabong is a still body of water that becomes linked to the main river or other wetlands by flood water. These water bodies are ecologically important as dry season refugia for many animals and also provide a permanent water supply for particular tree and grass species (see Finlayson *et al.* 1990). In the lowlands, the billabongs are formed by: i) in-channel erosion features; ii) damming of tributaries to the main ephemeral creeks by sediment deposited in these creeks; and iii) oxbow features of the rivers. The first two are characterised by sandy beds and muddy banks (Nanson *et al.* 1990), and are generally less than 4 m deep (Hart & McGregor 1980). While they usually retain water throughout the dry season, a series of low rainfall years can cause them to dry out. The oxbow lakes are deeper, reaching 7 m in depth (Walker & Tyler 1984, Hart & McGregor 1980) and retain water all year round.

3.4. Flood plain area

The flood plains of the Kakadu region are flat and lie predominantly on the Koolpinyah surface. Cracking clays overlie estuarine deposits and have high infiltration rates that are attributed to shrinkage cracks (Kellett *et al.* 1979). However, the wetting of the flood plains by the initial rains of the wet season tends to seal the soil surface, reducing the infiltration rate and allowing water to pond. Kingston (1981) estimated that the surface sealed within the first 150 mm of wet season rainfall. The abundant and permanent water supports a rich diversity of flora and fauna (see Finlayson *et al.* 1988,1990).

Water begins to enter the flood plains through ephemeral creeks about January and is immediately impeded by dense vegetation, which, in combination with the flat topography of the plains, results in water velocities < 0.1 m s^{-1} (Northern Territory Power and Water Authority unpublished data). Flow from the creeks and surrounding lowlands provides the majority of fresh water to the flood plains rather than direct rainfall (Vardavas 1989). The creeks flow through a network of billabongs and, at times, the channels become ill-defined, with the water reaching depths of several metres above the plain (Roos & Williams 1992). The flood plain billabongs are characterised by muddy bottoms and steep banks, and vary in depth from 1–4 m. As water enters these billabongs its velocity is greatly reduced and particulate matter deposited.

The velocity of the water through the billabongs has been further affected in recent years by the occurrence of the introduced weed *Salvinia molesta* (Finlayson *et al.* 1994b). A similar effect has also occurred on the flood plains where the eradication of feral buffaloes has allowed the re-establishment of thick mats of grasses in areas previously denuded of vegetation (Finlayson *et al.* 1994a).

The occurrence of plant species in different habitats on the flood plains during the wet season is determined primarily by the water depth and period of inundation by floodwaters (Finlayson *et al.* 1989). Growth of the many grass, sedge and herb species increases as soon as the plains become inundated, often keeping pace with the rising water level (Finlayson *et al.* 1989) and greatly increasing the amount of plant biomass (Finlayson 1991b). The vegetation growth on the flood plain was found by Vardavas (1989) to be an important factor in the flow of water during the period of inundation of the flood plain. In an extension of a previously developed simple water balance model (Vardavas 1988) it was shown that the surface roughness of the flood plain was affected by the vegetation growth in any given year and was an important factor in estimating the flood flow across the plain. Roos & Williams (1992) demonstrated that the surface roughness (described by a variable known as Manning's n) varied with the growth of vegetation and depth of water and was significantly different between years (Table IV). Such variability greatly reduces the accuracy of hydrological modelling of the flood plain (Gurnell & Midgley 1994).

McDonald & McAlpine (1990) modelled the seasonal water balance for flood plain soils near Gunbalanya and estimated the mean weekly soil moisture storage to range between 100 mm (late-wet season) and zero (mid-dry season). The maximum amount of excess water (excluding the amount consumed by evapo-transpiration) available for runoff and groundwater recharge is about 11 mm d^{-1}, with excess water occurring in about 20 weeks of the year. As a consequence of this pattern, plants with annual life cycles are common on the flood plains (Finlayson *et al.* 1989). Perennial species also occur, but as the available water supply decreases these senesce and shed leaves and stems (Finlayson *et al.* 1989, Finlayson 1991b).

Table IV. The variability of Manning's N.

Year	Manning's n	Year	Manning's n
1978	0.72 –0.130	1980	0.058–0.096
1981	0.086–0.165	1984	0.106–0.139
1985	0.167–0.244		

Depending on the annual rainfall and the length of the wet season the number of billabongs on the flood plains that hold water throughout the dry season varies markedly from year to year (see above). During the dry season the plains are generally dry with mainly senesced plants on the cracked soil (Finlayson *et al.* 1989). At irregular intervals fires sweep across the dry plains.

Woodroffe *et al.*(1985) suggested that the vegetation communities of the flood plains in the region have undergone considerable change as a consequence of changes in water level and salinity. Further increases in sea level could result in a shift to a salt water environment with corresponding change in the vegetation communities. Whilst the South Alligator and East Alligator River systems appear to be currently stable they have recently been affected by salt water intrusion as a consequence of buffaloes damaging the natural levees that separate the fresh and saline wetlands in the region (Woodroffe *et al.* 1986, 1989). Chin *et al.* (1992) have investigated the sub-surface hydrology of the Mary River flood plain, to the west of the Kakadu region and concluded that since about 1940 this flood plain has been converting from a fresh water wetland system to an estuarine system. This has been caused by the erosion of the natural flood plain barriers by the tidal outflow of the Mary River, possibly initiated by a human-induced event.

4. Acknowledgement

JTA and IB express their gratitude to the Director, Bureau of Meteorology for permission to contribute to this chapter.

5. References

ANPWS 1991. Kakadu national park plan of management. AGPS, Canberra. 171 pp.

Boughton, W.C. 1966. A mathematical model for relating run-off to rainfall with daily data. Civil Engineers Trans. CE8: 83–93.

Bowman, D.M.J.S. 1992. A burnt out case? Reply to Lonsdale and Braithwaite (1991). Aust. J. Ecol. 17: 103–106.

Brown, T.E, Morley, A.W. & Koontz, D.V. 1985. The limnology of a naturally acidic tropical water system in Australia II. dry season characteristics. Verh. Inter. Verein. Limnol. 22: 2131–2135.

Bureau of Meteorology 1988. Climatic Atlas of Australia, Map Set 3. AGPS, Canberra.

Bureau of Meteorology 1989. Climate of Australia. AGPS, Canberra. 49 pp.

Butterworth, I.J. & Arthur, J.T. 1993. Rainfall trends over the Northern Territory Top End. In: Research Papers Volume II, Bureau of Meteorology, Darwin. pp. 111–116.

Chapman, T.G. 1988. Groundwater recharge; and discharge in the Magela Creek Catchment, Northern Territory Department of Mines and Energy internal progress report. 15 pp.

Chartres, P.J., Walker, P.H., Willet, I.R., East, T.J., Cull, R.F., Talsma, T. & Bond, W.J. 1991. Soils and hydrology of Ranger Uranium Mine sites in relation to application of retention pond water. Supervising Scientist for the Alligator Rivers Region Technical Memorandum 34. 61 pp.

Chin, D., Martin, K., Schwartz, T. & Jolly, P. 1992. Mary River flood plain sub-surface hydrology study 1992. Northern Territory Power and Water Authority Report No. 20/93. 15 pp.

Finlayson, C.M. 1991a. Plant ecology and management of an internationally important wetland in monsoonal Australia. In: Kusler, J. & Day, S. (eds) Proceedings of an International Symposium on Wetlands and River Corridor Management. pp. 90–98. Association of State Wetland Managers, New York.

Finlayson, C.M. 1991b. Production and major nutrient composition of grass species on the Magela floodplain, Northern Territory, Australia. Aq. Bot. 44: 263–280.

Finlayson, C.M., Bailey, B.J. & Cowie, I.D. 1989. Macrophytic vegetation of the Magela flood plain, northern Australia. Supervising Scientist for the Alligator Rivers Region Research Report No. 5. AGPS, Canberra. 38 pp.

Finlayson, C.M., Cowie, I.D. & Bailey, B.J. 1990. Characteristics of a seasonally inundated freshwater system in monsoonal Australia. In: Wigham, D.F., Goode, D.F. & Kvet, J. (eds) Wetland Ecology and Management – Case Studies. pp. 141–162. Kluwer Academic Publishers, Dordrecht.

Finlayson, C.M., Bailey, B.J., Freeland, W.J. & Fleming, M. 1988. Wetlands of the Northern Territory. In: McComb, A.J. & Lake, P.S. (eds) The Conservation of Australian Wetlands. pp. 103–126. Surrey Beatty & Sons, Sydney.

Finlayson, C.M., Thompson, K., von Oertzen, I.A.M-L & Cowie, I.D. 1994a. Macrophytes of five Magela Creek billabongs, Alligator Rivers Region, northern Australia. Supervising Scientist for the Alligator Rivers Region Technical Memorandum (in press).

Finlayson, C.M., Julian, M.H., Russell-Smith, J. & Storrs, M.J. 1994b. Summary of a workshop on *Salvinia molesta* in Kakadu National Park. Plant Protect. Quart. (in press).

Gaffney, D.O. 1971. Seasonal rainfall zones in Australia. Working Paper 141, Bureau of Meteorology, Australia. 9 pp.

Gurnell, A.M. & Midgley, P. 1994. Aquatic weed growth and flow resistance: influence on the relationship between discharge and stage over a 25 year gauging station record. Hydrol. Process. 8: 63–74.

Hart, B.T. & McGregor, J.M. 1980. Limnological survey of eight billabongs in the Magela Creek catchment, Northern Territory. Aust. J. Mar. Freshw. Res. 31: 611–626.

Henderson-Sellers, A. & Blong, R. 1989. The Greenhouse Effect. New South Wales University Press, Sydney. 211 pp.

Holland, G.J. 1986. Interannual variability of the Australian summer monsoon at Darwin: 1952/82. Monthly Weather Review 114: 594–604.

Hooper, A.D.L. 1969. Soils of the Adelaide-Alligator Area. In: Story.R., Williams, M.A.J., O'Ferral, R.E. & McAlpine, J.R. (eds) Lands of the Adelaide-Alligator Area, Northern Territory. pp. 95–113. CSIRO Land Research Series 25. CSIRO, Melbourne.

Johnston, A. 1991. Water management in the Alligator Rivers Region: a research view. In: Hyne, R.V. (coordinator), Proceedings of the 29th Congress of the Australian Society of Limnology, Jabiru, NT 1990. Supervising Scientist for the Alligator Rivers Region, AGPS, Canberra. pp. 10–33.

Kellett, J.R., Evans, W.R., Jones, B. & Guy, D.B. 1979. Progress report on hydrogeological studies in the Magela and Cooper Creeks catchments, Alligator Rivers Region, NT. Bureau of Mineral Resource, Geology and Geophysics, Open File Record No. 1979/88. 31 pp.

Kingston, D. 1981. Northern Territory Coastal Plains Hydrology. Unpublished report to the Department of Transport and Works 49/81, Darwin, Northern Territory. 16 pp.

Lonsdale, W.M. & Braithwaite, R.W. 1991. Assessing the effects of fire on vegetation in tropical savannas. Aust. J. Ecol. 16: 363–374.

Lonsdale, W.M. & Braithwaite, R.W. 1992. Fiddling while Rome burns: Nero much misunderstood: reply to Bowman (1992). Aust. J. Ecol. 17: 106–108.

Lourensz, R.S. 1981. Tropical cyclones in the Australian region July 1909 to June 1980. Meteorological Summary. AGPS, Canberra. 94 pp.

Luke, R.H. & McArthur, A.G. 1978. Bushfires in Australia. AGPS, Canberra. 359 pp.

Madden, R.A. & Julian, P.R. 1971. Detection of a 40 to 50 day oscillation in the zonal wind in the tropical Pacific. J. Atmos. Sci. 28: 702–708.

McDonald, N.S. & McAlpine, J.R. 1990. Floods and drought: the northern climate. In: Haynes, C.D., Ridpath, M.G. & Williams, M.A.J. (eds) Monsoonal Australia: Landscape, Ecology and Man in the Northern Lowland. pp. 19–29. Balkema, Rotterdam.

McQuade, C.V. 1993. Probabilistic design for risk analysis in mine water resources management systems. Unpublished PhD thesis, The University of Queensland, Brisbane. 288 pp.

Meteorological Office. 1991. Meteorological Glossary. HMSO Publications Centre, London, England. 335 pp.

Morton, F.I. 1983. Operational estimates of areal evapotranspiration and their significance to the science and practice of hydrology. J. Hydrol. 66: 1–76.

Murphy, K. 1984. Big Blow up North – a history of tropical cyclones in Australia's Northern Territory. Government Printer of the Northern Territory, Darwin. 94 pp.

Nanson, G.C., East, T.J., Roberts, R.G., Clark, R.L. & Murray, A.S. 1990. Quaternary evolution and landform stability of Magela Creek catchment near the Ranger Uranium Mine, Northern Territory. Supervising Scientist for the Alligator Rivers Region Open File Record 63. 119 pp.

Needham, K.S. 1988. Geology of the Alligator Rivers uranium field, Northern Territory. Bureau of Mineral Resources, Geology & Geophysics Bulletin 224. 96 pp.

Oliver, J.E. 1972. Climate and Man's Environment. John Wiley and Sons Inc, Toronto. 515 pp.

Oliver, J.E. & Fairbridge, R.W. 1987. In: The Encyclopedia of Climatology. pp. 451–452. Van Norstrand Reinhold Company, New York.

Priestly, C.H.B. & Taylor, R.J. 1972. On the assessment of surface heat flux and evaporation using large-scale parameters. Monthly Weather Review 100: 81–92.

Roberts, R.G. 1991. Sediment budgets and Quaternary history of the Magela Creek catchment, tropical Northern Australia. Supervising Scientist for the Alligator Rivers Region Open File Record 80. 569 pp.

Roos, R.J. & Williams, D.K. 1992. Seasonal variations of Manning's n on the Magela Creek flood plain, N.T., Australia. In: Proceedings of the 8th Australasian Hydrographic Workshop, November 1992, Hosted by the Australian Capital Teritory Department of Electricity and Water, Canberra. 25 pp.

Supervising Scientist for the Alligator Rivers Region, 1992. Annual Report 1991–92. AGPS, Canberra. 113 pp.

Suppich, R., Pittock, A.P., Evans, J.L. & Mitchell, C.D. 1992. Regional impact of the enhanced Greenhouse effects on the Northern Territory. Annual Report 1991–92, CSIRO Division of Atmospheric Research, Melbourne. 71 pp.

Taylor, J.A. & Tulloch, D. 1985. Rainfall in the wet-dry tropics: extreme events at Darwin and similarities between years during the period 1870–1983 inclusive. Aust. J. Ecol. 10: 281–295.

Vardavas, I.M. 1987. Modelling the seasonal variation of net all-wave radiation flux and evaporation in a tropical wet-dry region. Ecol. Model. 39: 247–268.

Vardavas, I.M. 1988. A simple water balance daily rainfall-runoff model with application to the tropical Magela Creek catchment. Ecol. Model. 42: 245–264.

Vardavas, I.M. 1989. A water budget model for the tropical Magela floodplain. Ecol. Model. 46: 165–194.

Walker, T.D. & Tyler, P.A. 1985. Tropical Australia, a dynamic limnological environment. Verh. Inter. Verein. Limnol. 22: 1727–1734.

Webb, E.K. 1960. A pan-lake evaporation relationship. J. Hydrol. 4: 1–11.

Wiesner, C.J. 1970. Hydrometeorology. Chapman & Hall Ltd, London. 232 pp.

Wigley, T.M.L. & Raper, S.C.B. 1992. Implications for climate and sea level of revised IPCC emissions scenarios. Nat. 357: 293–300

Winter, T.C. & Llamas, M.R. 1993. Introduction to the 28th international geological congress symposium on the hydrogeology of wetlands. J. Hydrol. 141: 1–3.

Woodroffe, C.D., Thom, B.G. & Chappell, J. 1985. Development of widespread mangrove swamps in mid-Holocene times in northern Australia. Nat. 317: 711–713.

Woodroffe, C.D., Chappell, J., Thom, B.G. & Wallensky, E. 1986. Geomorphological Dynamics and Evolution of the South Alligator River Tidal River and Plains, Northern Territory. North Australian Research Unit Mangrove Monograph 3. ANU Press, Canberra. 190 pp.

Woodroffe, C.D., Chappell, J., Thom, B.G. & Wallensky, E. 1989. Depositional model of a macrotidal estuary and flood plain, South Alligator River, Northern Australia. Sedimentology 36: 737–756.

CHAPTER 3

Landform evolution

T. JON EAST

Abstract. The Kakadu region contains the catchments of the northerly-flowing Wildman, East Alligator, West Alligator and South Alligator Rivers. The rivers have an upstream highly seasonal freshwater reach and a meandering tidal estuarine reach. Erosion rates and stream suspended sediment and solute loads in the fresh water reach are low by world standards, but can be greatly increased in some catchments degraded by introduced cattle, feral buffalo and feral pigs. Landform development in the region commenced almost 2000 million years ago in the early Proterozoic with the development of the Pine Creek Geosyncline. The present day landforms reflect the influences of lithology, geological structure, past climate and sea level changes, and human activities throughout this long period. Nowadays, the landscape comprises six major physiographic units: a dissected sandstone plateau and escarpment; a stable erosion plain (the Koolpinyah Surface); flood plains of dominantly Pleistocene age; seasonally inundated flood basins of Holocene age formed in response to glacial and post-glacial climatic and sea level influences; estuarine flood plains; and a narrow coastal plain with saline mudflats and chenier ridges of shelly sand. In Aboriginal and European historical times the landscape has been modified by various burning practices, pastoral land use, mining, tourism and the introduction of exotic flora and fauna. The region is mineralogically rich with large reserves of uranium and smaller reserves of gold and platinate minerals.

1. Introduction

The Kakadu region broadly corresponds to an area defined by the catchments of the East Alligator, West Alligator, South Alligator and the Wildman Rivers east of Darwin (Fig. 1). The region is bounded to the east by the Aboriginal lands of Arnhem Land, and to the west and south by the catchments of the Mary and Katherine Rivers respectively; the headwaters of these rivers actually drain a very small part of the southern area of Kakadu National Park which is otherwise located entirely within the area known as the Alligator Rivers or the Kakadu region. All rivers drain northerly into van Diemen Gulf (Fig. 1).

The origins of the physical landscape of the region can be traced back almost 2000 million years, to early Proterozoic times, when epeirogenic tilting and orogenic upheaval in the Pine Creek Geosyncline set in place the geological foundations of the present landscape (Needham 1988). Subsequent geological and geomorphological events, including major erosional and depositional phases, extended over the vast span of time from the Proterozoic to the Holocene. Periods of intense weathering during the Tertiary (Williams 1969), and more recently the impact of Quaternary sea level changes (Nanson *et al.* 1993), left their imprint on the landscape. In Aboriginal and European historical times, the landscape was subjected to a major marine transgression (Wasson 1992) and was modified by Aboriginal burning practices, pastoral enterprises, mining, tourism and the introduction of exotic flora and fauna (Haynes *et al.* 1991). The present day landforms and soils reflect this long history of the interaction of lithology, geological structure, past climates and associated sea level changes, and human activities.

2. Geology

The regional geology of the Kakadu region is described by Ferguson & Goleby (1980) and Needham (1988); the key element being dominance by the mineralised metasediments and igneous rocks of the Pine Creek Geosyncline, and the younger sandstones of the Kombolgie Formation in the east of the region (Fig. 2). The Pine Creek Geosyncline extends from near Darwin in the west to Arnhem Land in the east. The geological sequence presented by Ferguson & Goleby (1980) and Needham (1988) is summarised in the following text.

C.M. Finlayson and I. von Oertzen (eds), Landscape and Vegetation Ecology of the Kakadu Region, Northern Australia, 37-55.

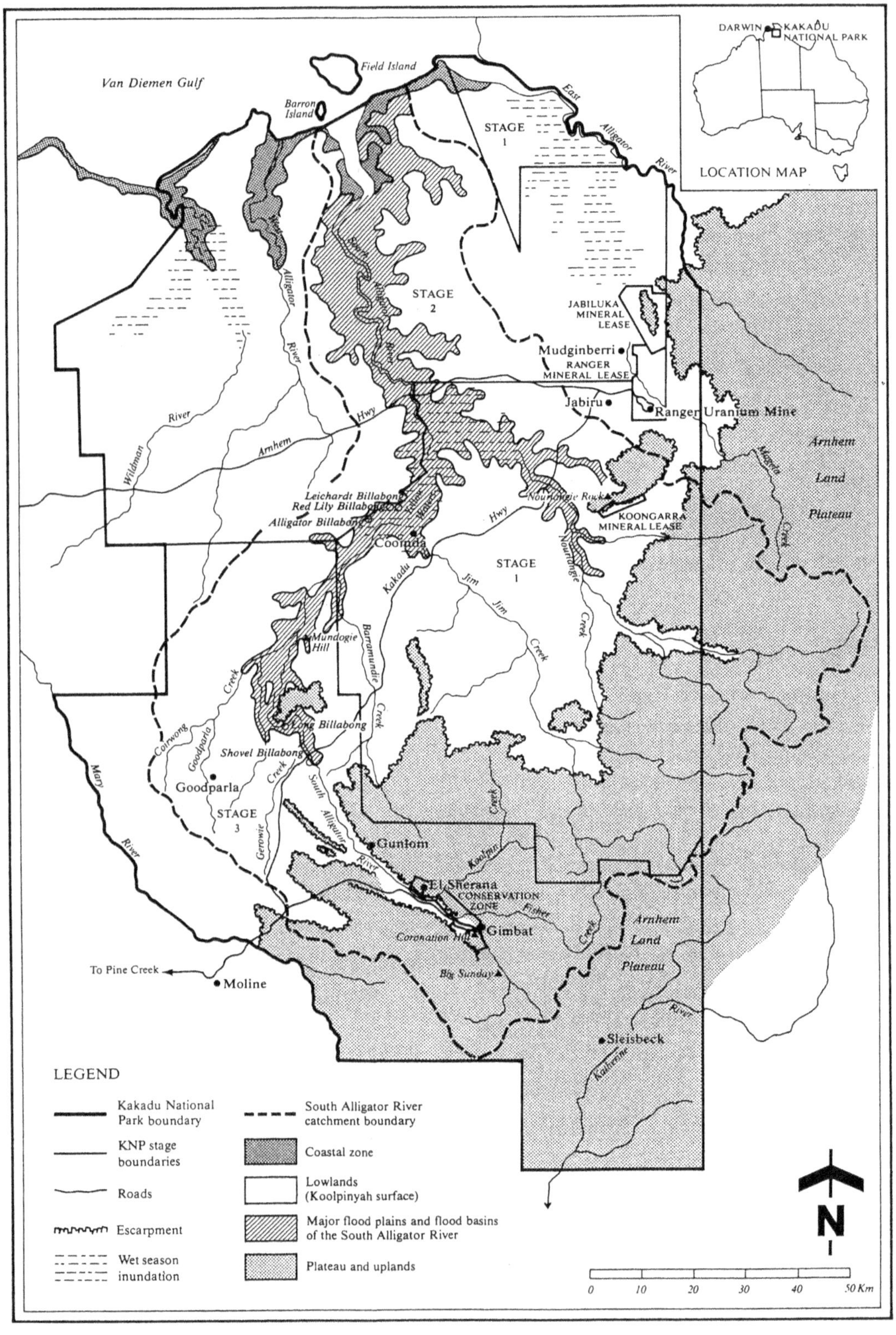

Fig. 1. Landforms and surface drainage of Kakadu National Park (adapted from East 1990).

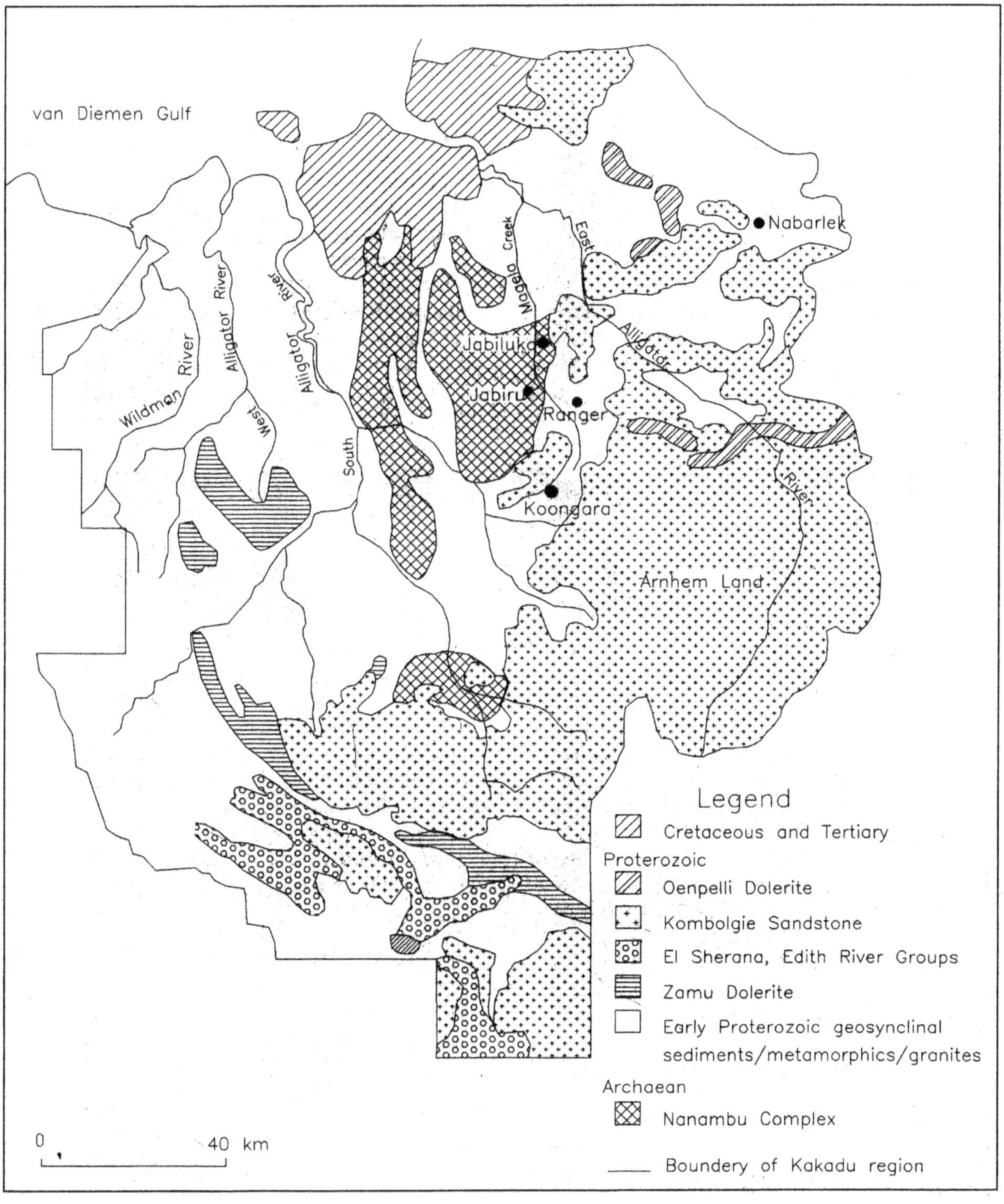

Fig. 2. Surface geology of the Kakadu region and north-eastern area of the Pine Creek Geosyncline (adapted from Needham 1988).

In early Proterozoic times arkosic sands were deposited on Late Archaean basement rocks of the Nanambu Complex and followed by deposition of carbonate rich sediments of the Lower Cahill Formation. The sediments underwent folding, with the subsequent deposition of the South Alligator Group and the later intrusion of the Zamu Dolerite.

The orogenic upheaval and folding processes in the region commenced about 1870 million years ago and spanned approximately 70 million years. Granites were intruded and migmatised during regional deformation and metamorphism to form the Nimbuwah Complex. The region underwent denudation and weathering in the remainder of the Early Proterozoic, during which lopoliths of the Oenpelli Dolerite were emplaced.

At the onset of the Middle Proterozoic, about 1650 million years ago, medium and coarse grained siliceous sands of the Kombolgie Formation were deposited in a braided alluvial fan environment on the planated surface of the Lower Proterozoic metasediments (Needham 1988). With the exception of localised downfaulting of Kombolgie Formation sandstone blocks, the region has been largely tectonically stable since the Late Proterozoic. The present Arnhem Land Plateau in the east and south of the region is formed dominantly in Kombolgie Formation sandstones.

In the Cretaceous, terrestrial sedimentation on the plateau surface and offshore marine sedimentation were accompanied by the formation of sea cliffs in Kombolgie Formation sandstone (Needham 1988). The terrestrial sediments were strongly lateritised by intense weathering in early to mid-Tertiary times (Williams 1969).

Since the mid-Tertiary, erosion has removed and redistributed much of the Cretaceous strata to form the Koolpinyah Surface, an extensive erosion plain of low relief. Denudation on the Koolpinyah Surface continued throughout the Quaternary, with fluvial incision, and valley, channel and estuarine aggradation accompanying at least two major Quaternary climate changes and sea level fluctuations (Nanson *et al.* 1993).

The Pine Creek Geosyncline is an area of high mineral prospectivity, with large reserves of uranium, and smaller reserves of gold and platinate minerals. All uranium mineralisation in the so-called Alligator Rivers Uranium Field (encompassing the Ranger, Nabarlek, Jabiluka and Koongarra deposits) occurs in the Cahill Formation, and is concentrated in the lower member of dominantly carbonaceous and carbonate rocks.

Current uranium reserves in the Pine Creek Geosyncline are approximately 360 000 t of uranium oxide, all of which has been discovered since 1969 (ARRRI 1988). The Ranger uranium deposit is presently being mined, whilst that at Nabarlek has been mined and rehabilitation of the minesite is now underway (Supervising Scientist for the Alligator Rivers Region 1994). A recent government decision has ruled against the development of the Coronation Hill gold and platinum deposit in the south of the region (Supervising Scientist for the Alligator Rivers Region 1994).

3. Landform development

The land surface of the Kakadu region consists of six major physiographic units, described by Williams (1969), Galloway (1976), East (1990) and Wasson *et al.* (1991) and shown in Fig. 1:

i) a plateau (Arnhem Land) and outliers of mainly resistant quartzose sandstone;
ii) an erosion plain (the Koolpinyah Surface) and isolated hills and strike ridges;
iii) intermittently inundated flood plains of predominantly Pleistocene age;
iv) low lying seasonally inundated flood basins of Holocene age;
v) deltaic estuarine flood plains; and
vi) a narrow coastal plain.

3.1. Plateau and outliers

The Arnhem Land Plateau is located in the east and south of the region (Fig. 1). The South Alligator and East Alligator Rivers, the two largest rivers, rise on the plateau and flow initially across the plateau surface in shallow channels, then in ravines and gorges of increasing depth to the plateau escarpment and then to the lowlands in the west and north of the region.

The plateau is formed predominantly in cross-bedded quartzose sandstones of the Kombolgie Formation of Middle Proterozoic age, and to a lesser extent in conglomerates and volcanics of the

Kombolgie Formation and in Cretaceous mudstones and siltstones. It has an overall height of about 300 m with hills rising to 570 m. The plateau sandstone has well developed horizontal bedding and vertical jointing, which control surface drainage through the formation of a generally deeply incised, trellised drainage pattern (Fig. 3).

The western margin of the plateau is marked by a spectacular escarpment, over 500 km long with an overall north-south orientation (Fig. 4). During the wet season (November-April) large and spectacular waterfalls, such as Twin Falls, Jim Jim Falls and Magela Falls, cascade 200 m or more over the escarpment edge onto adjacent lowlands (Fig. 5). Where an escarpment is not present, the rivers flow onto the lowlands through valleys bounded by plateau outliers, hills and strike ridges, as in the case of the South Alligator River between Coronation Hill and Gunlom (Fig. 1).

The formation of the escarpment has been largely controlled by the geological structure of the sandstone and to a lesser extent interstratified volcanics, and the erosion and removal of the

Fig. 3. Trellised drainage pattern, Arnhem Land plateau (photo T.J. East).

Fig. 4. Escarpment, Arnhem Land plateau (photo T.J. East).

Fig. 5. Magela Falls, East Alligator River in the wet season (photo T.J. East).

Fig. 6. Recent rockfall, Mt Brockman, East Alligator River catchment (photo T.J. East).

underlying and less resistant Lower Proterozoic rocks (Williams 1976). In places, the stability and rate of retreat of the escarpment depend to a large degree on the exposure and erosion of the underlying weaker rocks. Erosion of the underlying strata removes the basal support of the sandstone which fractures along vertical joints and bedding planes. Sandstone blocks are detached and accumulate as talus slopes and boulder fields at the base of cliffs. The development of a talus slope inhibits further erosion of the underlying rocks until the talus and blocks have been comminuted to sand by weathering, and are removed by surface wash and colluvial processes. The sand is redistributed on the lowlands, forming low angle sand sheets and sandy soils.

The dominant erosional processes along the plateau margins are mass movement processes – rock falls and debris slides. Scarp retreat is episodic, with phases of retreat alternating with periods of stability that may have a duration of some millions of years. The rate of escarpment retreat is consequently highly variable; a rate of 100 cm $(1000 \text{ years})^{-1}$ has been suggested by

Galloway (1976), and rates of 2–20 cm (1000 years)$^{-1}$ calculated by (Roberts 1991).

The escarpment exhibits considerable variation in form, and in the types and rates of retreat. Active scarps with precipitous slopes and talus slopes and boulder fields are developed at Mt Brockman in the north-east of the region (Fig. 6). Stable continuous cliffs abut low angle sandsheets, as in the case of the imposing 200 m high sandstone ramparts between Namargon, Deaf Adder and Jim Jim Creeks. This section of the escarpment was last active during the latter half of the Cainozoic (the last 25 million years) with retreat of the order of 25 km (Galloway 1976). Valleys and basins within the Arnhem Land Plateau, such as the Tin Camp Creek basin, were enlarged at this time.

Scarps have indented and irregular margins, where mass movement (rockfall) processes have been replaced by down-wearing by sand grain detachment and fluvial incision, yielding steps or terraces controlled by horizontal bedding planes. This scarp type is often associated with a sandstone pediment with a cover of sand which increases distally to a depth of 10 m (Roberts 1991). It is well developed in the north-east of the region where Magela Creek flows from the plateau onto the lowlands.

The plateau margin in the southern part of the region is discontinuous and indented by deep narrow valleys of the South Alligator River and its tributaries, Barramundie and Deaf Adder Creeks. The course of the main channel of the river here is determined by the north-west to south-east alignment of strike ridges and the regional fault systems (Fig. 2). The valleys are bounded by escarpments, strike ridges and hills. Sandy and silty deposits fill the valley and gorge bottoms forming narrow flood plains. Channels are commonly bounded by well developed sandy levees of Holocene age (East 1990).

The Arnhem Land plateau is deeply dissected by a dense network of intersecting linear gorges and ravines (Fig. 3). Extreme dissection has yielded locally a karstic landscape of pillars, blocks and caves (Fig. 7). Areal lowering of the plateau surface proceeds by weathering, solution and detachment of discrete sand grains. Liberated sand is transported by overland flow to a network of incised channels, some of which accumulate in broad depressions on the plateau surface, and in fans and flood plains of plateau streams. Following fluvial incision and erosion of the plateau deposits, sand is mobilised and transported by plateau streams onto the lowlands. A late Cainozoic denudation (surface lowering) rate of approximately 5 mm (1000 years)$^{-1}$ for the plateau surface has been inferred from the rate of infilling of an enclosed seasonal lake on the plateau top (Roberts 1991).

Caves in the escarpment and gorges vary in size from shallow overhangs to networks of passages

Fig. 7. Karstic sandstone landscape, Arnhem Land plateau (photo T.J. East).

which may exceed 100 m in length. The larger cave systems have formed initially through dissolution of siliceous sandstone along joints and bedding planes, with subsequent enlargement by spalling of the cave roof and walls. Elsewhere, erosion of a stratum of relatively more erodible Kombolgie Formation conglomerate has formed caves up to a hectare in area, the cave roof and floor forming in strata of resistant sandstone.

Erosion rates for sandstone plateau-dominated catchments are low for tropical catchments by world standards (Gibbs 1967). Annual erosion rates for the plateau-dominated upper Magela Creek catchment are between 3.6 and 4.9 t km^{-2} (Hart *et al.* 1986a), and are at the extreme low end of the range 5–15 t km^{-2} obtained for tropical northern Queensland by Douglas (1967), and the range 4–83 t km^{-2} for Amazon River basins reported in Gibbs (1967).

3.2. Lowlands

The lowlands (Fig. 8) are an extensive erosion plain, the Koolpinyah Surface, formed by the planation, weathering and duricrusting of folded geosynclinal sedimentary, metamorphic and intrusive rocks of Lower Proterozoic age during late Tertiary and early Pleistocene times (Williams 1969). Planation and dissection have yielded an almost flat landscape of broad valleys, long low-angle (1–4 %) slopes, and isolated hills and ridges of resistant rock. The lowlands are the most extensive of the landform types in the Kakadu region.

The main rivers in their lowland tracts have an upstream freshwater flow regime and a downstream tidal estuarine reach (Fig. 9). In their upper reaches they flow in braided sand channels in broad shallow valleys with long, low-gradient valley sides. Here, the rivers are flanked by extensive sandy and silty flood plains. Channels are both shallow and deep. Flow in the freshwater reaches is strongly seasonal in response to the markedly seasonal climate, and in the dry season flow ceases throughout most of the lowlands. Declining flow isolates permanent water bodies, locally called billabongs, in deeper sections of channels and on the flood plains.

Below the sand channel tract, the main rivers and main tributaries are commonly incompetent, and in the wet season abandon a defined channel and flood into extensive shallow flood basins. The freshwater flood basins merge downstream with broad flood plains which flank the tidal river channels and extend to the coast. The tidal channels are separated from the upstream flood basins by broad, low levees of variable width. The tidal rivers have a single large meandering channel (Fig. 9), which in the case of the South Alligator River, widens to about 10 km at its mouth.

The soils of the lowlands have developed on Lower Proterozoic metasediments and on colluvial quartzose sand derived by breakdown of Kombolgie Formation sandstone. The most widespread soils are red and yellow earths (sandy loams, silty loams) which have gradational profiles and high (> 30%) coarse fractions of quartz and ferricrete

Fig. 8. Lowlands (Koolpinyah Surface), Kakadu region; Ranger Uranium Mine is in foreground (photo T.J. East).

Fig. 9. Tidal meanders of the East Alligator River (photo T.J. East).

nodules and gravel (Hooper 1969, Aldrick 1976). Their properties and distribution have been strongly influenced by lithology and the history of weathering and erosion. As a consequence, the soils are acid, highly leached and deficient (< 1%) in organic matter. They have low soluble salt contents and low chemical activities, which are associated with low (< 20%) clay contents and a dominance of kaolinite in the clay fraction (Chartres *et al.* 1991). These soils are typically highly permeable and well drained.

The marked seasonality of rainfall and fluctuating water tables promoted lateritisation of the soils of the Koolpinyah Surface during the late Tertiary and intermittently through the Pleistocene. Lateritisation involves the removal of bases and the mobilisation and accumulation of iron and manganese sesquioxides producing red-yellow mottled profiles and ironstone nodules and concretions (Chartres *et al.* 1991). Denudation of the lowland surface has resulted in the development of a lag of quartz gravel and ferricrete nodules at the soil surface.

The Koolpinyah Surface was preceded by a series of erosion surfaces, the earliest having its origins in the Middle Proterozoic. During the vast span of time between the Middle Proterozoic when the sands of the Kombolgie Formation were deposited, and the Mesozoic when the Mullaman Beds (Petrel Formation) were laid down, erosion removed thousands of metres of rock to produce the northerly-sloping Pre-Cretaceous Surface (Fig. 10) (Galloway 1976). The Petrel Formation was deposited on the plateau part of this early surface, and the Bathurst Island Formation on the lowlands to the north and west (Needham 1988), forming the Post-Cretaceous Surface. These sediments were subsequently deeply weathered and

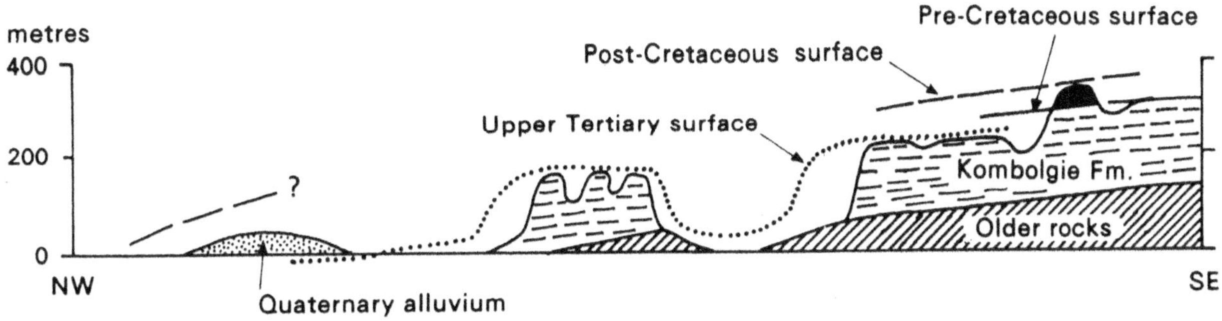

Fig. 10. Section through Kakadu landsurface (adapted from Galloway 1976).

lateritised. Sustained erosion during the second half of the Tertiary largely removed the Mesozoic strata and the Post-Cretaceous Surface to form a younger weathered surface, the Upper Tertiary Surface, which is still extensively preserved and is equivalent to the Koolpinyah Surface. Highs and horsts in the Lower Proterozoic basement rocks promoted localised removal of the Kombolgie Formation, forming basins such as those of Tin Camp Creek and upper Jim Jim Creek (Galloway 1976). Erosion was accompanied by weathering and lateritisation, which was particularly pronounced in the north of the region. The retreat of the escarpment in the southern part of the region during the late Tertiary resulted in the more resistant siliceous formations surviving as hills above the lowlands, as in the case of Mt Partridge, Mt Cahill and Mt Basedow.

Many of the lowland ridges and hills of resistant Lower Proterozoic rock have elevations higher than the basal beds of the younger Kombolgie Formation which forms the plateau. This suggests that these hills may have existed as islands (possibly submerged) in the Middle Proterozoic sea or lake in which the sands of the Kombolgie Formation were deposited. They were later exhumed by scarp retreat of the Kombolgie Formation. The Koolpinyah Surface, therefore, may approximate the ancient surface of unconformity between the Lower Proterozoic metamorphics and the overlying Kombolgie Formation.

The Koolpinyah Surface has a deeply weathered ferruginous surface zone which extends up to 60 m into the bedrock surface. There is less evidence of deep weathering and a sparseness of ferricrete on the lowlands in the south of the region. In the north of the region, near Mt Brockman, the intensity of deep weathering increases with distance from the escarpment because the more distant sites were the first to be exposed by scarp retreat (Galloway 1976).

The present lowlands have a surficial cover of Cainozoic soil and sands. Soils on the residual hills and strike ridges in the south-east of the region are generally clayey or loamy textured and skeletal (lithosols). They typically have high gravel contents and a surface scatter of boulders. Colluvial soils with smaller coarse fractions and fewer boulders have formed at the base of slopes. Red gravelly clay soils occur on rounded dolerite hills in the north and south of the region. Soil catenas are common and well developed; coarse gravelly soils characterise the upper slopes, sandy leached soils the mid-slopes, and texture contrast soils and earths the lower slopes and alluvial plains (Aldrick 1976, Cull & East 1987, Chartres *et al.* 1991).

Erosion has reduced the Cainozoic duricrusted layer to remnant ferricrete pavements, benches and gravel. Long, low-angle sand fans extend from the base of the plateau escarpment or from scree slopes at the escarpment base, coalescing to form a continuous sand plain. The present sand fans began to accumulate at the base of the escarpment at 230–220 and 120–100 thousand years ago; coinciding with the start of the penultimate and last interglacials respectively (Roberts 1991). Since then, the fans have accumulated at a fairly constant rate of 30–70 mm (1000 years)$^{-1}$. Surface wash processes dominate the lowland slopes, selective removal of the fine soil particles producing a residual surface lag of rounded quartz and ferricrete gravel.

Undisturbed lowland slopes have a high erosional stability because of low gradients, the high permeability of the sandy soils, the vegetation cover and the surface lag of gravel which protects the soil against rainsplash erosion (East 1990). The relative weighting of these factors is largely unknown, with the exception of vegetation cover which, because of the prominence of the other factors, may not be as important in ameliorating erosion as reported elsewhere (Duggan 1988). Mean denudation (surface lowering) rates for the East Alligator and South Alligator River catchments with minor areas of sandstone plateau are between 0.01 and 0.02 mm y^{-1} (Table I). These rates are lower, but broadly consistent with a denudation rate of 0.06 $\pm$0.01 mm y^{-1} obtained for the South Alligator River catchment using flood plain stratigraphic data (Chappell 1985).

Annual denudation (and erosion) rates are variable because of the high variability of annual rainfall and stream flow. Catchments with the highest proportions of resistant Kombolgie Formation sandstone outcrop tend to have the lowest denudation rates respectively (Table I). Annual solute yields, corrected for rainfall inputs, account for 12–35% of the total catchment load (Duggan 1988). The bedload component of the denudation in undisturbed catchments is 0.002

Table I. Suspended sediment concentrations and erosion rates for East Alligator and South Alligator River catchments. Suspended sediment concentrations are corrected for bedload and the number of years of records is given in [ ].

Catchment	Lowlands/ Uplands (%)	Area (km^2)	Suspended solids ($mg\ L^{-1}$)	Erosion ($t\ km^{-1}\ y^{-1}$)	Denudation ($mm\ y^{-1}$)
East Alligator River (Duggan 1988)					
7J Creek	60/40	53.5	22 [2]	15	0.006
Georgetown Creek	89/11	7.8	47 [1]	46	0.018
Georgetown Creek	82/18	4.8	83 [2]	9	0.004
Gulungul Creek	72/28	61.9	79 [2]	41	0.016
South Alligator River (Duggan 1988)					
Koongarra Creek	76/24	15.4	49 [2]	38	0.015
East Alligator River (Hart *et al.* 1986b)					
Magela Creek	c. 20/80	600	13 [1]	3.6	
Magela Creek above GS821009				4.96	
South Alligator River (Dames & Moore 1988)					
Koolpin Creek		477		13.5	
Fisher Creek		324		149.5	
South Alligator River above Fisher Creek		384		25.2	

mm y^{-1}, which is a minor component of the total surface lowering (Duggan 1988).

Regional denudation rates are low by comparison with rates for the seasonally wet tropics elsewhere in the world. Denudation rates for the seasonally wet savanna range from 0.1–0.5 mm y^{-1} (Saunders & Young 1983), which are about an order of magnitude greater than for the East Alligator River catchments. The seasonally wet savanna generally is characterised by high erosion rates because of a combination of high intensity rainfall and depleted vegetation cover at the end of the dry season. The lower denudation rates measured for the Kakadu region can be attributed to the high permeability of the sandy soils and the soil surface lag of gravel (East 1990).

In contrast to the high stability of undisturbed natural landforms, disturbed or engineered slopes have the potential for high rates of erosion (Silburn *et al.* 1990) with some reaching 1.25 mm y^{-1} (Duggan 1988). Erosion rates are elevated considerably above background rates in catchments in which cattle and feral animals (buffalo, pigs) have degraded soils and vegetation (Graetz 1989, East 1990). Some sub-catchments of the South Alligator River catchment, such as that of Fisher Creek, have greatly increased erosion rates (Table I) because of a combination of overgrazing by cattle and feral buffalo, and the presence of erodible duplex soils (East 1990). Measured erosion rates for the Fisher Creek catchment are two orders of magnitude greater than the adjoining but relatively undisturbed Koolpin Creek catchment (Table I).

3.3. Flood plains

The flood plains in the freshwater reaches of the regional rivers are of two distinct types:

i) those of mainly Pleistocene sandy and clayey-sand sediments which flank braided sand channels in the upstream reaches of rivers; and
ii) seasonally inundated flood basins of Holocene organic clay sediments which occur in the lower reaches of rivers and which merge with the upstream flood plains and tidal flats of the estuaries.

3.3.1. Pleistocene flood plains

Flood plains of clayey, silty and sandy sediments of predominantly Pleistocene age flank braided sand channel in the upstream reaches of the regional rivers and their major tributaries (Nanson *et al.* 1990, 1993, Wasson 1992).

The floodplain soils include silty brown earths, yellow earths, gradational red earths, texture contrast (duplex) soils and cracking clays (Aldrick 1976, Hooper 1969, Wells 1979). The duplex soils have hard setting sandy and clayey-sand A horizons and clayey mottled subsoils; the subsoil is either a massive alkaline calcic clay or an alkaline columnar-structured dispersive clay B horizon. They are prone to piping and gully erosion. In the valley reaches of rivers, as in the case of the South Alligator River between Coronation hill and Gunlom (Fig. 1), flood plains are generally narrow with deeply incised channels. In the downstream lowland tract, they fill shallow valleys with long low-gradient valley sides. The lowland flood plains are covered intermittently by wet season overbank floods and may be waterlogged for much of the wet season. They cover extensive areas; the flood plain at the confluence of the South Alligator River, Buffalo Creek and Coirwong Creek, for example, has an area of about 80 km^2 (Fig. 1). Flood plains include both treeless grassy plains and thickly wooded alluvial flats. Permanent billabongs occur in lower-lying areas of flood plain.

The lowland flood plains vary from one to more than 10 km in width, eventually merging downstream, as in the cases of the South Alligator River and Magela Creek, the major west bank tributary of the East Alligator River. The presence of well developed texture contrast profiles in the flood plain soils suggests that they are of early Holocene or Pleistocene age. Buried palaeosols from a flood plain in the upper reaches of Magela Creek have similar texture contrast profiles, and have been dated (by thermoluminescence (TL)) as middle to early Holocene (T.J. East unpublished data). The flood plains presently have low rates of accretion; the more elevated parts are above present flood levels (Nanson *et al.* 1990, 1993).

The formation of the Quaternary sand channels and flood plains of Magela Creek is described in Nanson *et al.* (1990, 1993). The valley of Magela Creek between the Arnhem Land plateau escarpment and Mudginberri Billabong is shallowly incised into the lowland surface, and is infilled by a chronosequence of concentrically arranged and inwardly younger sedimentary units (Fig. 11). The oldest outermost units are elevated above the adjoining younger flood plain, and are the remnants of a former valley fill. Erosion has reduced the former terraces to poorly differentiated rounded ridges forming interfluves between tributary valleys. The terrace sediments are highly weathered (based on scanning electron microscopy of sand grain surface textures; Nanson *et al.* 1990) and have red and yellow mottled sandy profiles and abundant ferricrete nodules. Remnant platforms and low scarps of ferricrete occur with these deposits. Uranium-thorium dating of ferricrete nodules in the terrace sediments showed that they were deposited in late Tertiary to early Pleistocene time (Nanson *et al.* 1993).

The Tertiary-Pleistocene valley fill was extensively excavated during periods of low sea level in the mid-Pleistocene. Predominantly siliceous sands, derived from erosion of the Arnhem Land plateau sandstones, were deposited in the excavated trench to form the contemporary flood plain. Thermoluminescence (TL) dating of the siliceous sand fraction of these deposits showed that deposition commenced about 300 thousand years ago (Nanson *et al.* 1990, 1993). These sediments have generally lower fines contents (mean silt+clay 19%) than the older early Pleistocene/Tertiary deposits (32%), indicating that the Arnhem Land plateau was the primary sediment provenance. Ages in the range of 220–300 thousand years ago for the basal units of the present flood plain (Nanson *et al.* 1990, 1993) are broadly consistent with a vertical accretion chronology for the colluvial sand fans that flank the Arnhem Land escarpment and merge with the Magela Creek flood plain (Roberts 1991). These similar chronologies are evidence for the commencement of a regional depositional phase about this time. Over the ensuing 300 000 years, alluvium progressively infilled the excavated valley, with glaciations and low sea levels resulting in localised incision and infilling of palaeo-channels in the flood plain deposits (Fig. 11). Extensive basal gravel beds and the sandy texture of the valley fill suggest that aggradation occurred through the lateral migration and infilling of the ancestral Magela Creek channel.

The youngest palaeo-channel had incised to bedrock by the time of the last glacial maximum (about 18 000 years ago) when sea level was about 120 m lower than at present. At this time, the coastline of the Kakadu region was some 300 km

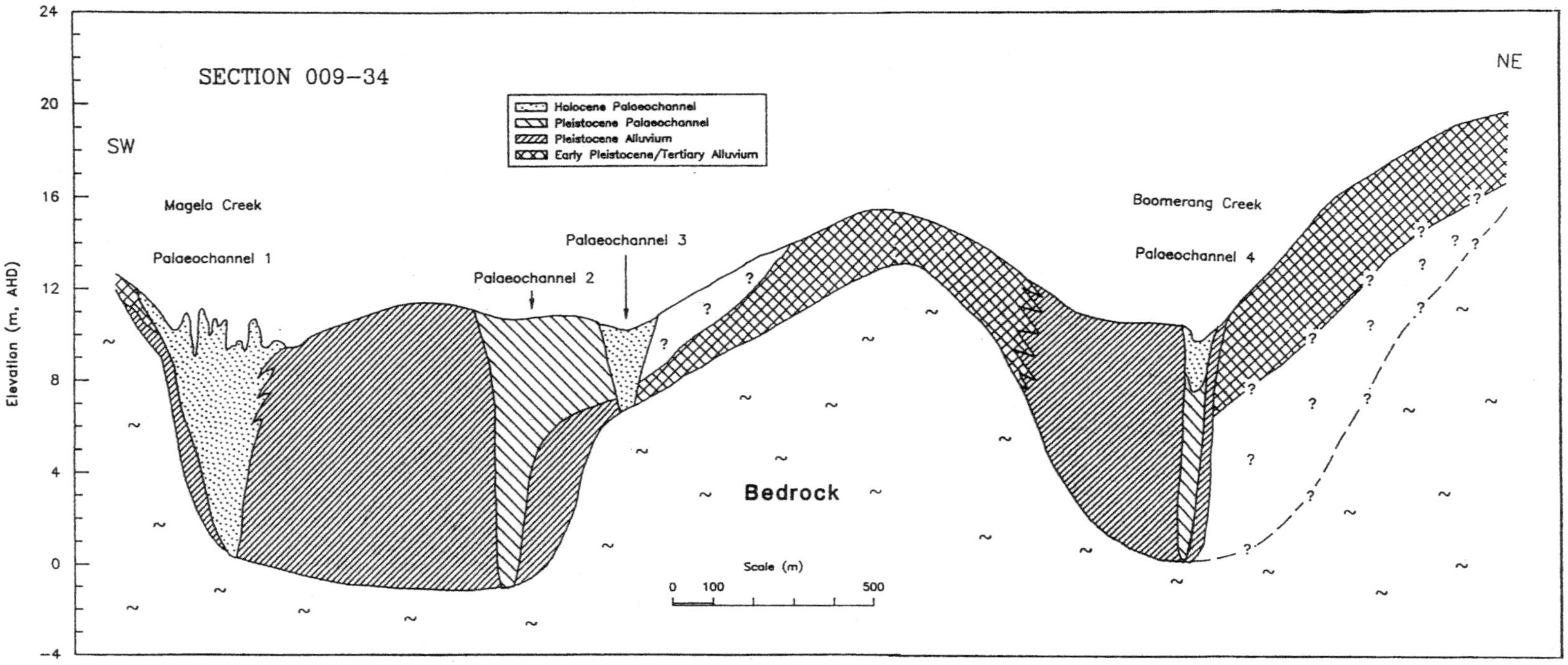

Fig. 11. Section through Magela Creek valley fill showing generalised chrono-stratigraphic units (adapted from Nanson *et al.* 1990).

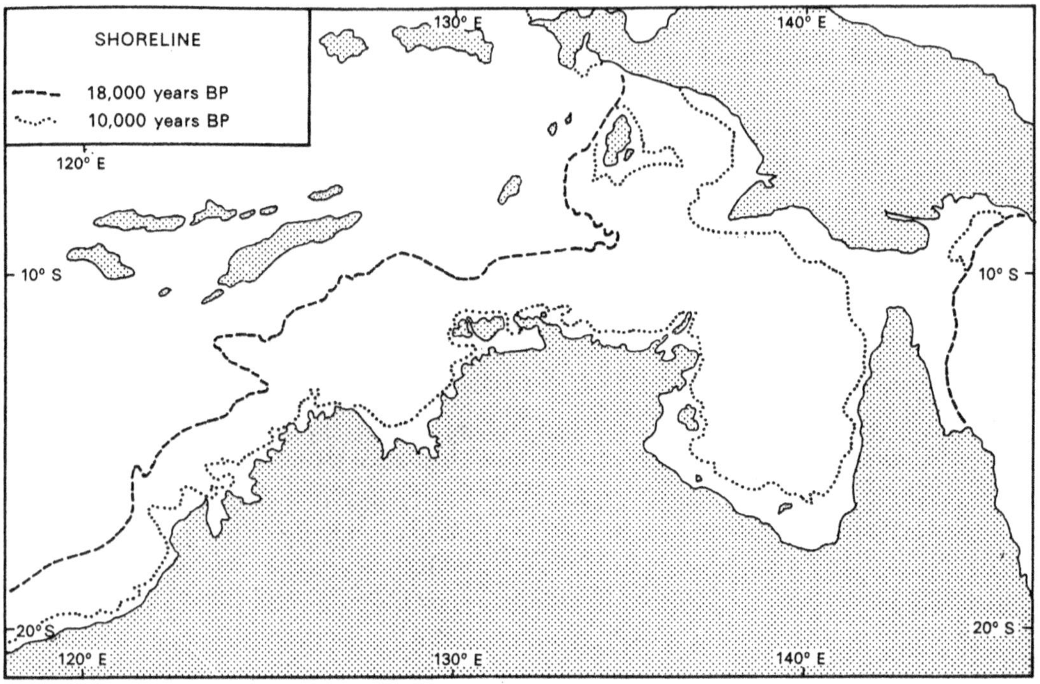

Fig. 12. Approximate location of the shoreline of northern Australia 18 000 and 10 000 years ago (adapted from Woodroffe *et al.* 1986).

to the north of the present coastline (Fig. 12). The palaeo-channel commenced infilling with clean siliceous fluvial sand (silt+clay 9%) at about 5–7000 years ago (radiocarbon dating of wood/charcoal; Roberts 1991), and accreted at a constant rate throughout the Holocene. The sand grains forming this deposit are polished and smooth with an absence of the textures such as etching, crusts, precipitations that characterise the older Pleistocene and Tertiary sands (Nanson *et al.* 1990).

The present Magela Creek flows in a braided sand channel on the Holocene sand fill. Its present bedload transport rate is 1800 $m^3 y^{-1}$ (Nanson *et al.* 1991). As the ancestral Magela Creek channel infilled, plugs of bedload sand formed alluvial-dammed lakes (known locally as backflow billabongs) at the mouths of tributaries. The beds of the present billabongs lie 1–2 m below the bed of the main channel of Magela Creek, indicating that Magela Creek is accreting vertically at a rate higher than that for its tributaries. Whereas Magela Creek downcut vertically in response to the last sea level fall, its tributaries tended to erode their flood plains laterally; in the case of Gulungul Creek, to a depth of about 1–2 m. The Magela Creek palaeo-channel is presently almost filled with sand. Excess sand is infilling perennial lakes, such as Mudginberri Billabong and the tributary backflow billabongs, and is spilling from the channel onto the adjacent flood plain. The present rate of infilling of Mudginberri Billabong has been estimated from aerial photography and bathometric surveys at between 1400–2100 $m^3 y^{-1}$ (Roberts 1991).

Downstream of the sand channel reach of Magela Creek, basal Pleistocene fluvial sands in the Magela Creek flood basin were deposited contemporaneously with the upstream Pleistocene sediments, and were later covered by Holocene marine and freshwater sediments (Wasson 1992).

The Quaternary alluvial record preserved in the Magela Creek flood plain is replicated to varying degrees in other tropical river valleys elsewhere in the region and across northern Australia (Nanson *et al.* 1993). The East Alligator River, for example, flows in a braided sand channel which is incised in older gorge bottom sediments of similar texture, age and degree of alteration to the Quaternary flood plain sediments of Magela Creek (Murray *et al.* 1992).

In the South Alligator River catchment, flood

plains in the sand channel reach have soils with well differentiated texture contrast profiles and are generally of Pleistocene age (East 1990, Wasson *et al.* 1991). The flood plains are formed in silty and fine-sandy alluvium indicating overbank depositional processes (Wasson *et al.* 1991). Deposits of Holocene-age form levees along channel margins and have either little or no pedological development (East 1990).

The sand channels are transit zones for fine suspended sediment because of the high wet season flow velocities. The suspended load is deposited in areas of low flow energy, including: i) the broad silty-fine sand flood plains of the lowlands; ii) the clay flood plains in the downstream tidal estuarine reaches; and iii) the clay flood basins.

3.3.2. Flood basins

The second type of flood plain are seasonally inundated flood basins. In the Kakadu region they occur in the lower reaches of the main rivers and their major tributaries. Major flood basins include the approximately 100 km^2 basin in the vicinity of Cooinda in the South Alligator River catchment (equivalent to the 'alluvial plain' of Woodroffe *et al.* 1986), and the 200 km^2 flood basin in the lower reaches of Magela Creek (the 'backwater plain' of Wasson 1992). The Holocene evolution and the contemporary geomorphological processes of the South Alligator River flood basin and estuarine flood plain have been described in Woodroffe *et al.* (1985a,b, 1986) and the formation of the Magela Creek flood basin in Wasson (1992).

Flood basins have markedly different geomorphological, hydrological, sedimentological and biological characteristics to the upstream, essentially Pleistocene flood plains. They are formed in organic clay sediments of mainly Holocene age, and merge upstream with the Pleistocene flood plains and downstream with the estuarine flood plains.

The flood basins have formed in response to climatic and sea level influences following the last post-glacial rise in sea level. Deposition of clayey sediments on the estuarine flood plain and the flood basins commenced following the flooding of the shallow basin of van Diemen Gulf and the stabilisation of sea levels at about 7000 years ago (Woodroffe *et al.* 1986, Wasson 1992). The channels of the tidal rivers and creeks changed position as the flood basins were formed, leaving abandoned channel segments which infilled with organic (mangrove) tidal muds.

In the case of the South Alligator River, the river changes from a defined sand channel to an approximately 100 km^2 flood basin (the 'alluvial plain' of Woodroffe *et al.* 1986) in the vicinity of Cooinda (Fig. 1). The Cooinda and Magela flood basins have similar geomorphological characteristics. Both basins are flooded during the wet season to a depth of about a metre, and are major deposition sites for suspended sediments (Wasson 1992, Wasson *et al.* 1991). Permanent lakes or billabongs are located in sections of prior river channel and include Yellow Water, Red Lily, Alligator, Leichhardt Billabongs in the South Alligator River basin, and Jabiluka, Nankeen and Island Billabongs in the Magela Creek basin (Fig. 1). Billabongs in the flood basins typically have levees which are largest at their downstream ends. They carry flow in the wet season at which time the levees may be submerged. In the dry season, with the exception of localised low lying areas, the clayey sediments in the basins dry out to form a hard surface fissured with numerous desiccation cracks.

The basins mark the limit of the freshwater reaches of the rivers of the region. Downstream of the basins, the rivers have a meandering tidal estuary form, and are flanked by deltaic-estuarine plains with a Holocene alluvial and marine history (Woodroffe *et al.* 1986).

Flood basin sediments are fine grained, typically clays with some silt and fine sand. They have variable but generally high organic and biogenic silica contents (Wasson 1992). Some solutes are deposited in flood basins as ponded flood waters infiltrate and evaporate, although high solute concentrations (approximately five times inflow concentrations) in outflow water from the Magela flood basin (Hart *et al.* 1986b) indicate that a significant proportion of the total solute load is transported from the basin. In contrast, almost the total suspended sediment load is deposited in the flood basin; in the case of Magela Creek flood basin in excess of 90% is deposited in the upstream (inflow) part of the basin (Johnston *et al.* 1987).

3.4. *Deltaic-estuarine plain*

The South Alligator, West Alligator and East Alligator Rivers have deltaic-estuarine plains. These are extensive flood plains which extend inland from the coastal plain to the tidal limit (Fig. 13). They are characterised by both estuarine and fluvial processes. The deltaic-estuarine plain of the South Alligator River has formed in the South Alligator Trough, a pre-Tertiary secondary tectonic basin associated with the Lower Proterozoic Pine Creek Geosyncline (Woodroffe *et al.* 1986). The estuarine flood plains of the East Alligator and West Alligator Rivers are morphologically similar to the South Alligator River deltaic-estuarine plain and may have similar origins.

Channel morphology changes markedly within the estuarine reach. In the South Alligator River, three distinct channel types have been recognised by Woodroffe *et al.* (1986), on the basis of meander morphology and the relation between channel width and distance from mouth; they are an estuarine funnel, a meandering segment of sinuous and cuspate elements, and an upstream segment.

The estuarine funnel of the South Alligator River has a broad mouth which narrows negative-exponentially upstream from six to one kilometres. There are a few angular bends and mangroves line the banks. The meandering reach is highly sinuous with steep undercut banks and opposing muddy point bars; fringing mangroves are generally absent. The cuspate meandering channel is much less sinuous and is characterised by pointed inside meander bends. Shoals commonly form in the middle of the channel and adjacent to banks. In the upstream tidal channel, individual meanders are separated by long straights. These segments are present in most of the large macrotidal rivers in northern Australia, including the other rivers of the Kakadu region, although the combination of channel types varies considerably.

The deltaic-estuarine plain has an approximately one metre thick surface layer of black organic freshwater cracking clays which overlies a freshwater-marine transitional zone and blue-grey, estuarine saline mangrove muds at depth. The surface of the plain is elevated above the level of the highest spring tides. It has numerous infilled palaeo-channels and palaeo-creeks. The plain slopes away from levees lining the river channel, the lowest lying land being located along the plain's margins. The East Alligator and West Alligator Rivers have estuarine flood plains which have broadly similar morphologies and processes. In an average wet season of about 1500 mm rainfall, rain water ponds on the plain which is inundated for 3–6 months or more depending on local variations in elevation. A number of morphologic units have been defined for deltaic-estuarine plains on the basis of elevation and duration of wet season flooding; they are the Upper and Lower Flood plain Groups, and the Backwater Swamps (Woodroffe *et al.* 1986).

The topographically lowest parts of the plain are the backwater swamps, which have standing water for six or more months of the year, generally drying out only along their margins (Woodroffe *et al.* 1986). Backwater swamps occur along the plain's margins, filling re-entrants in the lowlands. They are dominated by tall paperbark trees (*Melaleuca* species). Similar swamps occur along the margins of the flood plains of the other rivers in the region (Finlayson *et al.* 1988). Areas of the Lower Flood plain Group include infilled palaeo-channels and palaeo-creeks with surface depressions as well as depressions along the margins between the plain and backwater swamps. They have a seasonal herbaceous vegetation or are dominated by flood tolerant sedges (*Eleocharis* species). Areas of the Upper Flood plain Group are the most extensive on the estuarine plain, and include the higher land which is flooded for 3–4 months of the year. The surface sediments are typically black organic cracking clays with some gilgai. These areas are characterised by numerous infilled palaeo-channels and palaeo-creeks and a vegetation of grasses and sedges. The surfaces of infilled palaeo-channels are topographically continuous with the surface of the plain.

3.5. *Coastal plain*

The coastal plain at the mouth of the South Alligator River is a progradational landform, 4–5 kn wide, which has formed in sediments that become younger seawards. Fig. 13 shows that the plain has a linear seawards zonation of coastal plain (with upper and lower components), two

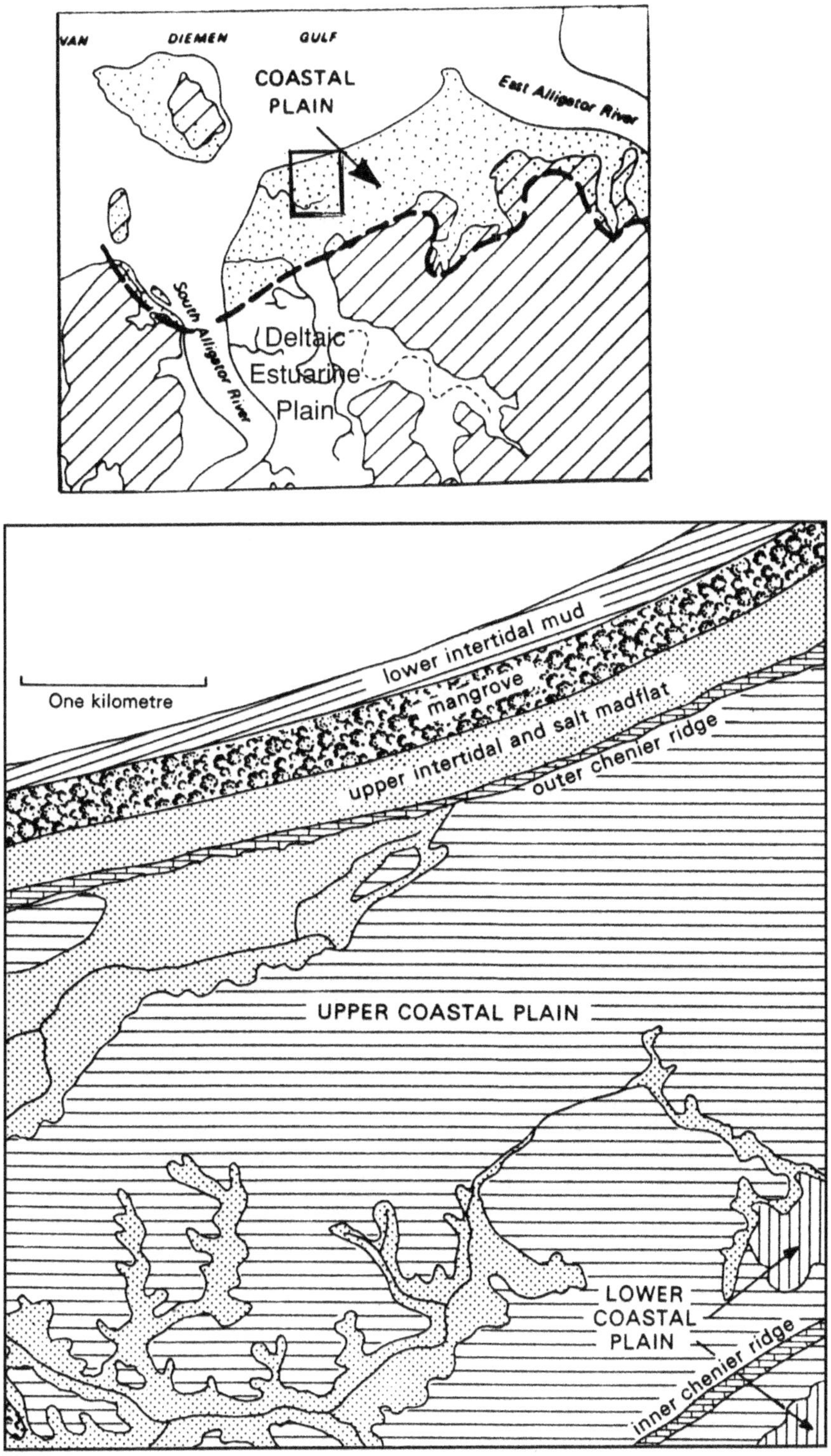

Fig. 13. Morphologic provinces of the estuarine South Alligator River flood plain (adapted from Woodroffe *et al.* 1986).

chenier ridges of shelly sand, an upper intertidal saline mudflat, mangroves, and a lower intertidal mudflat which extends into the shallow, 5–15 m deep south-eastern corner of van Diemen Gulf (Woodroffe *et al.* 1986).

The lower coastal plain occurs in the inland lower lying areas of Coastal Plain; it is most extensive at the mouth of the East Alligator River. The upper plain has an elevation of 2.5 m and is formed in black cracking freshwater clays which overlie sands and marine mangrove muds. The chenier ridges are discontinuous with an elevation of about one metre and widths of 25–60 m for the inland ridges. The seaward chenier was deposited about 1700 to 1500 years ago, and the inland chenier about 5000 years ago (Woodroffe *et al.* 1986). The upper intertidal mudflat is saline, bare or with localised samphire, and is partly covered by spring tides. The mangrove forest is strongly zoned with a seaward fringe of *Sonneratia* and *Camptostemon*, a middle zone of *Rhizophora* and a landward zone of *Avicennia*. Mudflats at the shoreline extend seawards of the mangrove. The position of the shoreline has changed little since about 2000 years ago. Similar coastal plains with multiple chenier ridges extend along the coast of van Diemen Gulf west of the Alligator Rivers region (Clarke *et al.* 1979).

4. References

Aldrick, J.M. 1976. Soils of the Alligator Rivers area. In: Story, R., Galloway, R.W., McAlpine, J.R., Aldrick, J.M. & Williams, M.A.J. (eds) Lands of the Alligator Rivers Region, Northern Territory, pp. 71–88. CSIRO Land Research Series No. 38, CSIRO, Melbourne.

ARRRI. 1988. Alligator Rivers Region Research Institute Annual Research Summary 1987/88. Australian Government Publishing Service, Canberra. 160 pp.

Chappell, J.M.A. 1985. Denudation and sedimentation in some Northern Territory river basins. In: Loughran, R.J. (compiler), Drainage Basin Erosion and Sedimentation, University of Newcastle Conference and Review Papers 2: 69–76.

Chartres, C.J., Walker, P.H., Willett, I.R., East, T.J., Cull, R.F., Talsma, T. & Bond, W.J. 1991. Soils and hydrology of Ranger Uranium Mine sites in relation to the application of retention pond water. Supervising Scientist for the Alligator Rivers Region Technical Memorandum 34. 61 pp.

Clarke, M.F., Wasson, R.J. & Williams, M.A.J. 1979. Point Stuart chenier and Holocene sea levels in Northern Australia. Search 10: 90–93.

Cull, R.F. & East, T.J. 1987. Geomorphic factors in the rehabilitation of earthen structures (with reference to uranium mining in the Northern Territory). Australian Mining Industry Council Environmental Workshop Papers, Adelaide. pp. 63–76.

Dames & Moore. 1989. Coronation Hill Gold, Platinum and Palladium Project, Draft Environmental Impact Statement, Prepared for Coronation Hill Joint Venture, Darwin. Unpaginated.

Douglas, I. 1967. Man, vegetation and the sediment yields of rivers. Nature 215: 925–928.

Duggan, K. 1988. Mining and erosion in the Alligator Rivers Region of Northern Australia. Ph D thesis, School of Earth Sciences, Macquarie University, Sydney. 317 pp.

East, T.J. 1990. Erosion and sediment processes in the Kakadu Conservation Zone, South Alligator River catchment, Northern Territory. Resource Assessment Commission Kakadu Conservation Zone Inquiry Consultantcy Series, Australian Government Publishing Service, Canberra. 97 pp.

Ferguson, J. & Goleby, A.B. (eds) 1980. Uranium in the Pine Creek Geosyncline. Proceedings Series, International Atomic Energy Agency, Vienna. 760 pp.

Finlayson, C.M., Bailey, B.J., Freeland, W.J. & Fleming, M.R. 1988. Wetlands of the Northern Territory. In: McComb, A.J. & Lake, P.S. (eds) The Conservation of Australian Wetlands. pp. 103–126. Surrey Beatty & Sons Pty Limited, Norton, Australia.

Galloway, R.W. 1976. Geomorphology of the Alligator Rivers area. In: Story, R., Galloway, R.W., McAlpine, J.R., Aldrick, J.M. & Williams, M.A.J. (eds) Lands of the Alligator Rivers Area, Northern Territory. pp. 52–70. CSIRO Land Research Series 38, CSIRO, Melbourne.

Gibbs, R.J. 1967. The geochemistry of the Amazon River system I; the factors that control the salinity and the composition and concentrations of the suspended solids. Geol. Soc. Amer. Bull. 78: 1203–1232.

Graetz, R.D. 1989. Analysis of land degradation in Stage III of Kakadu National park, including the Conservation Zone: Phase I. Unpublished report to the Australian National Parks and Wildlife Service, Canberra. 126 pp.

Hart, B.T., Ottaway, E.M. & Noller, B.N. 1986a. Nutrient and trace metal fluxes in the Magela Creek system, northern Australia. Ecol. Model. 31: 249–265.

Hart, B.T., Ottaway, E.M., Beckett, R. & Noller, B.N. 1986b. Materials transport in the Magela Creek system: 1982/83 wet season. Water Studies Centre Technical Report No. 27, Chisholm Institute of Technology, Melbourne. 14 pp.

Haynes, C.D., Ridpath, M.G. & Williams, M.A.J. 1991. Monsoonal Australia: Landscape, ecology and man in the northern lowlands. A.A. Balkema, Rotterdam. 231 pp.

Hooper, A.D.L. 1969. Soils of the Adelaide-Alligator Area. In: Story, R., Williams, M.A.J., Hooper, A.D.L., O'Ferrall, R.E. & McAlpine, J.R. (eds) Lands of the Adelaide-Alligator Area, Northern Territory. pp. 95–113. CSIRO Land Research Series 25, CSIRO, Melbourne.

Johnston, A., Murray, A.S., Marten, R., Martin, P. & Hancock, G. 1987. Radionuclide distributions in sediments and macrophytes. Alligator Rivers Region Research Institute Annual Research Summary 1986–87, pp. 62–64. Australian Government Publishing Service, Canberra.

Murray, A.S., Wohl, E. & East, T.J. 1992. Thermoluminescence and excess ^{226}Ra decay dating of late Quaternary fluvial sands, East Alligator River, Australia. Quatern. Res. 37: 29–41.

Nanson, G.C., East, T.J., Roberts, R.G., Clark, R.L. & Murray, A.S. 1990. Quaternary evolution and landform stability of Magela Creek catchment, near Ranger Mine, Northern Australia. Supervising Scientist for the Alligator Rivers Region Open File Record 63. 119 pp.

Nanson, G.C., East, T.J. & Roberts, R.G. 1993. Quaternary stratigraphy, geochronology and evolution of the Magela Creek catchment in the monsoon tropics of northern Australia. Sediment. Geol. 83: 277–302.

Nanson, G.C., Roberts, R.G. & East, T.J. 1991. Bedload sediment (sand) transport in Magela Creek near Ranger Uranium Mine. Supervising Scientist for the Alligator Rivers Region Open File Record 64. 15 pp.

Needham, R.S. 1988. Geology of the Alligator Rivers Uranium Field, Northern Territory. Bureau of Mineral Resources Bulletin 224, Australian Government Publishing Service, Canberra. 96 pp.

Roberts, R.G. 1991. Sediment budgets and Quaternary history of the Magela Creek catchment, tropical northern Australia. Ph D thesis, University of Wollongong, Wollongong. 569 pp.

Saunders, I. & Young, A. 1983. Rates of surface processes on slopes, slope retreat and denudation. Earth Surf. Proces. Land. 8: 473–501.

Silburn, D.M. Loch, R.J., Connolly, R.D. & Smith, G.D. 1990. Erosional stability of waste rock dumps at the proposed Coronation Hill Mine. Resource Assessment Commission Kakadu Conservation Zone Inquiry Consultancy Series. Australian Government Publishing Service, Canberra. 52 pp.

Supervising Scientist for the Alligator Rivers Region 1994. Annual Report 1993–94. Australian Government Publishing Service, Canberra. 98 pp.

Wasson, R.J. (ed.) 1992. Modern sedimentation and late Quaternary evolution of the Magela Creek Plain. Supervising Scientist for the Alligator Rivers Region Research Report 6, 322 pp.

Wasson, R.J., Caitcheon, G.G., East, T.J. & Murray, A.S. 1991. Prediction of sediment deposition sites in the South Alligator River Valley Downstream of the Proposed Coronation Hill Mine. Resource Assessment Commission Kakadu Conservation Zone Inquiry Consultantcy Series, Australian Government Publishing Service, Canberra. 60 pp.

Wells, M.R. 1979. Soil studies in the Magela Creek catchment 1978, Part I. Land Conservation Unit, Territory Parks and Wildlife Commission, Darwin. 101 pp.

Williams, M.A.J. 1969. Geomorphology of the Adelaide-Alligator area. In: Story, R., Williams, M.A.J., Hooper, A.D.L., O'Ferral, R.E. & McAlpine, J.R. (eds) Lands of the Adelaide-Alligator Area, Northern Territory. pp. 71–94. CSIRO Land Research Series No. 25. CSIRO, Melbourne.

Williams, M.A.J. 1976. Erosion in the Alligator Rivers area. In: Story, R., Galloway, R.W., McAlpine, J.R., Aldrick, J.M. & Williams, M.A.J. (eds) Lands of the Alligator Rivers Area, Northern Territory. pp. 112–125. CSIRO Land Research Series 38. CSIRO, Melbourne.

Woodroffe, C.D., Chappell, J.M.A., Thom, B.G. & Wallensky, E. 1986. Geomorphological Dynamics and Evolution of the South Alligator Tidal River and Plains, Northern Territory. Australian National University North Australian Research Unit, Darwin. 190 pp.

Woodroffe, C.D, Chappell, J.M.A., Thom, B.G. & Wallensky, E. 1985a. Geomorphology of the South Alligator tidal river and plains, Northern Territory. In: Bardsley, K.N., Davie, J.D.S. & Woodroffe, C.D. (eds) Coasts and Tidal Wetlands of the Australian Monsoon Region. pp. 3–15. Australian National University North Australian Research Unit, Mangrove Monograph No. 1, Darwin.

Woodroffe, C.D, Chappell, J.M.A., Thom, B.G. & Wallensky, E. 1985b. Stratigraphy of the South Alligator tidal river and plains, Northern Territory. In: Bardsley, K.N., Davie, J.D.S. & Woodroffe, C.D. (eds) Coasts and Tidal Wetlands of the Australian Monsoon Region. pp. 17–30. Australian National University North Australian Research Unit, Mangrove Monograph No. 1, Darwin.

CHAPTER 4

Terrestrial vegetation

BRUCE A. WILSON, JEREMY RUSSELL-SMITH and RICHARD WILLIAMS

Abstract. The Kakadu region is considered one of the most foristically diverse areas of monsoonal northern Australia. There are 1847 vascular species from over 500 genera and 180 families recorded from the region. Genera with pantropic origins are an important component of all plant communities. The extensive lateritic peneplains support eucalypt open-forest, woodlands and open-woodlands, with tall, mainly Andropogonoid, grass dominated understoreys; often referred to as 'savanna' vegetation. Closed canopy monsoon rain forests, heaths and open-woodlands with sparse Andropogonoid-Eragrostoid grass understoreys are also widespread. They occur on a variety of landforms including coastal dunes, drainage lines, river levees, rocky hills of varying geology, and sandstone escarpments and plateaux.

The annual drought is regarded as one of the most important climatic factors affecting the terrestrial plants of the region. Variability of the onset and duration of the annual wet season plays an important role, particularly with regards to plant regeneration strategies. Winds associated with cyclones and storms can have a dramatic, localised impact on vegetation adjacent to the coast. Vegetation distribution is best correlated with edaphic conditions associated with topographic position.

Life-forms and functioning of terrestrial plant species are described in terms of phenological responses to annual drought. The wet season is the major period for growth and flowering of the herbaceous species and small shrubs of the flood plains and the understorey of the open savanna forests and woodlands. However, there is considerable variation between different woody species with respect to monthly and yearly patterns in deciduousness, flowering and shoot growth. The vast majority of perennial species have the capacity to regenerate vegetatively following disturbance. In contrast to the copious vegetative sprouts, the incidence of seedlings, especially those of the eucalypts, is low. Competition and edaphic conditions probably interact with other factors, such as fire and feral animals to determine population dynamics, but the long-term consequences of these complex interactions are not known.

1. Introduction

The Kakadu region is considered one of the most floristically diverse areas of monsoonal northern Australia (Lazarides *et al.* 1988). While some of this diversity is a reflection of the intense survey effort the region has received (Dickinson & Dunlop 1988), it is mainly due to the diversity of terrestrial landforms and associated plant habitats which occur. Coastal dunes, extensive lateritic peneplains, rocky hills of varying geology, and sandstone escarpments and plateaux provide habitats for most of the broad vegetation types found across monsoonal northern Australia. Hence, the region (c 20 000 km^2) contains 80% of all plant communities identified within monsoonal Northern Territory (c 400 000 km^2) by Wilson *et al.* (1990). At finer scales, microhabitat diversity is associated with a variety of localised vegetation patterning. This is particularly evident within the rocky hills and sandstone escarpments which form the western edge of the Arnhem Land plateau.

The vegetation of the area is the most intensively surveyed in northern Australia. Although there is no written flora for the region, Cowie & Finlayson (1986), Lazarides *et al.* (1988) and Brennan (1992) have compiled comprehensive plant species checklists. General vegetation descriptions, in conjunction with various landscape feature mapping, have been provided by Christian & Stewart (1953), Story (1969, 1973, 1976), Schodde *et al.* (1987) and Wilson *et al.* (1990). Numerous surveys at various scales of intensity have been carried out to describe the vegetation of more localised areas or covering particular vegetation types (Specht 1958, Webb & Tracey 1979, Bell 1981, Burgman & Thompson 1982, Taylor & Dunlop 1985, Rice & Westoby 1985, Cowie *et al.* 1987, Russell-Smith 1984, 1986, 1991, Wilson *et al.* 1989, Bowman *et al.* 1988b, Bowman *et al.* 1990, Orr *et al.* 1990, Woinarski *et al.* 1989, Wilson *et al.* 1991, Duff *et al.* 1991, Russell-Smith *et al.* 1993).

C.M. Finlayson and I. von Oertzen (eds), Landscape and Vegetation Ecology of the Kakadu Region, Northern Australia, 57-79.

Fig. 1. Tropical savanna vegetation that is typical of about 55 % of the Kakadu region (photo B. McKaige).

Despite this historic effort, botanical survey and taxonomic revision are constantly adding to the floristic knowledge of the area. For example, Duff *et al.* (1991) found specimens of four species and one genus which were new to science, and many more communities and taxa new to the region during recent investigation of the southern part of the Kakadu region.

The terrestrial vegetation of the region is predominantly eucalypt open-forest, woodlands and open-woodlands (*sensu* Specht 1981a), with tall (1–2 m), mainly Andropogonoid grass dominated understoreys (see references above). These communities cover about 55% of the region, mainly on the lowland plains and lateritic plateaux. Known as 'tropical tallgrass savannas' (Fig. 1) within Australia, (Mott *et al.* 1985) these formations are often referred to as 'savannas' in other wet-dry tropical parts of the world (Walker & Gillison 1982, Johnston & Tothill 1985). Heaths and open-woodlands (Fig. 2) with sparse Andropogonoid-Eragrostoid grass understoreys, are also widespread, covering, in total, some 30% of the region. Closed canopy monsoon rain forests are confined to less extensive habitats, such as lowland springs, rock outcrops, flood plain or beach levees and sandstone escarpments (Russell-Smith 1991).

There has been little investigation into the ecophysiological and dynamic processes of vegetation. Most hypotheses on the functioning of vegetation are based on the extrapolation from observations on the variation in vegetation composition and morphology, and of relevant studies from elsewhere in monsoonal northern Australia.

The following discussion includes a brief overview of the flora and its global affinities, the environmental correlates, structure and functioning of the vegetation and a description of the structure and composition of the major vegetation types. Plant communities of the freshwater and saline wetlands and the interactions of fire and animals with terrestrial vegetation are largely excluded, as they are discussed elsewhere in this volume.

2. Taxonomic composition

Lazarides *et al.* (1988) recorded 1531 macrophytic species from over 500 genera and 180 families from the region. A more recent plant checklist by Brennan (1992) lists over 1800 species in the area. Statistics of the gross composition of the flora of the Kakadu Region are presented in Table I. Bryophyta (liver-worts and mosses – 42 species) and Pteridophyta (ferns – 52 species) make small contributions. Spermatophyta (seed plants) are represented mainly by Angiosperms (1791 species) and only a few Gymnosperms (4 species).

Table II lists the 20 largest families and genera ranked by number of species in the Kakadu

Fig. 2. Monsoonal rain forest vegetation is generally confined to less extensive habitats in the region (photo C.M. Finlayson).

region. Generally, the entries on this table are similar to the largest families and genera for the Northern Territory as a whole, listed by Bowman *et al.* (1988b). Many of the larger genera and families characterise the open-forests, woodlands and shrublands of the region. As for other tropical parts of the Northern Territory, there are a high proportion of families and genera re-

Table I. Taxonomic composition of the flora of the Kakadu region (source: Lazarides *et al.* 1988 and Brennan 1992). Numbers in brackets are total numbers in that category in the Northern Territory (source: Bowman *et al.* 1988b).

	Families	Genera	Species
Bryophyta	18	24	42
Pteridophyta	20	27	52
	(27)	(63)	(97)
Spermatophyta			
Gymnospermae	3	3	3
	(3)	(3)	(3)
Angiospermae			
Dicotyledons	119	498	1322
	(159)	(732)	(2426)
Monocotyledons	33	153	469
	(44)	(248)	(882)
Total	193	675	1889

Table II. Twenty largest plant families and genera ranked by number of species in the Kakadu region (source: Brennan 1992).

Family	No. of genera	No. of species	Genus	No. of species
Poaceae	72	204	*Acacia*	46
Fabaceae	42	149	*Fimbristylis*	54
Cyperaceae	17	141	*Cyperus*	39
Myrtaceae	16	95	*Eucalyptus*	37
Euphorbiaceae	27	77	*Tephrosia*	25
Mimosaceae	4	60	*Hibbertia*	22
Rubiaceae	19	49	*Utricularia*	22
Scrophulariaceae	13	38	*Mitrasacme*	21
Asteraceae	27	38	*Eriachne*	19
Convolvulaceae	12	36	*Stylidium*	19
Verbenaceae	10	33	*Ipomoea*	17
Malvaceae	9	28	*Goodenia*	17
Dilleniaceae	2	29	*Grevillea*	17
Proteaceae	6	23	*Desmodium*	15
Asclepiadaceae	14	23	*Eriocaulon*	14
Sterculiaceae	9	23	*Pityrodia*	14
Loganiaceae	3	23	*Phyllanthus*	13
Goodeniaceae	4	22	*Sauropus*	13
Lentibulariaceae	1	22	*Hibiscus*	13
Stylidaceae	1	19	*Lindernia*	11

presented by a low (< 3) number of genera and species in the region (Bowman *et al.* 1988b). Many of these smaller taxa are associated with monsoon rain forest communities (Taylor & Dunlop 1985).

3. Global affinities

Table III shows the mean percent frequency of endemic, Australasian (Australia, New Zealand and near Pacific Islands) and pantropic elements in major community groups in the Kakadu region. Genera with pantropic origins are an important component (58–94%) of all plant communities in the region. The proportion of genera which occur in Australasia is smaller than that for vegetation from southern Australia (Taylor & Dunlop 1985), and for the tropical Northern Territory generally (*cf* Bowman *et al.* 1988b). The region is particularly rich in endemic species compared with other parts of northern Australia (Adams *et al.* 1973). However, there are few endemic genera relative to southern Australia (Taylor & Dunlop 1985).

There are significant between-community differences in global affinities of the genera (Taylor & Dunlop 1985). The sandstone open-woodland and shrublands (Table III, Group 1) have the smallest proportion of pantropical genera and the largest proportion of Australasian genera in the region. Of particular note is the large number of Australasian 'heath' (*sensu* Specht 1979) genera (e.g. *Calytrix*, *Grevillea*, *Hibbertia*, *Leptocarpus*, *Cyperus*) that are found in these communities (Specht 1958, 1981b, Dunlop 1988). Several endemic heath genera from the Fabaceae family, such as *Hovea*, *Templetonia*, *Bossiaea*, *Leptosema*, *Daviesia* and *Burtonia*, are restricted within the Kakadu region to sandstone habitats or sandy soils derived from them (Dunlop 1988). A strong relationship exists, even at species level, between the sandstone flora of the Kakadu region, monsoonal India and, to a lesser extent, Africa. Thus, these heath species may be derived from taxa that existed on the southern landmass Gondwana before its breakup, which probably began in the early Cretaceous (e.g. Veevers 1984).

The majority of rare species found in Kakadu National Park are associated with communities on the hills and sandstone plateaux (Leach 1987, Briggs & Leigh 1988, Leach 1988). Furthermore, a majority of the few genera endemic to the Northern Territory, such as *Allosyncarpia*, *Neobyrnesia* and *Symplectrodia*, are restricted to sandstone environments in the region (Bowman *et al.* 1988b). This endemism can be partly explained by the restricted and disjunct distribution of these habitats across northern Australia (Wilson *et al.* 1990), but is also due to the heterogeneity of substrates within these communities. Thus, the wide variety of niches present, combined with their disjunct distribution, has resulted in speciation (Specht 1958, Carlquist 1979). Many rare species found on sandstone habitats in the region, from genera such as *Stylidium*, *Hibiscus*, *Pityrodia* and *Micraira*, are known from only one collection site (Dunlop 1988).

The monsoon rain forests associated with the escarpment (Table III, Group 2) contain a large proportion of pantropical genera compared with the eucalypt communities in the region, but also a greater number of endemic and Australasian taxa than the lowland monsoon rain forests (Table III; Russell-Smith 1986). The dominant tree in these communities, *Allosyncarpia ternata*, is restricted to the actively eroding northern and western rims of the Arnhem Land escarpment. Unlike many other species in the monsoon rain forests of the Northern Territory, *Allosyncarpia* is poorly dispersed. The relatively large dry seeds fall 2–3 m at most beyond parental canopies. Its nearest relatives occur in isolated populations in northern Queensland, New Guinea, and New Caledonia, strongly suggesting that its origins lie with the ancient

Table III. Mean percentage frequency of endemic, Australasian and pantropic elements in major community groups in the Kakadu region (source: Taylor & Dunlop 1985).

Community Group	Endemic	Australasian	Pantropic
1. sandstone wood/shrubland	11	31	58
2. escarpment monsoon rain-forest	10	16	74
3. lowland monsoon rain-forest	1	5	94
4. eucalypt forest/woodland	5	21	74

super-continent of Gondwana (Johnson & Briggs 1984, Russell-Smith *et a*l. 1993). Several associated understorey species are restricted to, or most common in, these forests, including the small trees *Cryptocarya exfoliata*, *Glycosmis sapindoides*, *Rhodamnia australis*, *Xanthostemon psidioides*, *Xanthostemon umbrosus*, and the woody vines *Desmos wardianus* and *Gynochtodes australiensis*.

While sandstone rain forests dominated by *Allosyncarpia* are not particularly rich floristically (Russell-Smith 1991), they are highly significant because they comprise a forest type endemic to the region. Such forests cover 1138 km^2, and comprise 41% of all rain forest in northern and north-western Australia (Russell-Smith *et al.* 1993).

4. Biogeographic affinities

Much attention has been paid to the biogeographic affinities of the species found within the lowland monsoon rain forests of the region (Table III, Group 3). These communities exist across monsoonal Australia in small, disjunct patches of vegetation with dramatically different structural and floristic attributes from the surrounding savannas (Webb & Tracey 1981). Unlike the savannas, however, they are dominated by taxa with mainly extra-Australian, tropical affinities (Table III; Russell-Smith 1986).

Fossil evidence indicates that rain forest assemblages were distributed extensively across Australia following the retreat of seas covering large parts of the Australian-New Guinea land mass in the Late Cretaceous (ca. 110 million years ago) (Veevers 1984, Williams 1991), until at least the mid-Miocene (ca. 15 million years ago). This extensive distribution persisted in northern regions perhaps until the Pliocene (ca. 2.5 million years) (Truswell & Harris 1982, Truswell *et al.* 1987, Archer *et al.* 1989, Truswell 1990, Russell-Smith *et al.* 1993). Fossil pollen samples from sediment cores collected near the source of the South Alligator River indicate that rain forest containing *Podocarpus* and *Nothofagus* species occurred in the region possibly 50 million years ago (Truswell 1982). Such vegetation occurs elsewhere today under cooler, less seasonally arid conditions.

There are conflicting arguments as to the biogeographic origin of the vegetation in the region. While it has been suggested that Tertiary (ca. 65 to 2.5 million years) rain forests formed a floristic and structural continuum along a gradient of declining rainfall (Webb 1968, Bowman *et al.* 1991), biogeographical disjunctions in other vegetation types (e.g. sandstone heath floras) indicate considerable antiquity of a range of vegetation types adapted to different substrate drainage and nutrient conditions. The development of marked rainfall seasonality in the latter part of the Tertiary is likely to have resulted in increased frequency and impact of fire which would have caused fragmentation of these communities from that time.

Russell-Smith & Dunlop (1987) argue that many northern Australian monsoon rain forest patches are of relatively recent origin. For example, many modern-day rain forests occur on geologically young landforms which have developed only since sea-level attained its present level around 6000 years ago (e.g. beach dunes and flood plain levees). Such observations emphasise the capacity for many species to be dispersed widely; for example, over 60% of regional rain forest species possess fruits amenable for dispersal by frugivorous birds (especially fruit pigeons) and flying foxes (*Pteropus* spp.) (Russell-Smith & Dunlop 1987).

Observations that many monsoon rain forest genera are typically widespread in Malesia (south-easterly most parts of Asia and the islands of New Guinea and the Solomons) and, to a lesser extent, Africa (Barlow & Hyland 1988), and that over 40% of regional rain forest species are shared with geologically young islands in the Indonesian archipelago (Russell-Smith 1986), have been used to support the argument that these are readily and widely dispersed taxa. Similar findings have been used to support the notion that floristic similarities in rain forests from Africa, the Indian sub-continent, and certain parts of South-East Asia reflect ancient geological connections through the super-continent, Gondwana (e.g. Webb & Tracey 1981, Barlow & Hyland 1988).

Most of the dominant species of the lowland eucalypt forest and woodland savannas (Table III, Group 4), such as *Eucalyptus*, *Acacia*, and the grass genera *Chrysopogon*, *Eragrostis*

Sorghum, are considered Australasian savanna genera (Clifford & Simon 1981, Dunlop 1988). However, a number of species from Australasian endemic heath genera, such as *Pachynema*, *Hibbertia*, *Calytrix* and *Stylidium*, occur predominantly in eucalypt savannas on red earth soils, rather than in sandstone habitats (Dunlop 1988). Other 'obligate' sandstone species, such as *Callitris intratropica*, the palm *Gronophyllum ramsayi*, and various *Grevillea*, *Boronia* and *Hovea* occur within the lowland eucalypt communities on sands derived from the eroding, retreating escarpment.

It has been argued that poor soils and marked seasonal aridity have limited the immigration of large numbers of tropical elements from the floristically rich Indomalaysian flora into Australian eucalypt-dominated formations (van Steenis 1979). However, compared with eucalypt communities in southern Australia, the lowland eucalypt savannas of the region have a greater number of pantropical genera and a corresponding smaller proportion of endemic and Australasian genera (Taylor & Dunlop 1985). Thus, they possess many characteristic genera of grasses (e.g. *Eriachne*, *Thaumastochloa*, *Pseudopogonatherum* and others listed by Clifford & Simon 1981), forbs (e.g. *Tephrosia*) and woody plants (e.g. *Senna*) that appear to have immigrated from Malesian and other tropical floras (Dunlop 1988). Many of these woody and forb genera show drought-avoiding traits rather than the typical drought resisting properties of the Australasian trees and shrubs, suggesting they may have evolved under wet-dry climatic regimes (Dunlop 1988). There are also some smaller genera with extra-Australian centres of distribution that appear to have speciated under Australian conditions (Specht 1958). For example, *Terminalia*, *Planchonella* and *Buchanania* are represented by species that occur in monsoon rain forests and by different species that occur in eucalypt open-forests and woodlands.

5. Spatial patterning and environmental associations

Spatial patterning of the terrestrial vegetation distribution in the region is influenced, to some extent, by the rainfall gradient that exists from the moister coastal areas to the drier southern parts. However, the major environmental correlate of vegetation distribution is topographic position, chiefly through its effects on edaphic conditions (Specht 1958, Story 1969, 1973, 1976, Taylor & Dunlop 1985, Bowman 1988). Major vegetation types can be arranged along an idealised topographic gradient (Fig. 3). This gradient represents a transition from the escarpment and hill morphological units to the lowland plains which parallels increasing soil fertility, depth, moisture and drainage impedance.

The spatial distribution of the plant communities (Fig. 4) reflects the relationship with topography through the geomorphological patterning of the region. Thus, the shrublands/open-woodlands and closed-forests on the escarpments, are clearly differentiated from the open-forests and woodlands of the peneplains that dominate the lowlands. Scattered across these areas are drainage lines, coastal dunes, river levees and other features, many too small to map at this scale, which support a variety of forest, woodland and shrubland types.

The structure and composition of the canopy and the understorey within the major savanna vegetation types, are related to subtle changes in edaphic conditions. Understorey type may vary independently to the tree layer (Rice & Westoby 1985). In *Eucalyptus tetrodonta* forests near Darwin, the changes from shrubby to grassy understoreys may be correlated with decreasing moisture availability in the subsoil (Bowman 1986, Bowman & Minchin 1987). These patterns imply that the canopy and ground strata of these savannas may utilise resources from the lower and upper soil layers in a similar manner to the resource partitioning model used by Walker *et al.* (1981) to explain tree-grass ratios in African savannas. Alternatively, or additively, such changes in the understorey may be due to changes in fire frequency and/or intensity (Braithwaite & Estbergs 1985). The temporal and spatial variation in the population structure and understorey composition of savannas, especially in relation to soils and fire, and the role of seed and seedlings in forest regeneration, require much further investigation.

In assessing the role of climate in shaping the structure and composition of vegetation of northern Australia, it is generally considered that the

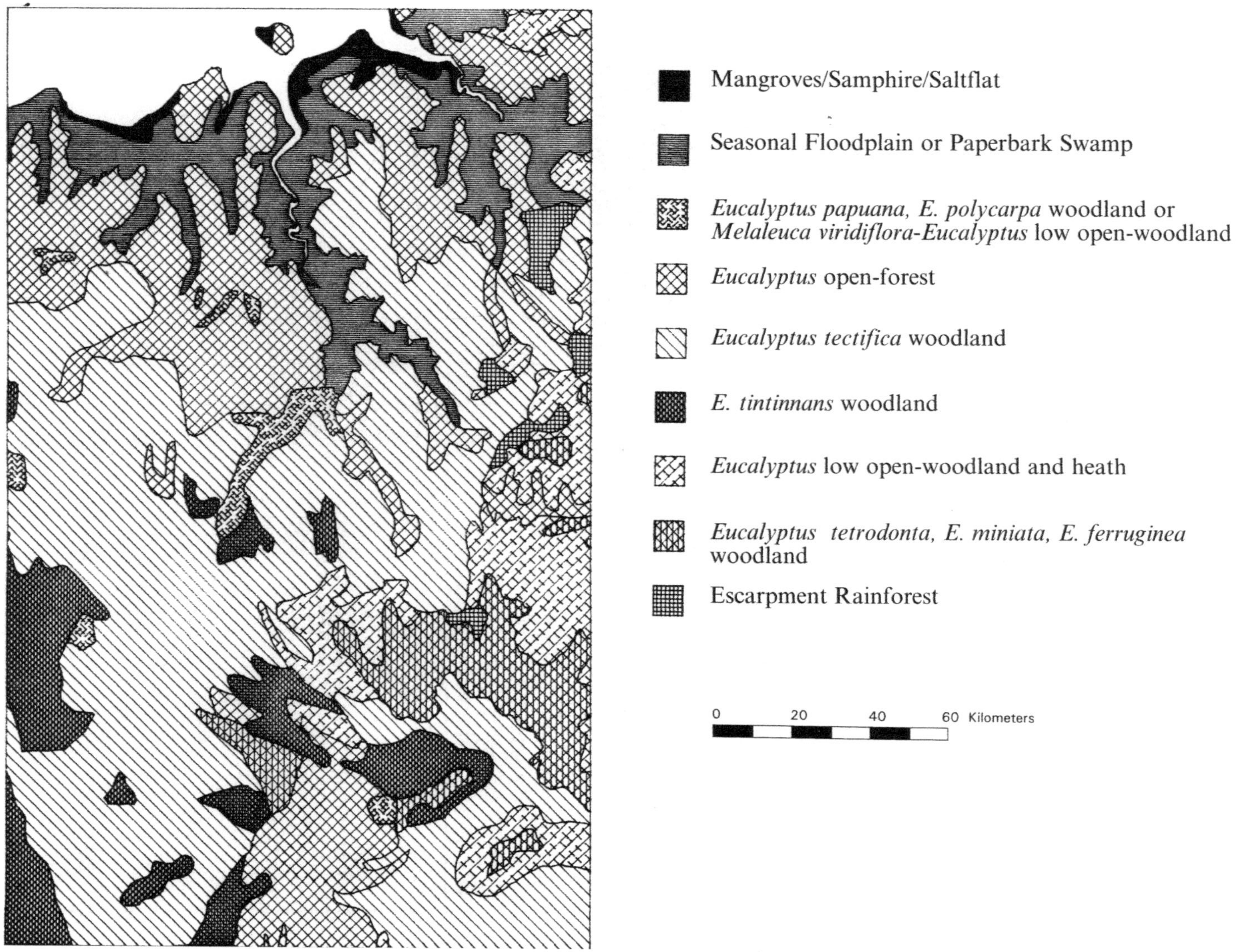

Fig. 3. Vegetation map of the Kakadu region (adapted from Wilson *et al.* 1990). Community labels are equivalent to those used in the text.

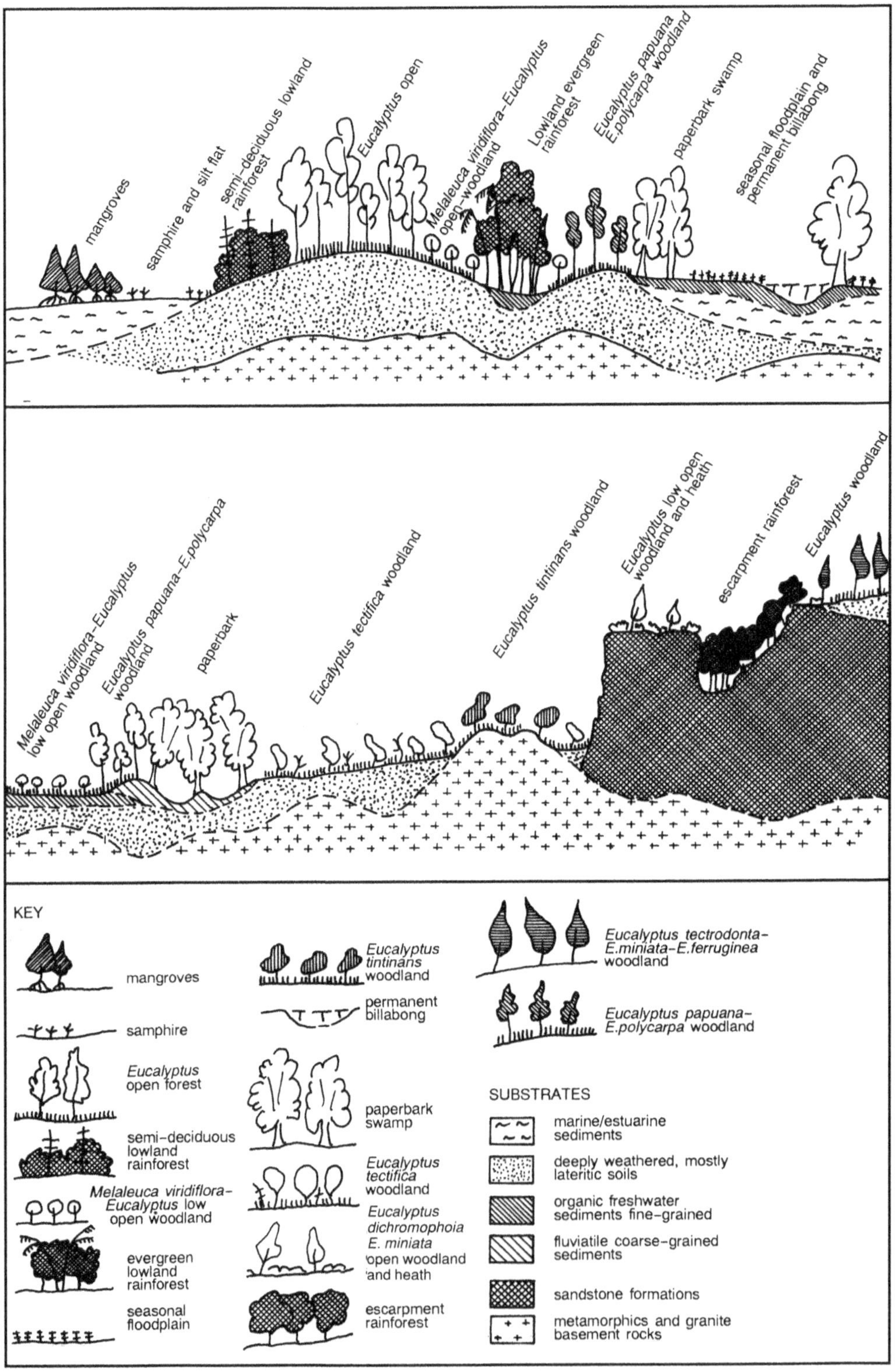

Fig. 4. Idealised topographic gradient of the Kakadu region along which the major vegetation types are arranged. Community labels are the same as those used in the text and Figure 1.

direct effects of temperature and light are non-limiting to biological growth, while the amount and distribution of rainfall is critical (Fitzpatrick & Nix 1970, Nix 1983). The annual drought is often regarded as one of the most important climatic stresses to which terrestrial plants of the region have had to adapt (Story 1973, Taylor & Dunlop 1985, Bowman 1988). The variability of the onset and duration of the annual wet season is regarded as another critical factor, particularly in relation to plant regeneration strategies (Taylor & Tulloch 1985). Winds associated with cyclones and storms can have a dramatic, localised impact on the structure and composition of vegetation adjacent to the coast (Wilson & Bowman 1987).

Temperature, rainfall and evaporation figures suggest that the potential growing season in the monsoonal tropics is limited to 20–22 weeks of the year (Mott *et al.* 1985). However, local variations in edaphic factors modify this figure considerably. The highly weathered relictual nature of most terrestrial soils in the region has resulted in extremely low soil nutrient status, low moisture storage capacity and high run-off rates (Aldrick 1976, Cole 1983). The deep red earth soils in the region, which support extensive open-forests, appear to have sufficient moisture supply at depth to support tree growth throughout the year (Wilson & Bowman 1994). Plant growth on these well drained soils is probably limited during the wet season by soil fertility (Mott *et al.* 1985). On low lying depressions or sites with water tables perched on shallow clay layers, lack of soil oxygen is a major limitation to wet season plant growth, while lack of effective soil depth often limits dry season water holding capacity (Fensham & Kirkpatrick 1992). Sandy soils on, or derived from, the Arnhem Land escarpment are of particularly low nutrient status and their moisture content is seasonally very low (Aldrick 1976, Russell-Smith 1986, Bowman *et al.* 1990). These latter soil types support distinctive woodlands, shrublands, grasslands and heaths (Story 1976, Specht 1958, Bowman *et al.* 1990).

Soils supporting lowland monsoon rain forests are generally considered to be more nutrient rich and have a higher moisture status than surrounding open-forests and woodlands (Bowman *et al.* 1991). Different rain forest community types in the region occupy landforms ranging from springs with nutrient-rich organic layers and a good year round supply of moisture, to rock outcrops with shallow soils that undergo severe moisture depletion towards the end of the dry season (Russell-Smith 1986). While the distribution of rain forest types associated with springs is clearly limited by water supply, the drier types are associated with various inter-correlated factors, such as soil moisture and nutrient status and topographic protection from fire (Russell-Smith 1986, Bowman & Dunlop 1986, Bowman *et al.* 1991, Russell-Smith *et al.* 1993). Favourable soil fertility and moisture relations probably become more important factors determining the distribution of the drier rain forest types, where fire protection from topographic features is lacking (Webb & Tracey 1981, Russell-Smith 1991).

6. Structure and function

Whilst the composition, distribution and environmental determinants of the plant communities of the region have been studied in detail, little work has been done on functional community processes such as growth and flowering phenology, seed dispersal, regeneration and nutrient cycling. However, major long-term studies of these processes within savannas were initiated in the early 1990s, both within the region and at sites with similar vegetation closer to Darwin. Much of the following discussion is based on preliminary, unpublished data from these studies.

6.1. *Lifeforms and species richness*

Within the region's vascular flora, both species richness and the diversity of life forms are high. The 1847 vascular species have been classified into 17 main life-forms (Table IV; Brennan 1992). The predominant life-forms are evergreen trees and shrubs (10% and 15%, respectively), annual forbs (22%), forbs with perennial roots but annual foliage (9%), perennial grasses (6%) and annual grasses (6%). Open-forest, woodland and shrubland communities in the Kakadu region generally show the highest ratios of species richness per area (Taylor & Dunlop 1985). The combination of an open canopy and ground stratum, and the layering of these communities into distinct tree and ground

Table IV. Life-forms of the 1847 vascular species of the Alligator Rivers Region (source: Brennan 1992).

Life form	No. of species	% of total
Trees – perennial, evergreen	185	10
Trees – deciduous	55	3
Palms	14	1
Shrubs – perennial, evergreen	284	15
Shrubs perennial, deciduous	17	1
Shrubs – roots perennial, foliage annual	36	2
Shrubs – annual	39	32
Forbs – perennial, evergreen	66	3
Forbs – perennial roots, foliage annual	164	9
Forbs – annual	411	22
Grasses – perennial	105	6
Grasses – annual	107	6
Sedges – perennial	67	3
Sedges – annual	88	5
Ferns	57	3
Vines – woody	46	3
Vines – herbaceous	106	6
Total	1847	100

strata allows a variety of woody and herbaceous life-forms to co-dominate. Thus, although the canopies are typically dominated by only a few (1–6) tree species from two or three families, the ground layer may comprise many life-forms and species from numerous families and genera. The number of species recorded in 80 x 20 m (0.16 ha) sites in open-forest and woodland is often 50–70 (Lazarides *et al.* 1988) and may be as high as 82 (Taylor & Dunlop 1985).

In contrast, the monsoon rain forests have a significantly lower species to area ratio, albeit a slightly higher family richness (Taylor & Dunlop 1985). The lifeforms and species that characterise the taller, denser, more complex wet tropical rain forests of the eastern seaboard (Webb 1978) are less abundant in the monsoon forests of the Kakadu region. The closed canopies of these forests are dominated mainly by trees, shrubs and, to a lesser extent, vines, and allow little herbaceous growth. Thus, the mean number of species recorded in 0.16 ha plots in monsoon rain forest was less than 40 (Lazarides *et al.* 1988) . Even in whole rain forest patches (mean size = 0.86 ha over the seven floristic groups recorded from the region by Russell-Smith (1991), the mean number of species was only 43, with few patches exceeding 60.

6.2. *Phenology*

There is a wide range of phenological behaviour, both vegetative and floral, exhibited within the major vegetation formations. Growth and flowering may occur in both the wet season and the dry season. The wet season is the main period for growth and flowering of the herbaceous species and small shrubs of the flood plains and the understorey of the open savanna forests and woodlands. In the dry season and the build-up to the wet season, there may be considerable growth and flowering in the woody vegetation, both deciduous and evergreen, in both the savannas and the monsoon forests.

The life-forms and functioning of terrestrial plant species of the Kakadu region may be described in terms of phenological responses to annual drought (Story 1976, Taylor & Dunlop 1985). For example, the herbaceous stratum of the savannas possess large numbers of annual forbs and grasses which behave as drought avoiders. These species germinate rapidly, grow, flower and set seed within each wet season. A good example is the tall, annual grass *Sorghum intrans*, a widespread dominant of the ground layer. It regenerates readily from seed, which is produced in copious amounts and stored within the soil over the dry season. The seed germinates rapidly with the onset of the wet season, and vegetative growth, flowering and seed set is complete before the end of the shortest known wet-season (Andrew & Mott 1983). The composition and abundance of the annual species may vary between years with the timing, duration and intensity of the wet season (Taylor & Dunlop 1985).

Many perennials, evergreen trees and shrubs possess drought-resistant traits (Walker & Gillison 1982). For example, the rain forest taxa *Allosyncarpia*, *Micromelum* and *Aglaia* have leaves that reduce water loss by efficient stomatal control or prolonged wilting (Russell-Smith 1986, Russell-Smith & Dunlop 1987, Fordyce 1992). *Allosyncarpia* can maintain a positive carbon balance at relatively low water potentials (–2.5 to –4 MPa at noon) during the late dry season, via osmoregulation and stomatal control of transpiration (Fordyce 1992). Evergreen trees develop deep taproot systems, which allow them to grow through the dry season by tapping moisture deep in the soil

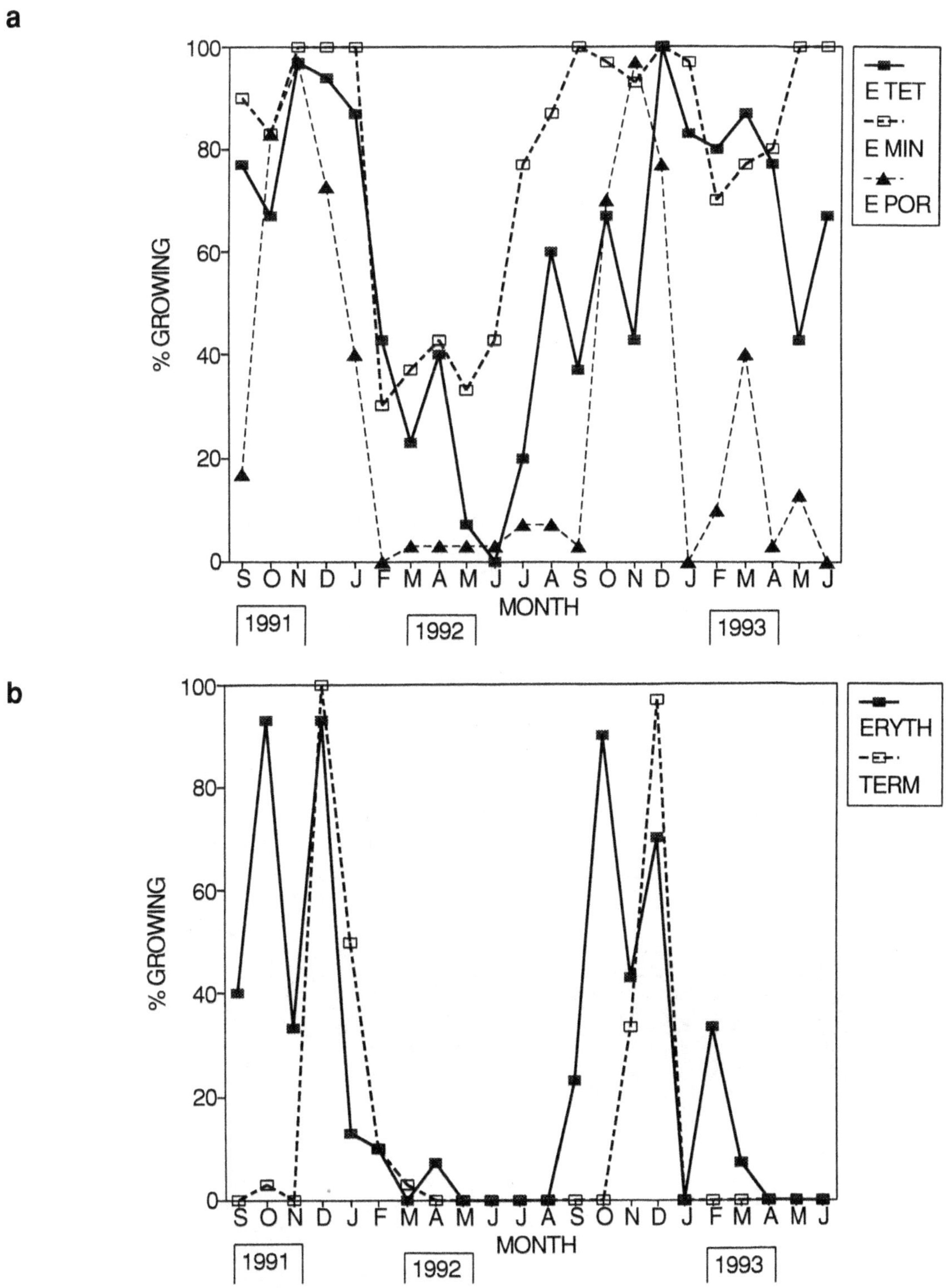

Fig. 5. Seasonal pattern of leaf growth in 5 species of canopy trees at the Kapalga Research Station from September 1991 to June 1993. Species are *Eucalyptus miniata*, *Eucalyptus tetrodonta*, *Eucalyptus porrecta*, *Erythrophleum chlorostachys* and *Terminalia ferdinandiana*. The survey was undertaken in three sub-catchments, each unburnt since 1988, with 10 individuals per species per sub-catchment per month; figures indicate the mean percentage of individuals producing new foliage (leaf buds and/or expanding foliage) at any one time, pooling subcatchments (n=30). (R.J. Williams unpublished data).

(Specht *et al.* 1976). In open-forest at Kapalga Research Station, 40–100% of individuals of *Eucalyptus miniata* produced leaf buds and/or expanding foliage during the 1992 dry season (Fig. 5). Characteristic morphological features of the heath taxa (Dunlop 1988) include evergreen, small scleromorphic foliage such as cladodes (*Pachynema*), phyllodes (*Acacia*), stiff, tough leaves (*Grevillea*) or reduced leaves (*Calytrix*, *Verticordia*). Although these heath traits clearly confer drought resistance, they are often thought to have evolved primarily in response to nutrient deficient soils (Beadle 1966, Specht 1979, Barlow 1988).

In contrast to the evergreen vegetation, numerous woody species evade drought by undergoing substantial or complete foliage loss during the dry season. Of 45 common species of tall shrubs and trees within savannas which are monitored at the Kapalga Researh Station and at Solar Village near Darwin, 17 are fully or partly deciduous (Brock 1988, Wilson *et al.* 1991, R.J. Williams unpublished data). Many of the deciduous species occur in rain forests or in the midstratum of the savannas, and have pantropical affinities (e.g. *Terminalia*, *Buchanania*). However, some of the eucalypts of the region are also completely or partially deciduous, (e.g. *Eucalyptus clavigera*, *Eucalyptus tintinnans*, *Eucalyptus alba*, *Eucalyptus porrecta*).

There is considerable range within the deciduous species with respect to both the degree of leaf loss and the timing of leaf fall. Some species, such as *Terminalia ferdinandiana* (Fig. 6), *Cochlospermum fraseri* and *Brachychiton megaphylla* are fully deciduous species, with leaf fall complete by the early-mid dry season (May-June). Others are fully deciduous, but with leaf fall complete by mid-late dry season (July-August); typical species are *Eucalyptus clavigera* and *Planchonia careya*. Species such as *Erythrophleum chlorostachys* and *Eucalyptus porrecta*, may be partly or fully deciduous, and lose their leaves late in the dry season (August-October).

Fig. 5 illustrates the monthly pattern of leaf growth for 5 common canopy species in open-forest (the evergreen dominants *Eucalyptus miniata* and *Eucalyptus tetrodonta*, the facultatively deciduous sub-dominants *Eucalyptus porrecta* and *Erythrophleum chlorostachys* and the fully deciduous *Terminalia ferdinandiana)* at the Kapalga Research Station for the period September 1991-June 1993. The data illustrate three strong patterns: i) the capacity of the two dominant species, *Eucalyptus miniata* and *Eucalyptus tetrodonta* to grow over the dry season; ii) the importance of both the latter part of the dry season (August-September) and the build-up to the wet season (October-November) for growth in all species; and iii) the relatively low level of growth during the latter part of the wet season and the transition period between wet season and dry season (February-May). This phenomenon of dry

Fig. 6. Terminalia ferninandiana is one of the fully deciduous trees in the region (photo B. McKaige).

season growth in evergreen dominant woody species is apparent in savannas elsewhere in the world (Medina 1982, Menaut & Cesar 1982, Sarmiento & Monasterio 1983, Mott *et al.* 1985). The rapid growth during the late-dry season/pre-wet season period has also been demonstrated for the open-forest savannas of Melville Island (Fensham 1990).

The timing of the major growth phases may differ between species and between years (Fig. 5). For example, *Eucalyptus miniata* commenced both the post-wet season and pre-wet season periods of growth approximately one month earlier than *Eucalyptus tetrodonta* in both years. The major period of growth in *Terminalia ferdinandiana* commenced in December in 1991, but in November in 1992. *Erythrophleum chlorostachys*, in comparison, displayed a relatively consistent pattern between years, with the initial growth phase in September, a relatively dormant period in November, and a second growth phase in December.

Flowering in woody species in open-forest savanna may occur throughout the year. Monthly surveys of 45 common woody savanna species since 1991/2 in a mixed *Eucalyptus miniata* and *Eucalyptus tetrodonta* forest near Darwin have shown that the number of species flowering is greatest during the dry season and build-up to the wet season, and least during the peak, monsoonal period of the wet season (Brock 1988, R.J. Williams, B.A. Myers, G.Duff & D. Eamus unpublished data). The factors affecting the initiation and maintenance of growth and flowering are likely to be both physiological and environmental, but the relative contribution of factors such as internal plant water status, day length, air temperature, relative humidity and soil moisture requires much further investigation.

6.3. Regeneration

The vast majority of perennial species, both herbaceous and woody, have the capacity to regenerate vegetatively following disturbance, either via epicormic sprouts, lignotubers, root suckering or rhizomes (Lacey 1974, Lacey & Whelan 1976, Taylor & Dunlop 1985). The height and density of the woody sprouts varies between years, and with fire regime (Fensham & Bowman 1992). As well as an adaptation to drought, these vegetative sprouts may also be an adaptation to low soil fertility (Beadle 1966) and frequent fire (Lacey & Whelan 1976, Braithwaite & Estbergs 1985).

Competition for water and nutrients may be important determinants of savanna regeneration and functioning. Removal of the overstorey may stimulate the development of understorey woody growth (Wilson & Bowman 1987), and overstorey density is inversely related to grass abundance in open-forests near Darwin (Bowman 1988). Thus, the very low density of intermediate-sized tree saplings, 2–4 m tall, in open-forests of the Kakadu region may be caused by suppression of the adult trees in the overstorey. However, frequent fire has also been suggested as a potential cause of this phenomenon (Braithwaite & Estbergs 1985, Werner 1986), which appears to be highly variable in space and may constitute a barrier to forest regeneration (Werner 1986).

Studies by Fensham & Bowman (1992) indicate that, in open-forest on Melville Island, Northern Territory, the current population structure is sufficient to sustain forest structure. However, there may be shifts in the composition of the dominants in these forests, as populations of juvenile and adult trees may be substantially different (Fensham & Bowman 1992). Age and physiological condition are important determinants for the successful recruitment of juvenile woody sprouts into the overstorey (Fensham & Bowman 1992). Only sprouts with sufficiently large underground parts have the potential to develop into sapling regrowth. However, sprouts that have been suppressed for long periods of time (> 5 years), may be physiologically incapable of vertical development and eventually become moribund and die (Fensham & Bowman 1992). Competition and edaphic conditions probably interact with other factors such as fire (Bowman & Wilson 1987, Bowman *et al.* 1988c) and feral animals (Werner 1986, Braithwaite *et al.* 1984) to determine population dynamics, but the long-term consequences of these complex interactions are not yet known.

In contrast to the copious vegetative sprouts, the incidence of seedlings, especially those of the eucalypts, within the population of juvenile trees is apparently low (Mott *et al.* 1985, Bowman 1986, Werner 1986, Wilson 1992, Bowman & Panton 1993a). This is a feature of neotropical savannas

(Medina 1982), and may be an adaptation to frequent fire. The regeneration of the dominant eucalypts from seed may require episodic events, such as the co-incidence of seed fall with early, heavy wet season rains. For example, none of the eucalypts hold their capsules beyond maturity (Dunlop & Webb 1991), and seed fall in both *Eucalyptus miniata* and *Eucalyptus tetrodonta* generally occurs 1–3 months prior to the onset of the wet season (Brock 1988, R.J. Williams unpublished data); seed set may also be small and sporadic (Wilson 1992). Seed which lies on the soil surface for the latter part of the dry season is likely to be eaten by ants (A.N. Andersen unpublished data); thus, there may be no reserve of soil-borne seed at the onset of the wet season. Seedlings may, however, be very common in some forests, although the reasons are unclear (Fensham 1993, Bowman & Panton 1993a).

Little work has been carried out on site specific factors that effect the germination of seed from tree species in the region. *Eucalyptus tetrodonta* seedlings have been observed growing in eucalypt open-forest leaf litter (Bowman 1986, Bowman & Panton 1993a, Fensham 1993). However, seed from monsoon rain forest species does not appear to readily germinate in eucalypt communities, possibly due to factors such as grass competition (Bowman 1993), unsuitable soil mycorrhiza (Bowman & Panton 1993b) and higher surface soil temperature due to a lack of shading. These latter species may require microclimate and soil conditions to be ameliorated by woody plant growth for successful germination and establishment of seedlings (Bowman & Panton 1993b).

7. Community descriptions

Following is a brief description of the major plant communities of the region, including dominant species, structure, distribution and major environmental associations. Community names are those presented in Figs 3 and/or 4 (with expanded or alternative names given in brackets). The descriptions are distilled from the studies listed in the introduction to this chapter and include the wide variation in floristic and structural composition that exists within each broad group (Rice & Westoby 1985). The structure and composition of plant communities of the plains of the 'lowlands' in the northern and central parts of the region contrast sharply with those of the rugged 'hill and escarpments' in the east and south (Fig. 3). The vegetation is described under these environments in turn.

7.1. Lowlands

A series of woodlands and open-forests occur on the generally low, undulating plains and ridges between the flood plains and the rocky hills and escarpments. Scattered throughout these areas are poorly drained depressions, drainage lines, rock outcrops and flood plain fringes, which support a variety of woodlands, shrublands, closed-forests and grasslands, dominated by a variety of genera.

7.1.1. Eucalyptus open-forest (Eucalyptus miniata, Eucalyptus tetrodonta open-forest)

The open-forests of the area are generally dominated by tall (16–25 m) mixed stands of *Eucalyptus miniata* and/or *Eucalyptus tetrodonta*. These forests are widespread across the northern coastal parts of the region and associated with undulating lateritic peneplains with deep, well drained, mainly red or yellow sandy loam soils. The height and density of the tree layer decreases with distance from the coast and on rockier or sandier soil types.

Associated canopy species include *Eucalyptus porrecta*, *Eucalyptus bleeseri* (on stony soils) and *Erythrophleum chlorostachys*. There is usually a low open-woodland/tall sparse-shrubland understorey present, dominated by eucalypts or other species such as *Acacia oncinocarpa*, *Acacia aulacocarpa*, the palm *Livistona humilis* and broadleaf species, such as *Buchanania obovata* and *Terminalia ferdinandiana*. The ground layer is dominated by seasonally tall (2–3 m) annual or perennial grasses such as *Heteropogon triticeus*, *Chrysopogon fallax* and *Sorghum* spp. Vines from the Fabaceae family and other forbs are common wet season components of the ground layer, although they are often obscured by grasses. Low (<1 m) shrubs, comprising eucalypt and broad leaved species, are prominent in the dry season when most of the herbaceous growth has died back or been removed by fire (Fig. 7). Open-forest on

Fig. 7. Woodlands and open forests often burn during the dry season leaving behind a low layer of resprouting shrubs (photo B. Wilson).

sand sheets in the region have similar composition to forests described by Bowman *et al.* (1988a) on Cobourg Peninsula with *Callitris intratropica* and the palm *Gronophyllum ramsayi* co-dominating the tree layer and 'heath' shrubs the understorey (B. Wilson personal observation).

7.1.2. Eucalyptus tectifica woodland (Eucalyptus foelscheana, Eucalyptus tectifica, Eucalyptus latifolia woodland)

These 10–12 m tall woodlands are dominated by a variety of eucalypt species and occur on the slopes of undulating plains with loam and clay loam soils. Soils supporting these communities are often subject to impeded drainage during the wet season due to the presence of a heavy, shallow clay layer. This community is the most widespread vegetation type in the region, dominating much of the lowland plains in the south and west.

Overstorey dominance and structure varies with edaphic conditions ranging from *Eucalyptus latifolia* on clay soils, to low (8 m) *Eucalyptus foelscheana* and *Eucalyptus confertiflora* trees on shallow imperfectly drained soils and, to *Eucalyptus tectifica* and *Eucalyptus tetrodonta* on better drained sites. An open-shrubby layer is generally present and can be tall (> 2 m) and dense (> 10% p.f.c.) in places. Common species of this stratum include various *Acacia* spp, *Grevillea decurrens*, the palm *Livistona humilis* and broad-leaf species such as *Buchanania obovata*, *Brachychiton paradoxum*, *Gardenia megasperma*, *Planchonia careya*, *Cochlospermum fraseri*, *Petalostigma quadriloculare*, and *Terminalia ferdinandiana*. The ground layer is dominated by tall grasses such as *Sorghum* spp., *Heteropogon contortus*, *Themeda triandra*, *Sehima nervosum*, *Eriachne avenacea* and *Chrysopogon fallax* (Fig. 8), with sedges and annual forbs conspicuous, particularly on more poorly drained sites and during the wet season.

7.1.3. Eucalyptus papuana, Eucalyptus polycarpa woodland (margin woodland)

This woodland is usually associated with the margins of the alluvial plains or the levees of larger rivers systems. Soils are generally clayey, poorly drained and may be subject to inundation for short periods during the wet season.

Eucalyptus papuana and *Eucalyptus polycarpa* are two common dominants of the tree layer that is predominantly woodland, although the vegetation spans low open-woodland to open-woodland classes. There is a variety of associated overstorey species including *Eucalyptus alba*, *Melaleuca viridiflora* and *Erythrophleum chlorostachys*. The shrub layer is variable, commonly including species such as *Pandanus spiralis*, *Planchonia careya*, *Flueggea virosa*, *Buchanania obovata* or *Melaleuca* spp. Grasses such as *Chrysopogon fallax*, *Sorghum* spp., *Sehima nervosum*, *Themeda avenacea* and *Heteropogon contortus* and sedges or forbs dominate the ground stratum. In many areas the exotic weeds *Hyptis suaveolens*, *Sida* spp. and *Senna* spp. form dense ground layer swards.

7.1.4. Melaleuca viridiflora-Eucalyptus low open-woodland

This vegetation type occurs throughout the region, generally on poorly drained, texture contrast, col-

Fig. 8. The woodlands and open forests develop tall grassy understoreys in the wet season (photo B. Wilson).

luvial or alluvial soils, fringing watercourses and drainage depressions. The overstorey is generally made up of low (8–10 m) scattered trees, interspersed with denser woodland or scattered trees among grasslands. *Melaleuca viridiflora* dominates this layer in association with various other species including *Eucalyptus polycarpa*, *Eucalyptus latifolia*, *Eucalyptus oligantha* and *Syzygium eucalyptoides*. A sparse-shrubland is generally present consisting of species such as *Pandanus spiralis*, *Livistona humilis*, *Planchonia careya*, *Grevillea pteridifolia* and a range of broadleaf species. The ground layer is dominated by the grasses *Chrysopogon fallax*, *Sehima nervosum*, *Eulalia aurea*, *Themeda avenacea*, *Eriachne* spp., *Sorghum* spp., forbs and sedges.

7.1.5. Grevillea, Banksia shrubland (heath)

Mixed species shrubland or heathland communities occur on sand-filled, poorly drained depressions on and adjacent to the escarpment and in depressions or drainage lines throughout the region. The upper layer is generally a low open-woodland to tall shrubland dominated by *Banksia dentata*, *Melaleuca nervosa* and *Grevillea pteridifolia* and smaller shrubs such as *Jacksonia dilatata*, *Melaleuca symphyocarpa*, *Verticordia cunninghamii* and *Acacia* spp. Emergent tree species such as *Eucalyptus polycarpa*, *Eucalyptus ptychocarpa*, *Melaleuca viridiflora* and *Lophostemon lactifluus* are common. The ground layer is dominated by a mixture of tall-mid height (0.5–2 m) grasses, such as *Sorghum*, *Eriachne triseta*, *Eriachne burkittii*, *Eriachne avenacea* and *Germainia grandiflora*. The rush *Leptocarpus spathaceus*, sedges and forbs are less common, but characteristic components of the ground layer.

7.1.6. Melaleuca argentea open-forests (riparian forest)

Riparian forest/woodland occurs on the sandy levee banks of major rivers, upstream from the flood plains. The rivers all flood to some extent each wet season and are typically fringed by a series of channels and levees with sandy soils that often overlay clays. The major river channels are often lined by *Melaleuca argentea* and/or *Melaleuca leucadendra* in association with a range of other species including the trees *Lophostemon grandiflorus*, *Barringtonia acutangula*, *Bambusa arnhemica*, *Ficus racemosa*, *Nauclea orientalis* and *Syzygium forte*, the shrub *Pandanus aquaticus* and the vines *Passiflora foetida* and *Flagellaria indica*. The ground layer is usually sparsely vegetated with grasses and sedges.

Some of the component species of this community are commonly associated with monsoon rain forests in the region and some riparian forests are classified as such by Russell-Smith (1991). Further from the river, on older levees, *Eucalyptus papuana*, *Eucalyptus polycarpa*, *Eucalyptus alba* woodlands (similar to the margin woodlands described above) often occur.

7.1.7. Lowland rain forest (Mixed species closed-forest)

Lowland monsoon rain forests occur as small habitat islands within the *Eucalyptus* or *Melaleuca* dominated vegetation. Russell-Smith (1991) has defined five types of lowland monsoon rain forests located within the Kakadu region (Groups 2, 4, 6, 9 and 10). These floristic groups are classified by the rain forest structural classification system of Webb (1978), as mesophyll, semi-deciduous mesophyll or semi-deciduous notophyll vine forests and deciduous vine thickets (Russell-Smith 1991). Lowland rain forests are broadly described below under either permanently or seasonally wet habitats.

Scattered throughout the region there are many small, usually less than 5 ha in extent, patches of monsoon rain forests associated with sites of perennial moisture availability at springs, seepages or situations where the water table remains close to the surface throughout the year. Canopies are dominated by tall (20–30 m), evergreen trees including species such as *Euodia elleryana*, *Syzygium* spp, *Ficus* spp, *Melaleuca leucadendra*, *Gmelina schlechteri*, *Fagraea racemosa* and *Calophyllum sil*. The tall feather palm *Carpentaria acuminata*, a monotypic genus endemic to the Northern Territory, is a conspicuous canopy component. Species with fern, shrub, vine and epiphytic growth forms are common, although usually less abundant, components of these communities.

Most lowland monsoon rain forests in the region occur on freely draining landforms, the surfaces of which dry out to varying degrees each year. Such habitats include coastal beach dunes, the upland margins of riverine flood plains, seasonal watercourses, and rock outcrops with skeletal soils. Patches of these monsoon rain forests are generally small, less than 10 ha, although larger patches are associated with the flood plain margins of the East Alligator River and along parts of the coastline. Many of the component species from seasonally dry monsoon rain forests are deciduous or partly deciduous and range in height from over 20 m on more favourable sites, to a mixture of vines and shrubs, often no more than 3 m in height and with an open canopy, on seasonally harsh sites such as rock outcrops. Typical tree species include *Acacia auriculiformis*, *Alstonia actinophylla*, *Bombax ceiba*, *Canarium australianum*, *Dysoxylum oppositifolium*, *Drypetes lasiogyna*, *Strychnos lucida*, *Peltophorum pterocarpum*, *Diospyros* spp, *Ficus virens*, *Grewia* spp., *Pouteria sericea*, *Hibiscus tiliaceus*, *Maranthes corymbosa* and *Sterculia quadrifida*.

7.1.8. Coastal dune complex

Narrow bands of unconsolidated beach sands on the coast support a variety of vegetation types. Many of the coastal dunes in the Kakadu region support lowland monsoon rain forest. Other less extensive, but still prominent and characteristic, communities include *Casuarina equisetifolia* open-woodland and a mixed grassland/forbland dominated by species such as *Sorghum* and *Spinifex* or the vine *Ipomoea pes-caprae*. These strand areas are not as extensive or diverse as in other parts of northern Australia, such as the more exposed beach dunes in eastern Arnhem Land (Wilson *et al.* 1990).

7.2. Hills and escarpments

The rugged terrain of the Arnhem Land escarpment, which is composed chiefly of quartz and sandstone, stands in sharp contrast to the surrounding lowlands. Soils here are generally shallow, coarse sands. However, the complex arrangement of sandstone pavements, escarpments and boulders, intermixed with drainage lines and lateritic plateaux, provides a wide range in edaphic conditions, particularly in relation to soil depth. The resulting array of habitats supports a complex, intergrading mosaic of forests, woodlands, shrublands, heaths and grasslands (Fig. 9).

7.2.1. Eucalyptus low open-woodland and heath (Eucalyptus dichromophloia, Eucalyptus miniata or sandstone low open-woodland/shrubland)

Spatial patterning in these communities is complex (Bowman *et al.* 1990), with dramatic changes in structure and composition occurring at a scale of metres. There is often little correlation between floristic composition of understorey and overstorey structure within this community (Rice & Westoby 1985), as the tree layer is often dis-

Fig. 9. The hills and escarpments support a complex mosaic of vegetation interspersed by bare rock (photo B. McKaige).

continuous, with areas of woodland intermixing with areas where the understorey shrubs and grasses become the dominant layers. Thus, these communities are often treated as a complex (e.g. Specht 1958) which can be subdivided on floristic and structural composition depending on scale of interest (e.g. Rice & Westoby 1985, Schodde *et al.* 1987, Bowman *et al.* 1990).

Eucalyptus dichromophloia, the closely related *Eucalyptus arnhemensis*, and *Eucalyptus miniata* are the most common dominants of the upper layer. A variety of other tree species such as *Eucalyptus koolpinensis*, *Eucalyptus kombolgiensis*, *Eucalyptus ferruginea*, *Eucalyptus brachyandra*, *Eucalyptus herbertiana* and *Eucalyptus phoenicea* are locally dominant. The gymnosperm *Callitris intratropica* (Fig. 10) and the palm *Gronophyllum ramsayi* occur on some sandy, boulder strewn areas where there appears to be some fire protection. The mid layer is dominated by a mixture of species including *Vitex acuminata*, *Terminalia canescens*, *Blepharocarya depauperata*, *Boronia lanuginosa*, *Owenia vernicosa*, *Grevillea*, *Calytrix*, *Jacksonia* and *Acacia* species. The ground layer is typically a hummock or tussock grassland dominated by *Plectrachne pungens*, *Triodia microstachys*, *Eriachne* and *Sorghum* spp. Annual forbs and sedges are common wet season components, although their abundance may be influenced by fire history (Bowman *et al.* 1990).

On still rockier sites the vegetation is reduced to scattered shrubs or heath genera such as *Acacia*, *Jacksonia*, *Grevillea*, *Melaleuca*, *Banksia* and *Calytrix* along with *Terminalia carpentariae* or the rock fig *Ficus leucotricha*. A scattering of tussock (*Eriachne bleeseri*, *Eriachne* spp.) or hummock (*Plectrachne pungens*, *Triodia microstachys*, *Triodia procera*) grasses also occur on the rockier sites. The "resurrection grasses" (which possess spirally arranged leaves) *Micraira* spp., are a distinctive component of this vegetation on some sandstone pavements. A myriad of seasonal forbs and small shrubs occur scattered throughout the range of habitats available. Common genera include *Mitrasacme*, *Hibiscus*, *Utricularia*, *Tephrosia*, *Goodenia*, *Pityrodia*, *Fimbristylis*, *Cyperus*, *Leptocarpus* and *Stylidium*. Duff *et al.* (1991) have described a rarer variant of a heathland, dominated by *Melaleuca magnifica* and *Leucopogon acuminata*, which occurs on the margins of the Tertiary laterite on the Marrawal Plateau in the south of the region.

7.2.2. Eucalyptus tetrodonta, Eucalyptus miniata, Eucalyptus ferruginea woodland

Remnants of Tertiary laterite that overlie the Protozoic sandstone on the Marrawal Plateau, support mainly *Eucalyptus miniata*, *Eucalyptus tetrodonta* dominated open-forests or woodlands. On more loamy soils this community may be similar in composition and structure to the lowland open-forests, although the typically sandy soils on

Fig. 10. Callitrus intratropica occurs on sandy, boulder strewn areas that are not regularly burnt (photo B. McKaige).

the plateau surface support tree canopies that are lower and more open and with a more variable understorey composition. Common subdominants include *Eucalyptus bleeseri*, *Erythrophleum chlorostachys*, *Xanthostemon paradoxus* and *Eucalyptus ferruginea*. There is generally a sparse-shrub/low open-woodland layer of species such as *Eucalyptus* spp, *Acacia* spp, *Bossiaea bossiaeoides*, *Terminalia canescens*, *Petalostigma quadriloculare*, *Grevillea* spp. and *Calytrix exstipulata*. The ground layer is dominated by *Plectrachne pungens*, *Chrysopogon fallax*, *Heteropogon triticeus*, *Sorghum* spp. and other tussock grasses.

7.2.3. Eucalyptus tintinnans woodland (hill woodland)

Towards the south of the region, particularly surrounding the headwaters of the South Alligator River, are rugged hills originating from a mixture of sedimentary, volcanic and metamorphic geologies. These hills support mainly low (8–10 m) woodlands, which merge into taller and denser lowland woodlands on flatter areas with deeper soils and sparser sandstone woodlands on rockier sites.

The upper layer of these communities is characterised by *Eucalyptus tintinnans* (formerly *Eucalyptus alba* var *alba*) with *Eucalyptus dichromophloia* on rockier sites. This community has the greatest species richness in the tree layer of woodlands and open-forests in the region (Woinarski *et al.* 1989). Thus, other species such as *Erythrophleum chlorostachys*, *Eucalyptus foelscheana*, *Eucalyptus setosa*, *Eucalyptus confertiflora*, *Eucalyptus latifolia*, *Eucalyptus tectifica* and *Xanthostemon paradoxus* are common canopy components depending on soil type. There is generally a sparse-shrub layer with species such as *Grevillea decurrens*, *Gardenia megasperma*, *Brachychiton paradoxus*, *Terminalia ferdinandiana*, *Petalostigma quadriloculare*, *Calytrix exstipulata* and *Owenia vernicosa* characteristic. Woody plants in this community (including the eucalypts) are deciduous or semi-deciduous. The ground layer is dominated by the tall grasses *Sorghum* spp, *Sehima nervosum*, *Rottboellia formosa*, *Themeda triandra*, *Eriachne avenacea* and *Heteropogon triticeus*. On some rockier sites *Plectrachne pungens* may be present.

7.2.4. Escarpment rain forest (Allosyncarpia or mixed species closed-forest)

The escarpment monsoon rain forests include a further two floristic groups defined by Russell-Smith (1991), in addition to the lowland rain forests. These groups are classified by the rain forest structural classification system of Webb (1978), as mesophyll, simple semi-evergreen notophyll, evergreen notophyll or simple notophyll evergreen vine forests (Russell-Smith 1991).

These communities occur on perennially moist sites such as those near springs, or more widely on a range of seasonally dry, often rugged, sandstone landforms with skeletal, sandy/rocky soils (Fig. 11). They are typically dominated by the large, evergreen, fire-tolerant tree *Allosyncarpia ternata*, especially on seasonally dry substrates (Russell-Smith *et al.* 1993). This species is endemic to the actively eroding northern and western rims of the

Fig. 11. Dense evergreen crowns of *Allosyncarpia ternata* at the base of the escarpment stand in sharp contrast to the woodlands etc. on the sandstone plateau (left) and the vast woodlands of the lowland plain (right) (photo J. Russell-Smith).

Arnhem Land escarpment and is often the sole component of the tree layer. It provides a protective canopy for other species, particularly in the seasonally dry situations. *Allosyncarpia* forests range in extent from small groves of scattered trees to large tracts of forest extending over hundreds of hectares.

In protected, moist sandstone gorges, canopies may be as tall as 35 m. As well as *Allosyncarpia*, other common tree species in such situations include *Calophyllum sil*, *Gmelina schlechteri*, *Horsfieldia australianum*, *Ilex arnhemensis*, *Melaleuca leucadendra*, *Syzygium angophoroides*, *Syzygium minutuliflorum*, *Xanthostemon eucalyptoides* and the palm *Carpentaria acuminata*. Some of these forests are similar in floristic composition to the lowland spring monsoon rain forests. In seasonally dry situations canopy height may be as low as 10 m, with an open understorey. Other hardy species associated with *Allosyncarpia* in these situations include *Notelaea microcarpa*, *Vitex acuminata*, *Drypetes lasiogyna*, *Myristica insipida*, *Maranthes corymbosa* and *Buchanania arborescens*.

8. Acknowledgement

Kim Brennan kindly extracted and supplied the data presented in Table IV.

9. References

Adams, L.G., Byrnes, N. & Lazarides, M. 1973. Floristics of the Alligator Rivers area. In: Alligator Rivers Region Environmental Fact Finding Study. Physical Features and Vegetation, Part II. Unpublished report, CSIRO, Melbourne. 76 pp.

Aldrick, J.M. 1976. Soils of the Alligator Rivers area. In: Story, R., Galloway, R.W., McAlpine, J.R., Aldrick, J.M., & Williams, M.A.J. (eds) Lands of the Alligator Rivers Area, Northern Territory. pp. 71–88, Land Use Research Series No. 38. CSIRO, Melbourne.

Andrew, M.H. & Mott, J.J. 1983. Annuals with transient seed banks: the population biology of indigenous sorghum species of tropical north-west Australia. Aust. J. Ecol. 8: 265–86.

Archer, M., Godthelp, H., Hand, S.J. & Megirian, D. 1989. Fossil mammals of Riversleigh, northwestern Queensland: preliminary overview of biostratigraphy, correlation, and environmental change. Aust. Zool. 25: 29–65.

Barlow, B.A. 1988. The Australian flora: Its origin and evolution. In: Flora of Australia 1. pp. 25–76. Bureau of Flora and Fauna, Canberra.

Barlow, B.A. & Hyland, B.P.M. 1988. The origins of the flora of Australia's wet tropics. Proc. Ecol. Soc. Aust. 15: 1–17

Beadle, N.C.W. 1966. Soil phosphate and its role in moulding segments of the Australian flora and vegetation, with special mention to xeromorphy and sclerophylly. Ecol. 47: 992–998.

Bell, A. 1981. Fire damages top end forests. Ecos 30: 18–20.

Bowman, D.M.J.S. 1986. Stand characteristics, understorey associates and environmental correlates of *Eucalyptus tetrodonta* F. Muell. forests on Gunn Point, Northern Australia. Vegetatio 65: 105–114.

Bowman, D.M.J.S. 1988. Stability amid turmoil?: towards an ecology of north Australian eucalypt forests. Proc. Ecol. Soc. Aust. 15: 149–158.

Bowman D.M.J.S. 1993. Establishment of two dry monsoon forest tree species on a fire protected monsoon forest-savanna boundary, Cobourg Peninsula, northern Australia. Aust. J. Ecol. 18: 235–238.

Bowman, D.M.J.S. & Dunlop, C.R. 1986. Vegetation pattern and environmental correlates in coastal forests of the Australian monsoon tropics. Vegetatio 65: 99–104.

Bowman D.M.J.S. & Minchin P.R. 1987. Environmental relationships of woody vegetation patterns in the Australian monsoon tropics. Aust. J. Bot. 35: 151–169.

Bowman D.M.J.S. & Panton, W.J. 1993a. Differences in the stand structure of *Eucalyptus tetrodonta* forests between Elcho Island and Gunn Point, northern Australia. Aust. J. Bot. 41: 211–215.

Bowman D.M.J.S. & Panton, W.J. 1993b. Factors that control monsoon-rainforest seedling establishment and growth in north Australian *Eucalyptus* savanna. J. Ecol. 81: 292–304.

Bowman, D.M.J.S. & Wilson, B.A. 1987. Fuel characteristic of coastal monsoon forests, Northern Territory, Australia. J. Biogeog. 15: 807–817.

Bowman, D.M.J.S., Wilson, B.A. & Davis, G.W. 1988a. Response of *Callitris intratropica* R.T Baker & H.G. Smith to fire protection, Murgenella, Northern Australia. Aust. J. Ecol. 13: 147–159.

Bowman, D.M.J.S., Wilson, B.A. & Dunlop, C.R. 1988b. Preliminary biogeographic analysis of the Northern Territory vascular flora. Aust. J. Bot. 36: 503–517.

Bowman, D.M.J.S., Wilson, B.A. & Hooper, R.J. 1988c. Response of *Eucalyptus* forest and woodland to four fire regimes at Munmalary, Northern Territory, Australia. Journal of Ecol. 70: 57–67.

Bowman, D.M.J.S., Wilson, B.A. & Fensham, B.S. 1990. Sandstone vegetation pattern in the Jim Jim Falls region, Northern Territory, Australia. Aust. J. Ecol. 15: 163–174.

Bowman, D.M.J.S., Wilson, B.A. & McDonough, L. 1991. Monsoon forests in northwestern Australia I. Vegetation classification and the environmental control of tree species. J. Biogeog. 18: 679–686.

Braithwaite, R.W., Dudzinski, M.L., Ridpath, M.G. & Parker, B.S. 1984. The impact of water buffalo on the monsoon forest ecosystem in Kakadu National Park. Aust. J. Ecol. 9: 309–322.

Braithwaite, R.W. & Estbergs, J.A. 1985. Fire patterns and woody vegetation trends in the Alligator Rivers Region of northern Australia. In: Tothill, J.C. & Mott, J.J. (eds) Ecology and Management of the Worlds Savanna. pp. 359–364. Australian Academy of Science, Canberra.

Brennan, K. 1992. Checklist of Vascular Plants of the Alligator Rivers Region, Northern Territory. Supervising Scientist for the Alligator Rivers Region, Second Edition. Open File Record No. 62. 44 pp.

Briggs, J.D. & Leigh, J.H. 1988. Rare or Threatened Australian Plants. Special Publication 14, Australian National Parks and Wildlife Service, Canberra. 278 pp.

Brock, J. 1988. Top End Native Plants. J.Brock, Darwin. 354 pp.

Burgman, M.A. & Thompson, B.S. 1982. Cluster analysis, ordination and dominance-structural classification applied to diverse tropical vegetation at Jabiluka, Northern Territory. Aust. J. Ecol. 7: 375–387.

Carlquist, S. 1979. *Stylidium* in Arnhem Land: new species, modes of speciation on the sandstone plateau, and comments on floral mimicry. Aliso 9: 411–461.

Christian, C.S. & Stewart, G.A. 1953. General Report on Survey of Katherine-Darwin Region, 1946. Australian Land Use Research Series No. 1, CSIRO, Canberra. 24 pp.

Clifford, H.T. & Simon, B.K. 1981. The biogeography of Australian Grasses. In: Keast, A.(ed.) Ecological Biogeography of Australia. pp. 471–498. W. Junk, The Hague-Boston-London.

Cole, M.M. 1983. The influence of soils, geomorphology and geology on the distribution of plant communities in savanna ecosystems. In: Huntly, B.J. & Walker, B.H. (eds) Ecology of Tropical Savannas. pp. 145–174. Springer-Verlag, Berlin.

Cowie, I.D. & Finlayson, C.M. 1986. Plants of the Alligator Rivers Region, Northern Territory, Australia. Supervising Scientist for the Alligator Rivers Region, Technical Memorandum 23. AGPS, Canberra. 26 pp.

Cowie, I.D., Finlayson, C.M. & Bailey, B.J. 1987. Alien plants in the Alligator Rivers Region, Northern Territory, Australia. Supervising Scientist for the Alligator Rivers Region, Technical Memorandum 17. AGPS, Canberra. 50 pp.

Dickinson, K.J.M. & Dunlop, C.R. 1988. Assessment of Kakadu Stage III and the Kakadu Conservation Zone for World Heritage listing on the basis of vegetation and floristics. In: Kakadu Stage 3: A Preliminary Assessment with Particular Reference to the Operational Guidelines for the Implementation of the World Heritage Convention, Technical Report No.45. pp. 71–86. Conservation Commission of the Northern Territory, Darwin.

Duff, G.A., Orr, T.M., Belbin, L. & Andersen, A. 1991. Flora Survey of Kakadu Stage III. Unpublished report to Australian National Parks and Wildlife Service. Canberra. 50 pp.

Dunlop, C.R. 1988. Sandstone habitats. In: Top End Native Plants. pp. 9–11. J.Brock, Darwin.

Dunlop, C.R. & Webb, L.J. 1991. Flora and vegetation. In: Haynes, C.D., Ridpath, M.G. & Williams, M.A.J. (eds) Monsoonal Australia: Landscape, Ecology and Man in Northern Lowlands. pp. 41–59. A.A. Balkema, Rotterdam.

Fensham, R.J. 1990. Regeneration in Tropical Eucalypt Forest on Melville Island. Unpublished PhD Thesis, Department of Geography, University of Tasmania. 266 pp.

Fensham, R.J. 1993. The establishment of eucalypt seedlings in tropical savanna forest. N.T. Natural. 13: 26–32.

Fensham, R.J. & Bowman, D.M.J.S. 1992. Stand structure and the influence of overwood on regeneration in tropical eucalypt forest on Melville Island. Aust. J. Bot. 40: 335–52.

Fensham, R.J. & Kirkpatrick, J.B. 1992. Soil characteristics and tree species distribution in the savannah of Melville Island, Northern Territory. Aust. J. Bot. 40: 311–333.

Fitzpatrick, E.A. & Nix, H.A. 1970. The climate factor in Australian grassland ecology. In: Moore, R.M. (ed.) Australian Grasslands. pp. 3–26. ANU Press, Canberra.

Fordyce, I.R. 1992. Some aspects of the ecology and physiology of *Allosyncarpia ternata* S.T. Blake. Graduate Diploma of Science Thesis, Northern Territory University, Darwin. 50 pp.

Johnson, L.A.S. & Briggs, B.C. 1984. Myrtales and Myrtaceae a phylogenetic analyses. Ann. Mo. Bot. Gard. 71: 700–56.

Johnston, R.W. & Tothill, J.C. 1985. Definition and broad geographic outline of savanna lands. In: Tothill, J.C. & Mott, J.J. (eds) Ecology and Management of the Worlds Savannas, pp 1–13. Australian Academy of Science, Canberra.

Lacey, C.J. 1974. Rhizomes in tropical eucalypts and their role in recovery from fire damage. Aust. J. Bot. 22: 29–38.

Lacey, C.J. & Whelan, P.I. 1976. Observations on the ecological significance of vegetative reproduction in the Katherine-Darwin region of the Northern Territory. Aust. J. Bot. 39: 131–139.

Lazarides, M., Craven, L.A., Dunlop, C.R., Adams, L.G. & Byrnes, N. 1988. A Checklist of the Flora of Kakadu National Park and Environs, Northern Territory, Australia, Occasional Paper No. 15. Australian National Parks and Wildlife Service, Canberra. 42 pp.

Leach, G.J. 1987. Rare and threatened plants of the monsoon tropics of the Northern Territory with special emphasis on Kakadu National Park. In: Kakadu Stage 2: A Preliminary Assessment with Particular Reference to the Operational Guidelines for the Implementation of the World Heritage Convention. Technical Report No. 37. pp 93–96. Conservation Commission of the Northern Territory, Darwin.

Leach, G.J. 1988. The rare and threatened flora of Kakadu National Park (Stage 3) in relation to world heritage listing. In: Kakadu Stage 3: A Preliminary Assessment with Particular Reference to the Operational Guidelines for the Implementation of the World Heritage Convention. Technical Report No.45. pp. 95–98. Conservation Commission of the Northern Territory, Darwin.

Medina, E. 1982. Physiological ecology of neotropical savannas. In: Huntley, B.J. & Walker, B.H. (eds) Ecology of Tropical Savannas. pp. 308–335. Springer-Verlag, Berlin.

Menaut, J.C. & Cesar, J. 1982. The structure and dynamics of a West African savanna. In: Huntley, B.J. & Walker, B.H. (eds) Ecology of Tropical Savannas. pp. 80–100. Springer-Verlag, Berlin.

Mott, J.J., Williams, J., Andrew, M.H. & Gillison, A.W. 1985. Australian savanna ecosystems. In: Tothill, J.C. & Mott, J.J. (eds) Ecology and Management of the Worlds Savannas. pp. 56–82. Australian Academy of Science, Canberra.

Nix, H.A. 1983. Climate of tropical savannas. In: Bouliere, F. (ed.) Tropical Savannas. Ecosystems of the World 13. pp. 1–17. Elsevier Scientific Publishing Company, Amsterdam.

Orr, T.M., Brennan, K. & Andersen, A.N. 1990. Vegetation of the Kakadu Conservation Zone. In: Woinarski, J.C.Z. & Braithwaite, R.W. (ed.) The Terrestrial Vertebrate Fauna and Vegetation of the Kakadu Conservation Zone: results of a field survey and interpretation of available data. pp. 7–14. Resource Assessment Commission, Kakadu Conservation Zone Inquiry Consultancy Series, Canberra.

Rice, B. & Westoby, M. 1985. Structure of local floristic variation and how well it correlates with existing classification schemes: vegetation at Koongarra, NT, Australia. Proc. Ecol. Soc. Aust. 13: 129–137.

Russell-Smith, J. 1984. The Status and Condition of Monsoon Vine-forests in the Kakadu Region: A Management Report. Unpublished report to Australian National Parks and Wildlife Service, Canberra. 37 pp.

Russell-Smith, J. 1986. The Forest in Motion: Exploratory Studies in Western Arnhem Land, Northern Territory. Unpublished PhD thesis, ANU, Canberra. 601 pp.

Russell-Smith, J. 1991. Classification, species richness, and environmental relations of monsoon forest in northern Australia. J. Veg. Sci. 2: 259–278.

Russell-Smith, J. & Dunlop, C.R. 1987. The status of monsoon vine forest in the Northern Territory: A perspective. In: Werren, G.L. & Kershaw, A.P. (eds) The Monsoon Forest Legacy: Australian National Monsoon Forest Study. Vol. 1. pp. 227–288. AGPS, Canberra.

Russell-Smith, J., Lucas, D.E., Brock, J. & Bowman, D.M.J.S. 1993. *Allosyncarpia*-dominated rain forest in monsoonal northern Australia. J. Veg. Sci. 4: 67–82.

Sarmiento, G. & Monasterio, M. 1983. Life forms and phenology. In: Bouliere, F. (ed.) Tropical Savannas. Ecosystems of the World 13. pp. 79–108. Elsevier Scientific Publishing Company, Amsterdam.

Schodde, R., Hedley, A.B., Mason, I.J. & Martensz, P.N. 1987. Vegetation Habitats in Kakadu National Park, Alligator Rivers Region, NT, Australia. Unpublished report to Australian National Parks and Wildlife Service, Canberra. 29 pp.

Specht, R.L. 1958. The geographical relationships of the flora of Arnhem Land. In: Specht, R.L. & Mountford, C.P. Records of the American-Australian Scientific Expedition to Arnhem Land. Botany and Plant Ecology. pp. 415–478. Melbourne University Press, Melbourne.

Specht, R.L. 1979. The sclerophyllous (heath) vegetation of Australia: The eastern and central states. In: Specht, R.L. (ed.) Heathlands and Related Shrublands. Ecosystems of the World 9. pp. 125–210. Elsevier Scientific Publishing Company, Amsterdam.

Specht, R.L. 1981a. Foliage projective cover and standing biomass. In: Gillison, A.N. & Anderson, D.J. (eds) Vegetation Classification in Australia. pp. 10–21. CSIRO and ANU, Canberra.

Specht, R.L. 1981b. Evolution of the Australian flora: some generalisations. In: Keast, A., (ed.) Ecological Biogeography of Australia. pp. 783–806. W. Junk, The Hague-Boston-London.

Specht, R.L., Salt, R.B. & Reynolds, S.T. 1976. Vegetation in the vicinity of Weipa, North Queensland. Proc. Roy. Soc. Qd 88: 17–38.

Story, R. 1969. Vegetation of the Adelaide-Alligator area. In: Story, R., Williams, M.A.J., Hooper, A.D.L., O'Ferrall, R.E. & McAlpine, J.R. (eds) Lands of the Adelaide-Alligator Rivers Area, Northern Territory. pp. 114–130. Land Use Research Series No.25. CSIRO, Melbourne.

Story, R. 1973. Vegetation of the Alligator Rivers area. In: Alligator Rivers Region Environmental Fact-finding Study: Physical Features & Vegetation, Part IX. Unpublished Report, CSIRO, Melbourne. 46 pp.

Story, R. 1976. Vegetation of the Alligator Rivers area. In: Story, R., Galloway, R.W., McAlpine, J.R., Aldrick, J.M., & Williams, M.A.J. (eds) Lands of the Alligator Rivers Area, Northern Territory. pp. 89–111. Land Use Research Series No. 38. CSIRO, Melbourne.

Taylor, J.A. & Dunlop, C.R. 1985. Plant communities of the wet-dry tropics of Australia. Proc. Ecol. Soc. Aust. 13: 83–127.

Taylor, J.A. & Tulloch, D. 1985. Rainfall in the wet-dry tropics: Extreme events at Darwin and similarities between years during the period 1870–1983 inclusive. Aust. J. Ecol. 10: 281–296.

Truswell, E.M. 1982. Palynology of a Tertiary Core Sample from South Alligator area, E.L. 1287. Bureau of Mineral Resources, Canberra: Professional Opinion. Unpaginated.

Truswell, E.M. 1990. Australian rain forests: the 100 million year record. In: Webb, L.J. & Kikkawa, J. (eds) Australian Tropical Rainforests: Science, Values, Meaning. pp. 7–22, CSIRO, Melbourne.

Truswell, E.M. & Harris, W.K. 1982. The Cainozoic palaeobotanical record in arid Australia: fossil evidence for the origins of an arid-adapted flora. In: Barker, W.R. & Greenslade, P.J.M. (eds) Evolution of the Flora and Fauna of Arid Australia. pp. 67–76. Peacock Publications, Adelaide.

Truswell, E.M., Kershaw, A.P. & Sluiter, I.R. 1987. The Australian-southeast Asia connection: evidence from the palaeobotanical record. In: Whitmore, T.C. (ed.) Biogeographical Evolution of the Malay Archipelago. pp. 32–49. Clarendon Press, Oxford.

van Steenis, C.G.G.J. 1979. Plant-geography of east Malesia. Bot. J. Linn. Soc. 79: 97–178.

Veevers, J.J. (ed.) 1984. Phanerozoic Earth History of Australia. Clarendon Press, Oxford. 418 pp.

Walker, B.H., Ludwig, D, Holling, C.S. & Peterman, R.M. 1981. Stability of semi-arid grazing systems. J. Ecol. 69: 473–498.

Walker, B.H. & Gillison, A.N. 1982. Australian savannas. In: Huntley, B.J. & Walker, B.H. (eds) Ecology of Tropical Savannas. pp. 5–24, Springer-Verlag, Berlin.

Webb, L.J. 1968. Environmental relationships of the structural types of Australian rain forest vegetation. Ecol. 49: 296–311.

Webb, L.J. 1978. A general classification of Australian rain forests. Aust. Plants 9: 349–63.

Webb, L.J. & Tracey, C.R. 1979. An Ecological Survey of the Monsoon Forests of the North-western Region of the NT. Unpublished report to Australian National Parks and Wildlife Service, Canberra. 70 pp.

Webb, L.J. & Tracey, C.R. 1981. Australian rainforests: patterns and change. In: Keast, A. (ed.) Ecological Biogeography of Australia. pp. 605–94. W. Junk, The Hague-Boston-London.

Werner, P.A. 1986. Population Dynamics and Productivity of Selected Forest Trees in Kakadu National Park. Report to Australian National Parks and Wildlife Service. Canberra. 63 pp.

Williams, M.A.J. 1991. Evolution of the landscape. In: Haynes, C.D., Ridpath, M.G. & Williams, M.A.J. (eds) Monsoonal Australia: Landscape, Ecology and Man in the Northern Lowlands. pp. 5–17, Balkema, Rotterdam.

Wilson B.A. 1992. The Open-Forest – 'Treeless Plains' Boundary on Melville Island, Northern Territory. Unpublished MSc thesis, Department of Geography, University of Tasmania, Hobart. 198 pp.

Wilson, B.A. & Bowman, D.M.J.S. 1987. Fire, storm, flood and drought: the vegetation ecology of Howards Peninsula, Northern Territory, Australia. Aust. J. Ecol. 12: 165–174.

Wilson, B.A. & Bowman, D.M.J.S. 1994. Factors influencing tree growth in tropical savanna: studies of an abrupt Eucalyptus boundary at Yapilika, Melville Island, northern Australia. J. Trop. Ecol. 10: 103-120.

Wilson, B.A., Dunlop, C.R., Brocklehurst, P.S., Clark, M.J. & Barritt, M.J. 1989. The Vegetation of an Area of Mt Bundy Station – Proposed Training Area for the Second Cavalry Regiment. Technical Memorandum 89/6. Conservation Commission of the Northern Territory, Darwin. 73 pp.

Wilson, B.A., Brocklehurst, P.S., Clark, M.J. & Dickinson, K.J.M. 1990. Vegetation Survey of the Northern Territory. Technical Report 49. Conservation Commission of the Northern Territory, Darwin. 222 pp.

Wilson B.A., Brocklehurst P.S. & Whitehead, P.J. 1991. Classification, Distribution and Environmental Relationships of the Vegetation of Major Flood plains, Northern Territory, Australia. Technical Memorandum 91/2, Conservation Commission of the Northern Territory. 114 pp.

Woinarski J.C.Z., Gambold N., Menkhorst K. & Braithwaite R.W. 1989. Wildlife of Kakadu National Park Stage III. Final Report to Australian National Parks and Wildlife Service. CSIRO, Darwin. 144 pp.

CHAPTER 5

Wetland vegetation

C. MAX FINLAYSON and COLIN D. WOODROFFE

Abstract. Wetlands in the Kakadu region include mangroves, freshwater flood plains, salt flats and small permanent lakes. One of the major environmental changes in the region over the last 6000 years has been the reduction in mangrove swamps as freshwater flood plains were formed. Mangroves are now restricted to a narrow band along the coast and the tidal reaches of the rivers. Zonation within the mangroves is greatly influenced by the high tidal range with some forests extending and others receding. The tall creek-side mangroves are relatively productive compared to mangroves elsewhere with litterfall values of around 1 kg m^{-2} y^{-1}. The freshwater wetlands are found along the major rivers and are greatly influenced by the seasonal hydrological cycle. During the wet season the flood plains are covered with 1–2 m of water and contain a multitude of plants. During the dry, in contrast, the plains are parched and the vegetation is sparsely distributed. The distribution of plants on the flood plains has been well documented over the past decade. Many of the plants on the flood plain are annuals and their relative abundance and distribution varies seasonally. Seasonal changes in the distribution and biomass dominance of the vegetation are greatly affected by the flooding patterns, although feral animals and exotic weeds have also exerted an influence. The widespread grasslands are very productive with dry weight biomass values ranging from 0.5–1.1 kg m^{-2} y^{-1}. The *Melaleuca* forests are similarly productive with litterfall values of 0.7–1.5 kg m^{-2} y^{-1}.

1. Introduction

Wetland classification and definition is a difficult and often contentious subject (see Finlayson & van der Valk 1995). The most widely accepted definition of wetlands is that adopted by the Ramsar Convention on Wetlands of International Importance – areas of marsh, fen, peatland or water, whether natural or artificial, permanent or temporary, with water that is static or flowing, fresh, brackish or salt, including areas of marine water the depth of which at high tide does not exceed six metres (Lyster 1985). Even though there has been widespread dissatisfaction with this definition in Australia (see McComb & Lake 1988) it has more recently been generally accepted (Phillips 1993). Accepting the Ramsar definition, then wetlands in the Kakadu region encompass coastal and riverine mangrove fringes and salt flats, freshwater seasonally inundated flood plains and billabongs, and small permanent lakes (Finlayson *et al.* 1988).

The concept of a wetland plant used here is also broad and is defined simply as a plant that is adapted to withstand periods of waterlogging or flooding for a period of time. Under this general definition plants found in the seasonally inundated areas (woodlands and grasslands) that fringe the billabongs and flood plains are included along with the wholly aquatic species. Thus, the broad approach to both marine and freshwater wetlands follows that of Finlayson & von Oertzen (1993).

The wetlands of Kakadu National Park have been designated as internationally important under criteria established by the Ramsar Convention. This recognition was due, in part, to their importance in a biogeographical context, the outstanding diversity of their plant communities and their role in conserving the large numbers of waterfowl that congregate on the flood plains during the dry season (Ramsar Convention Bureau 1990). World Heritage listing of Kakadu National Park also made reference to the natural heritage value of the tidal flats and flood plains, and the floristic diversity and endemism of the wetland vegetation (ANPWS 1980, 1987). Under the terms of the Ramsar Convention this recognition requires the managing authority (Australian Nature Conservation Agency – ANCA) to make wise use of the wetlands, and under the World Heritage Convention, to conserve and preserve them.

C.M. Finlayson and I. von Oertzen (eds), Landscape and Vegetation Ecology of the Kakadu Region, Northern Australia, 81-112.

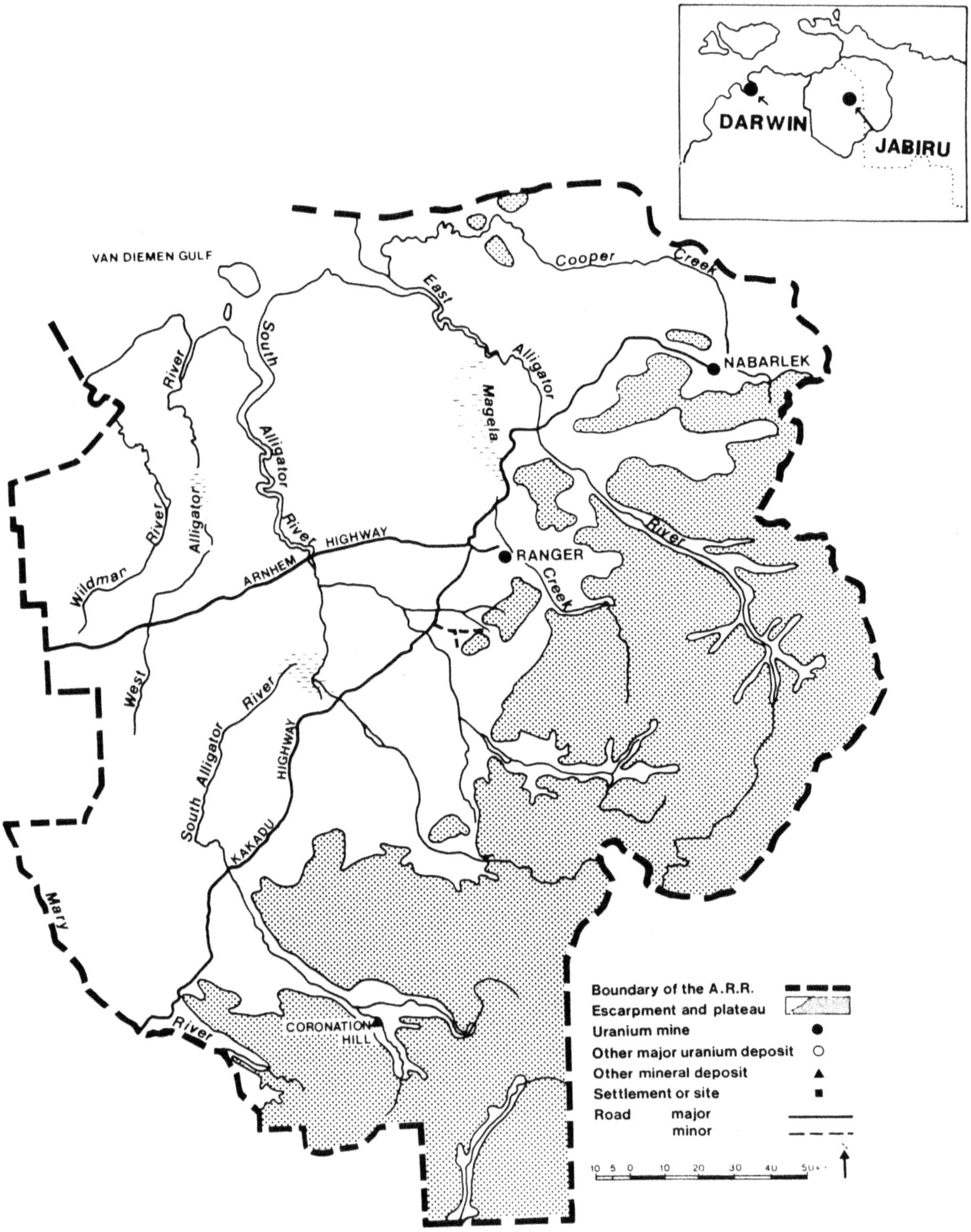

Fig. 1. Map of the Kakadu region showing the major rivers and creeks. The inset shows the location of the region in northern Australia.

Studies of the wetland vegetation in the region have centred on the mangroves of the South Alligator (e.g. Woodroffe *et al.* 1985a) and East Alligator Rivers (Hegerl *et al.* 1979), and the flood plains of the South Alligator (Taylor & Friend 1984) and Magela Creek (e.g. Sanderson *et al.* 1983, Finlayson *et al.* 1989). Little attention has been given to the vegetation of the salt flats and freshwater lakes. Other information on the vegetation is available from studies on feral buffalo (Williams & Ridpath 1982), alien plants (Cowie *et al.* 1988) and general plant taxonomy (Cowie & Finlayson 1986, Brennan 1990).

As a prelude to describing the compositional attributes of the vegetation of the mangrove and seasonally inundated freshwater wetlands an account of the evolutionary history of wetlands in the region is given. Threats and management problems associated with the wetland vegetation are also discussed (e.g. the occurrence of alien plant and exotic animal species). Locations mentioned in the text are shown in Fig. 1.

2. Wetland evolution

Stratigraphic studies on the wetlands of the South Alligator River have demonstrated that substantial environmental change has occurred over the last 20 000 years (Woodroffe *et al.* 1986, 1987). Many of these changes are attributed to changes in the sea-level over this period. At the peak of the last glaciation, about 18 000 years ago, the sea levels across northern Australia were some 150 m lower than today, the coastline was several hundred kilometres north of the present location (Fig. 2) and the rivers were fluvially dominated. The sea has since risen and around 8000 years ago tidal water flooded into the South Alligator River valley which is now filled to a depth of 14–16 m with fine muds and alluvium (Woodroffe *et al.* 1986, 1987). This sequence of events is depicted in a simplified sea-level envelope (Fig. 2).

The stratigraphy of the deltaic-estuarine plain of the South Alligator River, and probably the other river systems in the region, indicates that the wetlands developed in three major phases: i) transgressive; ii) big swamp; and iii) sinuous/cuspate. There has been a series of studies which generally support the notion of widespread vegetational changes on the wetlands over the past 8000 years or so (Chappell & Grindrod 1985, Woodroffe *et al.* 1985a, Hope *et al.* 1985, Russell-Smith 1985, Grindrod 1988, Clark & Guppy 1988). The major changes over this period are described below and are illustrated in Fig. 3.

The transgressive phase (8000–6800 years ago) was characterised by marine incursion into the prior valley and terrestrial ecosystems were displaced by mangrove forests as the sea level rose. The pollen record (Fig. 4) shows that by 6800 years ago a mangrove forest dominated by Rhizophoraceous species had become established.

The onset of the 'big swamp' phase (6800–5300 years ago) is indicated by the accumulation of estuarine muds containing abundant mangrove remains. During this period the sea stabilised around its present level (Woodroffe *et al.* 1985a, 1986) and mangrove forests established over most of the present-day flood plains for around 6000 years (Fig. 3). Pollen records show vegetation changes with *Rhizophora* slowly gaining dominance over *Bruguiera/Ceriops* species. A brief period of dominance by *Sonneratia* sp. possibly reflects the stabilisation of the present day sea-level. Subsequent succession from *Sonneratia* through Rhizoporaceous forest to stands of *Avicennia* indicates the replacement of mangroves normally found in seaward locations by species more characteristic of landward positions. The similarity of pollen diagrams from shallow drill holes in the Magela Creek area (Clark & Guppy 1988) and those from river systems elsewhere in the Northern Territory and in Western Australia (summarised in Woodroffe 1988) implies that the 'big swamp' phase was a widespread phenomena across north-western Australia.

The sinuous/cuspate phase began about 5300 years ago and represents a period during which the South Alligator River re-established a distinct channel meandering across the estuarine plains. This channel has changed during the last 2500 years from an entirely sinuous form to a predominantly cuspate form in the central part of the tidal section. Associated with the transition from the 'big swamp' to the 'sinuous' phase the pollen record shows widespread replacement of mangrove species with the grasses and sedges characteristic of the present day freshwater wetlands.

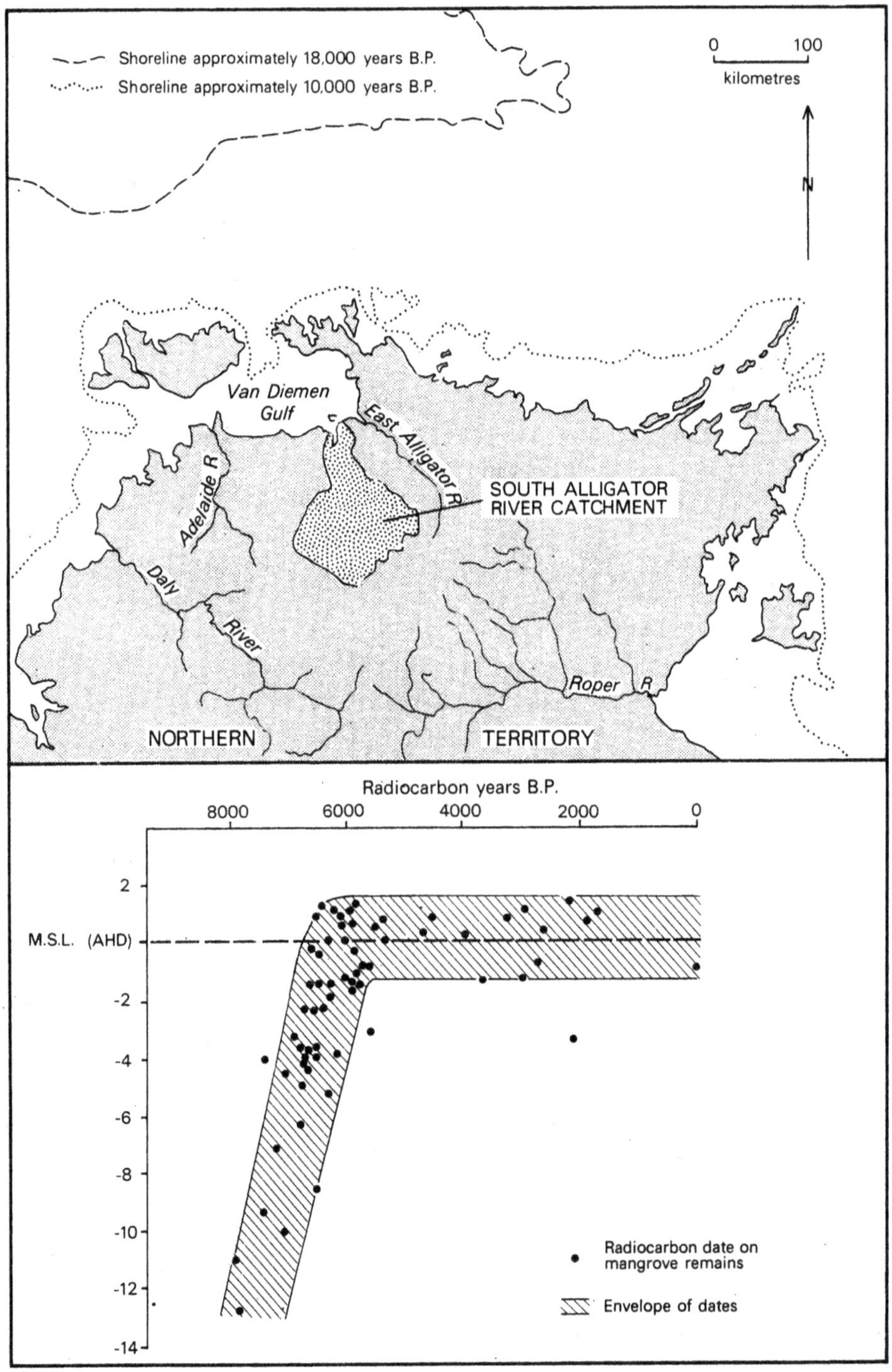

Fig. 2. Sea-level history in the Kakadu region: a) South Alligator River catchment and the shorelines of approximately 18 000 and 10 000 years ago; and b) Holocene sea-level envelope based on radiocarbon dates on mangrove sediments (after Woodroffe *et al.* 1987).

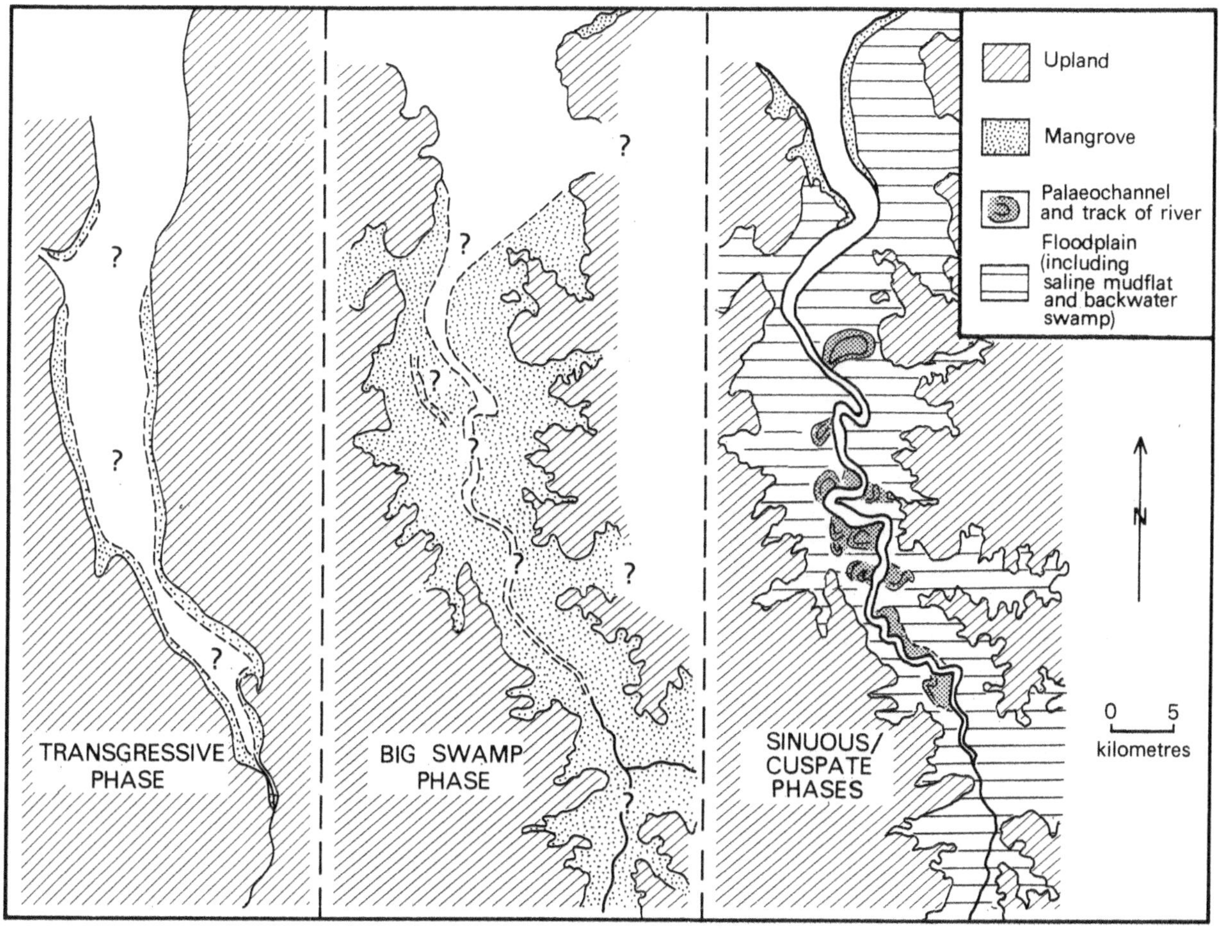

Fig. 3. Transgressive, big-swamp and sinuous/cuspate phases in the development of the South Alligator tidal river and plains (after Woodroffe 1988).

The timing of the transition from mangroves to freshwater wetlands probably varied throughout the region. Radiocarbon dates from middens on the mid-South Alligator flood plain indicate that mangroves had disappeared from these sites by about 4000 years ago (Woodroffe *et al.* 1986). Further upstream, the transition does not appear to have occurred until 1500 years ago (Jones 1985). Clark & Guppy (1988) also suggest that the upstream section of Magela Creek did not change from an estuarine swamp to a freshwater flood plain until about 1500 years ago.

3. Mangrove swamps

3.1. General description

Mangroves are halophytic trees or shrubs which dominate sheltered, muddy, intertidal environments along tropical and subtropical shorelines. The term 'mangrove' can refer either to the individual plants, or to the entire mangrove forest. In the Northern Territory, mangrove forests are found along river banks and the coastline and cover about 4120 km^2 (Galloway 1982) which represents about 35% of the total mangrove area in Australia. The most extensive mangrove areas in the Northern Territory are associated with large embayments, such as Arnhem Bay and

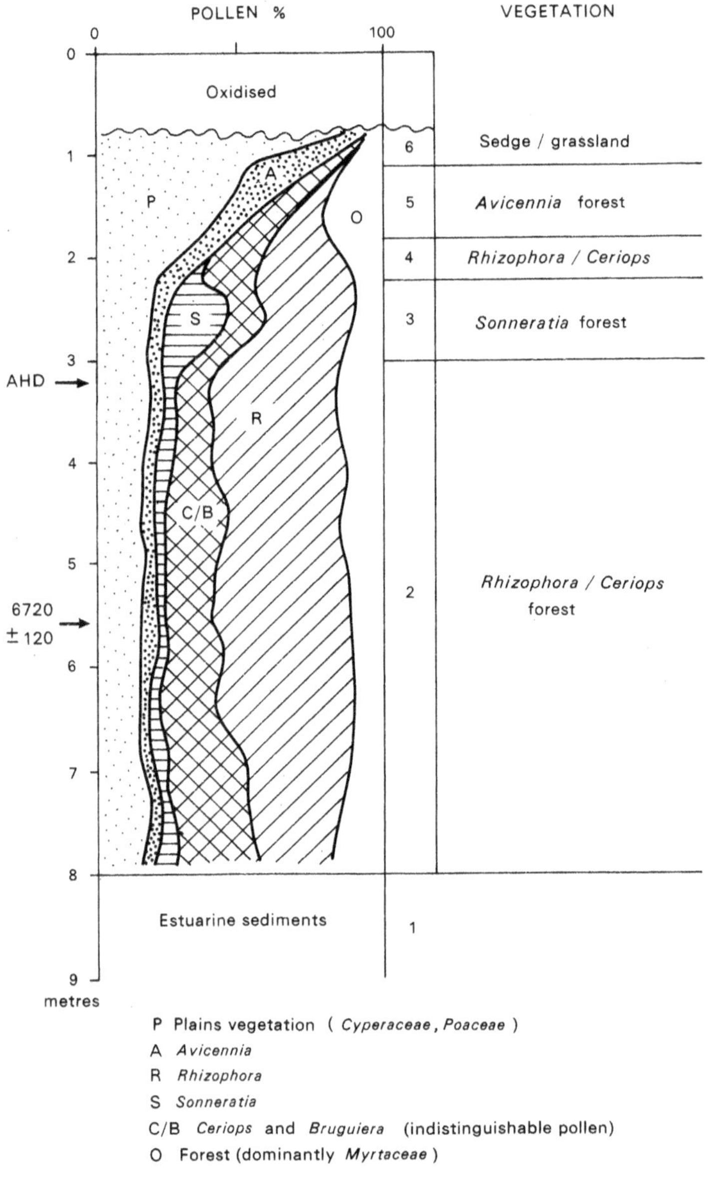

Fig. 4. Pollen diagram from (drillhole SAH 40) mid-plains South Alligator River (after Chappell & Grindrod 1985).

Darwin and Bynoe Harbours. These extensive mangrove areas are distinct from the mangrove forests that fringe the large macrotidal river systems, in terms of both their florisitic composition and their past development and successional trends (Wells 1985).

In the Kakadu region mangroves do not cover an especially large area, but they are floristically diverse (Dames & Moore 1984, Wightman 1989). Extensive stands occur along the Wildman, West Alligator, South Alligator and East Alligator Rivers with those of the South Alligator River being described in some detail. The broad distribution of mangrove on the West Alligator and South Alligator Rivers is shown in Fig. 5 (based on Story 1969, 1976).

Rivers in the Kakadu region have a pronounced seasonal flow pattern and drain into the van Diemen Gulf which has a spring tidal range of around 6 m. The high tidal range means that

Fig. 5. Vegetation types in the lower reaches of the South Alligator and West Alligator Rivers (based on Story 1969, 1976).

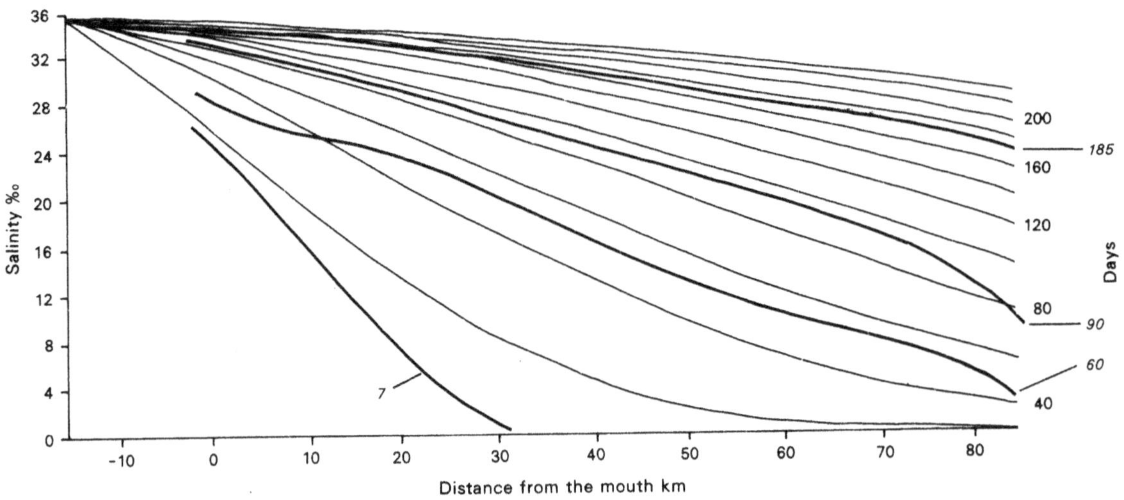

Fig. 6. Salinity changes with distance upstream in the South Alligator River through the dry season. Fine lines represent predicted salinity for number of days after the last wet season flood peak (prediction based on turbulent diffusion of salt). Solid lines represent observations. Agreement between model and observations is good (after Woodroffe *et al.* 1986).

strong tidal currents with velocities up to 2 m s^{-1} at spring tides occur in the rivers (Wolanski *et al.* 1988). The tidal reaches of the rivers are dominated by fluvial processes in the wet season, and by tidal processes in the dry season. The rivers are fresh during the wet season, although still tidally reversing throughout much of the channel, with a salt wedge at the mouth (Woodroffe *et al.* 1986). Saline water penetrates progressively upstream during the dry season (Fig. 6).

3.2. Vegetation description

There are 18 mangrove species commonly found in the Kakadu region. In addition, Wightman (1989) records *Rhizophora apiculata* and *Rhizophora lamarkii* (a hybrid between *Rhizophora stylosa* and *Rhizophora apiculata*), and a series of other plants frequently observed in the mangrove forest, including *Hibiscus tiliaceus, Thespesia populneoides, Diospyros littorea*, the succulents *Batis argillicola, Halosarcia indica* and *Sesuvium portulacastrum*, the grasses *Cynodon dactylon* and *Sporobolus virginicus*, the fern *Acrostichum speciosum*, the creeper *Derris trifoliata*, and the epiphytes *Amyema mackayense* and *Cynanchum carnosum.*

The distribution of the main mangrove species along the tidal reach of the South Alligator River (based on Wells 1984 and supplemented from Messel *et al.* 1979, Davie 1985 and C.D. Woodroffe personal observations) is shown in Fig. 7. Generally, the mangroves occupy a narrow fringe along the river banks. Near the mouth this may be more than 50 m wide, but it soon decreases to a discontinuous fringe little more than 10 m wide. Along much of the river this fringe is reduced to a line of scattered trees. In the upstream section of the river there is a discontinuous fringe of *Sonneratia lanceolata* and *Avicennia marina.*

Differences in the frequency of tidal inundation along the van Diemen Gulf coastline results in a clear zonation of mangrove species. Although the detailed zonation pattern has not received a great deal of attention it is considered to be similar to that in Darwin Harbour which is shown in Fig. 8 (Woodroffe & Bardsley 1988). Fig. 8 plots the elevational occurrence of a series of mangroves in Darwin Harbour, where the maximum tidal range is 7.2 m. The five distribution patterns shown by individual species are similar to those along the shore of the van Diemen Gulf, except that the lower tidal range restricts the landward limit.

While the zonation pattern in the mouth of the Alligator Rivers is similar to that on the coast – *Sonneratia, Rhizophora, Ceriops, Avicennia* – the occurrence and elevational range of species changes often, as commonly happens along the tidal length of tropical rivers (Bunt *et al.* 1982,

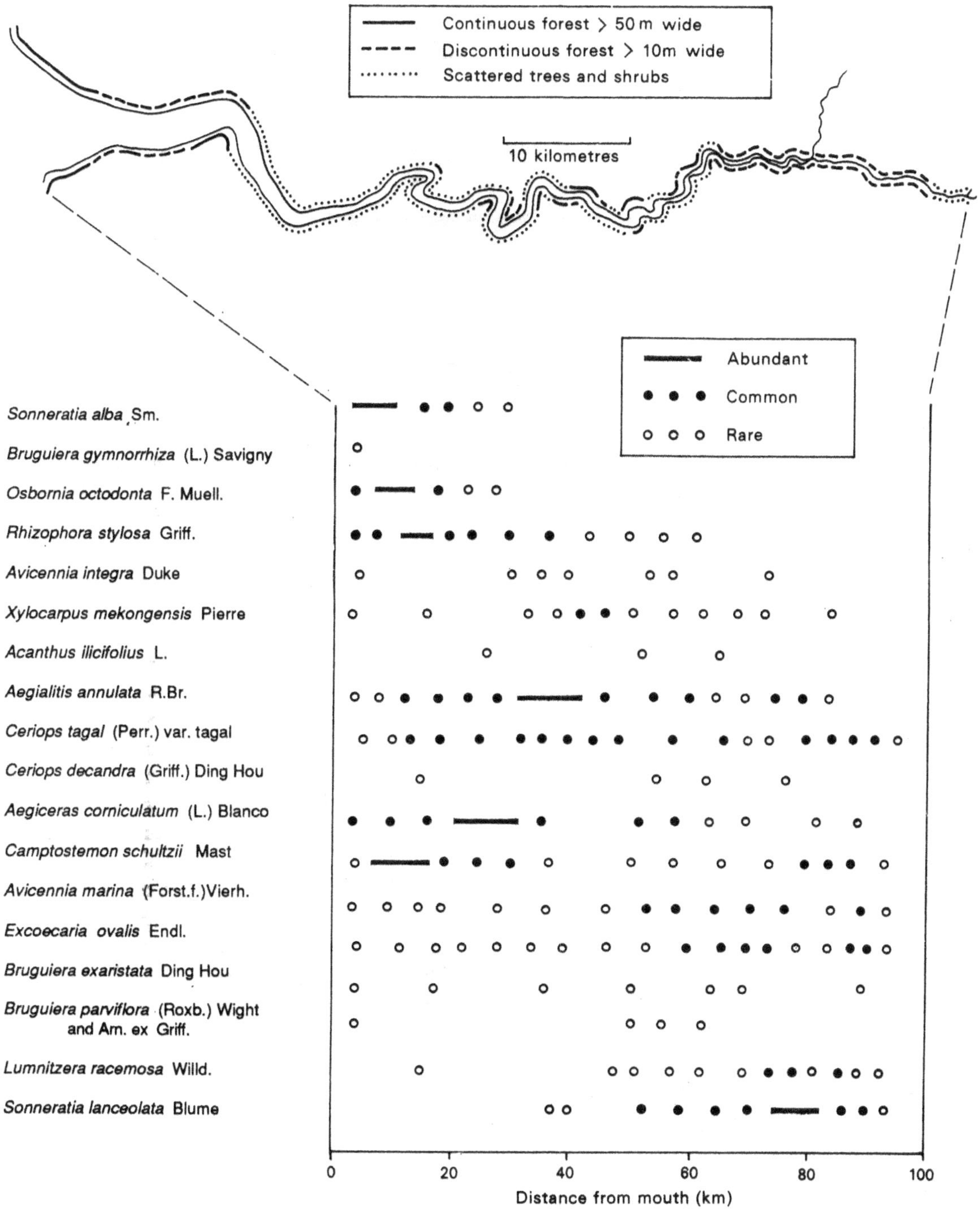

Fig. 7. Distribution and composition of mangrove forests along the South Alligator River (from Woodroffe *et al.* 1986, based on Wells 1984 and supplemented from Messel *et al.* 1979 and Davie 1985).

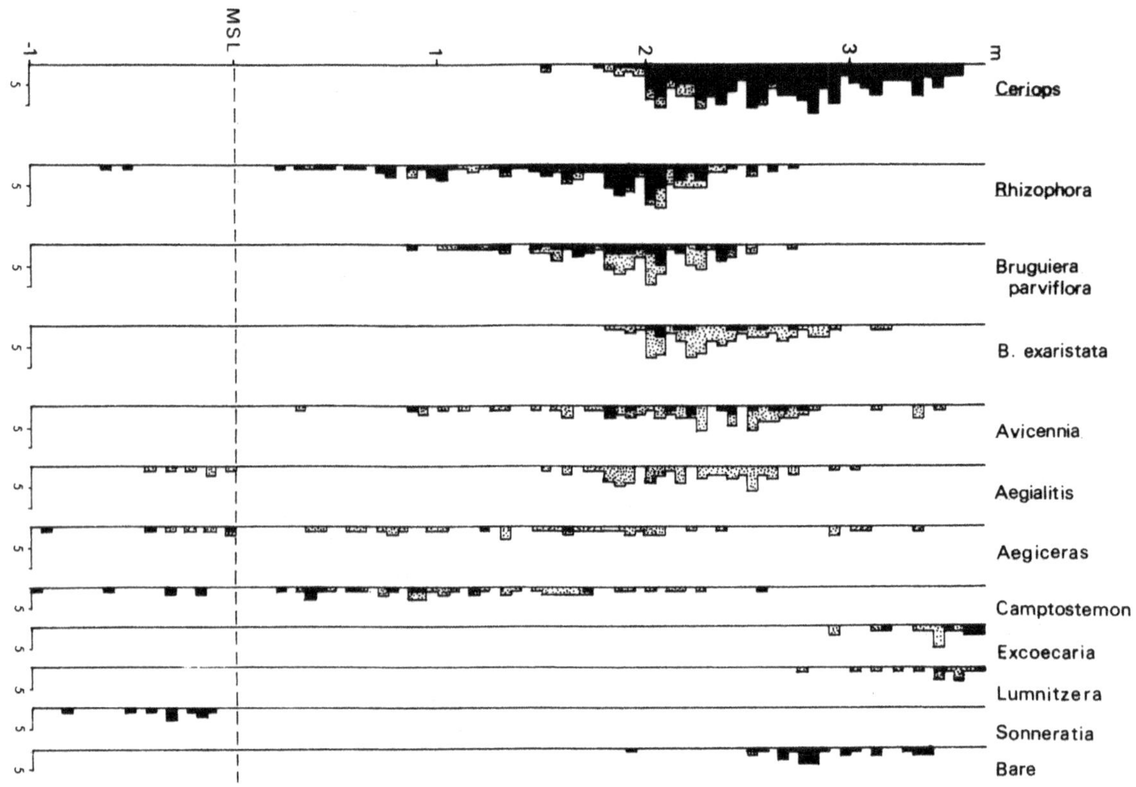

Fig. 8. Frequency of occurrence of mangrove species in elevational ranges in Darwin Harbour (black indicates species is dominant at a particular height; stiple indicates it is present; MSL represents mean sea level) basec on 10 transects through Creek H, East Arm (after Woodroffe & Bardsley 1988).

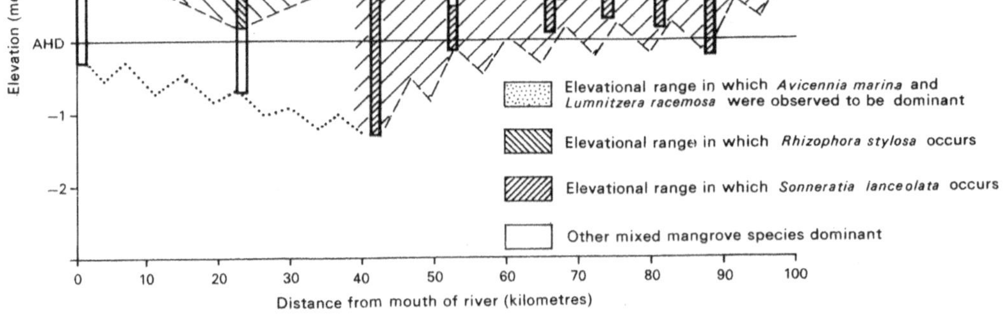

Fig. 9. Elevational range of mangroves along the South Alligator River (from Woodroffe *et al.* 1986).

1985). The most obvious changes along the South Alligator (Fig. 9) include *Rhizophora* not being found more than 68 km upstream, *Sonneratia alba* only dominating the seawardmost areas until 25 km upstream from the mouth, and *Sonneratia lanceolata* progressively occupying a larger part of the intertidal zone from 40 km to the tidal limit 105 km upstream from the mouth. The higher parts of the mangrove forests generally contain *Avicennia marina*, *Excoecaria ovalis* and *Lumnitzera racemosa*, and often *Clerodendrum inerme* and *Cathormion umbellatum*, and *Bombax ceiba* on the discontinuous silty levees. The salinity of the water is a major factor determining the distribution of mangrove species along the course of tropical rivers (Bunt *et al.* 1982, Wells 1984, Ball 1988, Ball & Pidsley 1988).

Along the tidal parts of the South Alligator there is evidence that the various channel segments recognised within the tidal reaches of the river (Fig. 10) are actively changing (Woodroffe *et al.* 1986). Stratigraphy and radiometric dating of sediments have demonstrated that sediments deposited by lateral accretion underlie the inner bends of sinuous meanders which are eroding their outer banks (see Fig. 10). These meanders are actively migrating. In many parts of the estuarine funnel and cuspate meandering segments the banks are undergoing erosion and bank slumping is common. Such changes can affect the distribution

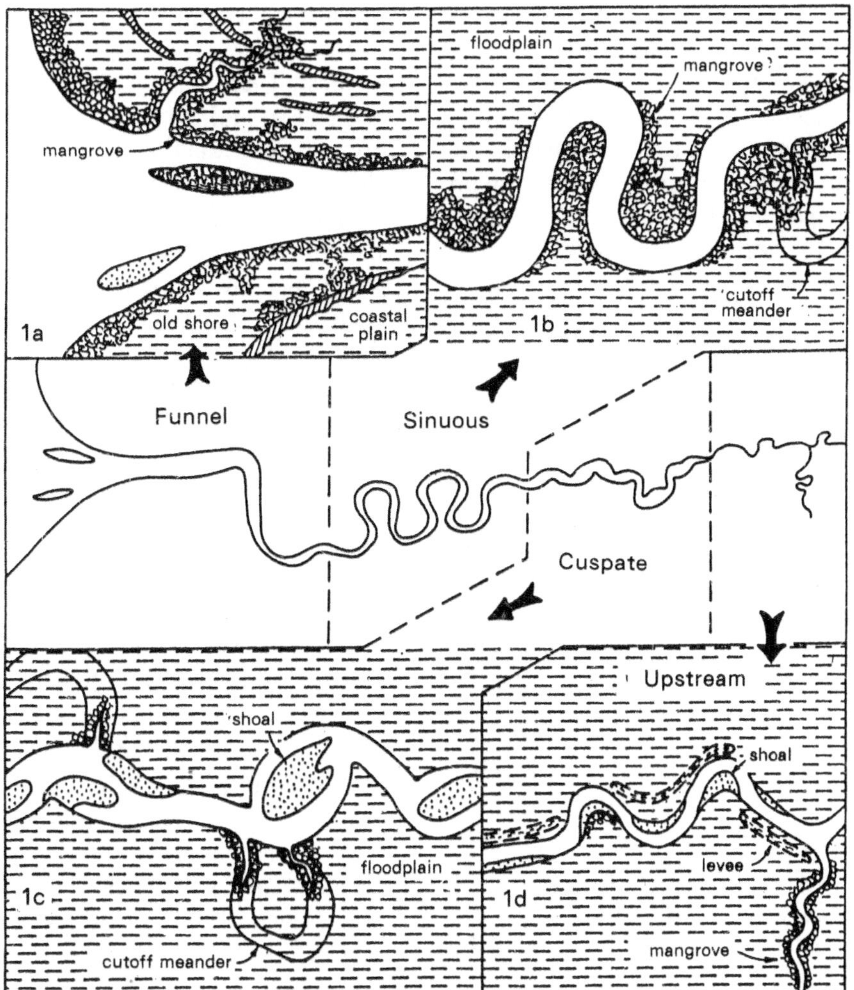

Fig. 10. Schematic tidal river showing typical channel segments (from Chappell & Woodroffe 1985).

of mangrove species along the river bank with the zonation pattern being truncated in places and more landward zones being exposed to the active erosional processes of the river.

3.3. Mangrove productivity

In recent decades it has been recognized that mangrove forests represent very productive ecosystems and play an indispensable role in the supply of above-ground primary production to adjacent nearshore communities. There are no data available on the rate of primary production of mangroves in the Kakadu region. However, productivity measurements in Darwin Harbour can be extrapolated to the region, with caution, in view of the broad similarity of the mangrove communities. Mangrove productivity in Darwin Harbour was measured as litterfall over a three year period, from 1984–87 (Woodroffe & Bardsley 1988, Woodroffe *et al.* 1988; Table I). These values are, in reality, a surrogate measure of productivity as they do not account for changes in below-ground biomass or changes in the trunk and branch biomass. Rather, they measure the contribution of the mangrove forest to the detrital-based food-web.

Litterfall under monospecific *Ceriops tagal* trees with an average height of 4.2 m was 0.69 ± 0.14 kg m^{-2} y^{-1} compared to 0.30 ± 0.04 kg m^{-2} y^{-1} beneath trees with an average height of 2.5 m. The *Ceriops* thicket is highest on the landward edge and decreases in stature until it reaches an area of high salinity, which is free of mangroves. Litterfall seems to parallel the trend in mangrove structure. In a mixed *Ceriops tagal* and *Bruguiera exaristata* stand, seaward of the bare area, with an average height of 3.5 m, litterfall was 0.75 ± 0.07 kg m^{-2} y^{-1}; each species contributing in roughly equal proportion. Mangrove trees on creek-bank sites had even higher litterfalls. At one creek-bank site, litterfall averaged 1.22 ± 0.19 kg m^{-2} y^{-1} and 1.10 ± 0.09 kg m^{-2} y^{-1} beneath 10.8 m and 10.0 m high *Rhizophora stylosa* trees respectively and 1.40 ± 0.19 kg m^{-2} y^{-1} beneath 13 m high *Avicennia marina* (Woodroffe & Bardsley 1988, Woodroffe *et al.* 1988).

Table I. Average annual dry weight (g m^{-2}) of litterfall (± standard deviation) beneath mangrove species in Creek H, Darwin Harbour from 1984–87. (from Woodroffe *et al.* 1988)

Location	Species	Leaves	Total
A	*Ceriops tagal*	502 ± 108	686 ± 140
B	*Ceriops tagal*	213 ± 37	300 ± 40
C	*Rhizophora stylosa*	852 ± 113	1158 ± 165
	Bruguiera parviflora	56 ± 56	63 ± 63
D	*Ceriops tagal*	290 ± 33	362 ± 44
	Bruguiera exaristata	319 ± 25	383 ± 30
E	*Avicennia marina*	847 ± 107	1251 ± 180
	Bruguiera parviflora	110 ± 68	132 ± 82
	Rhizophora stylosa	17 ± 10	20 ± 12
F	*Rhizophora stylosa*	686 ± 35	999 ± 45
	Avicennia marina	50 ± 33	68 ± 43
	Bruguiera parviflora	26 ± 16	30 ± 17

These estimates suggest a pattern in which litter production is greatest beneath the taller creek-bank stands of mangroves, where biomass is greatest. There is a peak in litter production in the wet season and leaf litter accounts for the largest proportion of the total (Fig. 11). The wet season is also the time of fruit production for most of the mangroves. The rate of litter production is spatially variable within these macrotidal, mangrove-lined embayments. However, the more productive tidal-creek mangroves are as productive as similar mangrove forests in more equatorial settings (Sasekumar & Loi 1983, Leach & Burgin 1985).

3.4. Mangroves and succession

Mangrove forests generally occupy coasts that are actively accreting mud and the vegetation pattern changes over both short- and long-term time scales. The concept of succession in mangroves was strongly perpetrated by Davis (1940) in south-western Florida. Many mangroves (especially the Rhizophoraceae) produce viviparous fruit that are able to establish in shallow sea water and it was assumed that stands of seedlings on the seaward margin of mangrove forests were pioneer individuals in shallow water habitats, and that they were stages of succession towards a climax terrestrial vegetation. This view gave rise to the notion of mangroves 'making land from the sea'.

Evidence for succession in mangroves is available from stratigraphic studies in the Kakadu region (Woodroffe *et al.* 1986, 1989) where the

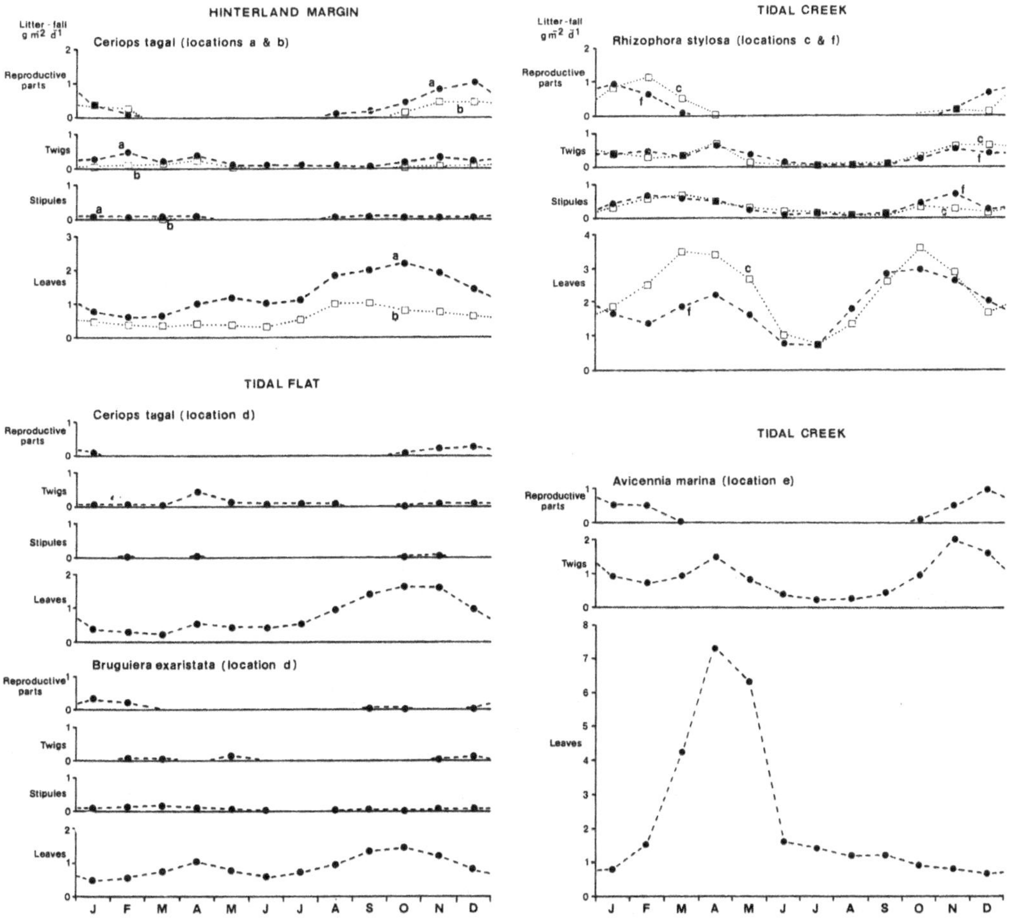

Fig. 11. Average rate of litterfall (g m^{-2} d^{-1}) beneath mangrove species in different habitats (hinterland margin, tidal flat and tidal creek) in Darwin Harbour (from Woodroffe *et al.* 1988).

clearest zonation is on the 4–5 km wide coastal plain that appears to have progressed seaward since sea-level stabilised around 6000 years ago (Woodroffe *et al.* 1986, 1989). The zoned mangrove forest at the seaward margin of the coastal plain may represent an important element within the progradation of the plain. Replacement of the seaward zones by progressively more landward species at the rear of the forest would be consistent with succession as a part of progradation and is supported by several observations. Firstly, the surface gradient through the forest is much steeper than that to seaward or that behind the mangroves towards dry land. This is especially noticeable through the *Rhizophora* zone, where a steep rise in elevation of 1 m or more is evident and suggests more rapid accretion of mud in this zone than in the more seaward *Avicennia* and *Sonneratia* open forest. Secondly, the landward margins of mangrove zones are often marked by dead trees. This is particularly true of the landward margin of the *Ceriops* zone where, in places, a band of dead trees several metres wide may be present.

However, it is not always the case that zonation of coastal vegetation indicates succession. Thom (1967) found that mangrove zonation was unrelated to succession in a deltaic setting. Furthermore, many mangrove forests may be zoned in a steady-state fashion in response to a static environmental gradient and there is no replacement of zones (Lugo 1980). It is also important to emphasise that zoned mangroves are probably a relatively weak force in the overall progradation processes which are usually driven by more powerful marine currents and tidal flows (see Wolanski *et al.* 1988, Woodroffe *et al.* 1986). Within the

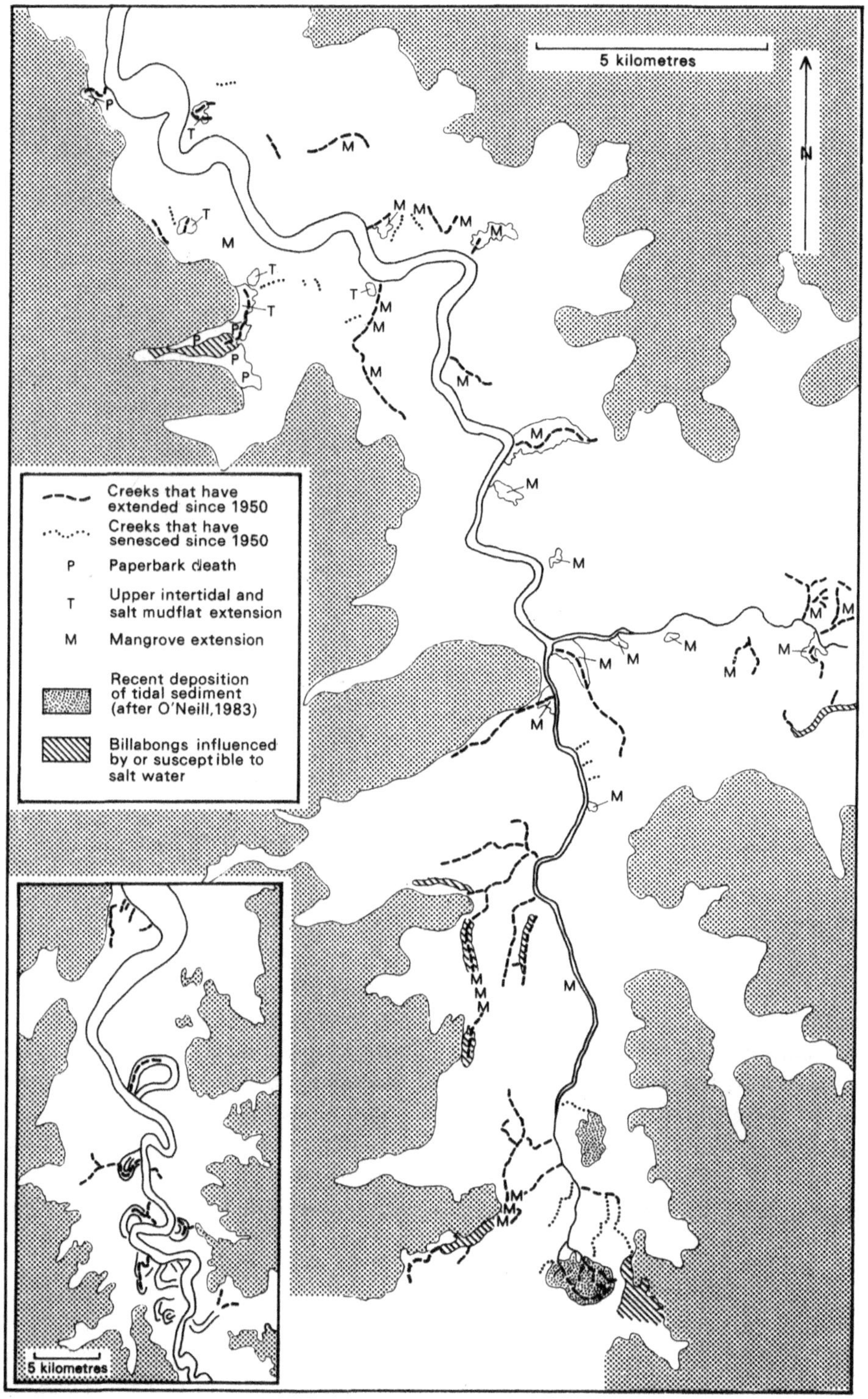

Fig. 12. Recent channel changes and mangrove extension on the South Alligator River plains derived from a comparison of 1950 and 1983 aerial photography.

estuarine funnel of the South Alligator River, bank erosion is a characteristic feature and the mangrove zonation has been, and is often continuing to be, truncated. Erosion has often removed the seaward-most *Sonneratia*, and in many cases, has also resulted in collapse of *Rhizophora* trees, in some cases exposing *Ceriops*, or the stunted, landward *Avicennia* and *Lumnitzera* trees on the riverbank. This process has been described on other mangrove-lined estuaries in northern Australia (Semeniuk 1980).

Elsewhere within the region there is evidence of mangrove extension. This has been particularly evident over the last three or four decades in response to the lengthening of a series of tidal creeks (Fig. 12). Extension of these creeks has almost always been along previous palaeochannels and has been accompanied by death of *Melaleuca* trees, as a result of saltwater intrusion, and replacement by the landward-most mangroves, *Avicennia marina* and *Sonneratia lanceolata*. Mortality of *Melaleuca* trees has also been recorded on the Magela flood plain (Williams 1984), and on the Mary River (Stocker 1970, Knighton *et al.* 1991). The latter have been ascribed to buffalo induced damage, but there is little direct evidence to implicate buffalo except in localised areas where their activity has accelerated the geomorphological processes of channel abandonment and reoccupation.

4. Freshwater wetlands

Freshwater wetlands in the Kakadu region comprise seasonally inundated flood plains/basins and billabongs, and small permanent lakes. The latter are small and not common. The areas generally referred to as flood plains are in fact more accurately described as organic clay flood basins and occur in the lower reaches of the rivers. However, throughout the region they are known as flood plains, although this does lead to confusion with the less widesread sandy plains that flank the braided streams. The following description refers to the large organic clay basins (and associated billabongs) that cover about 195 000 ha and are found along the lower reaches of the major rivers (West Alligator, South Alligator and East Alligator) and creeks (Magela Creek and Cooper Creek) throughout the region (Fig. 1). Published botanical studies have centred on the flood plain of Magela Creek (Fig. 13), a tributary of the East Alligator River.

4.1. Hydrological cycle

The hydrological cycle has been identified as an important factor in shaping the characteristics of the vegetation in these wetlands (Williams 1979, Taylor & Dunlop 1985, Finlayson *et al.* 1989). Williams (1979) even related vegetation pattern to depth of flooding on Magela Creek. The extremes of flood and drought have resulted in a diverse flora with characteristic wet and dry season plant communities.

Water flows in the creeks and rivers on a seasonal basis. Flow commences early in the wet season and lasts until after the end of the rains. In Magela Creek, the flow consists of a series of flood events superimposed on a 'base' flow. Sanderson *et al.* (1983) have presented the flow pattern for Magela Creek diagramatically (Fig. 14). In brief, there is a period of intermittent rain storms that saturate the soils, followed by more consistent rains and widespread flooding. As rainfall increases continuous flow occurs in the braided channels and eventually spills out onto the flood plain. Floodwaters cover the plain and connect the permanent billabongs and swamps. In some cases there is no continuous channel running through the flood plain areas (e.g. on the Magela flood plain) and these areas are more strictly referred to as flood basins or backwater plains.

At the end of the wet season the creeks stop flowing and the water on the flood plains evaporates; in most years the plains dry out except for isolated pockets of water in billabongs and a few permanent swamps. The drying-out phase can take many months. Except for the tidally influenced channels the creeks also dry out completely, or are reduced to a series of waterholes (billabongs).

The flooding sequence on the flood plains is partly influenced by overbank flooding. Woodroffe *et al.* (1985b) described the morphology of the South Alligator River. The area generally referred to as the flood plain includes a deltaic-estuarine area and an alluvial plain. The greater part of the deltaic-estuarine area consists of a blacksoil plain that is separated from the river by

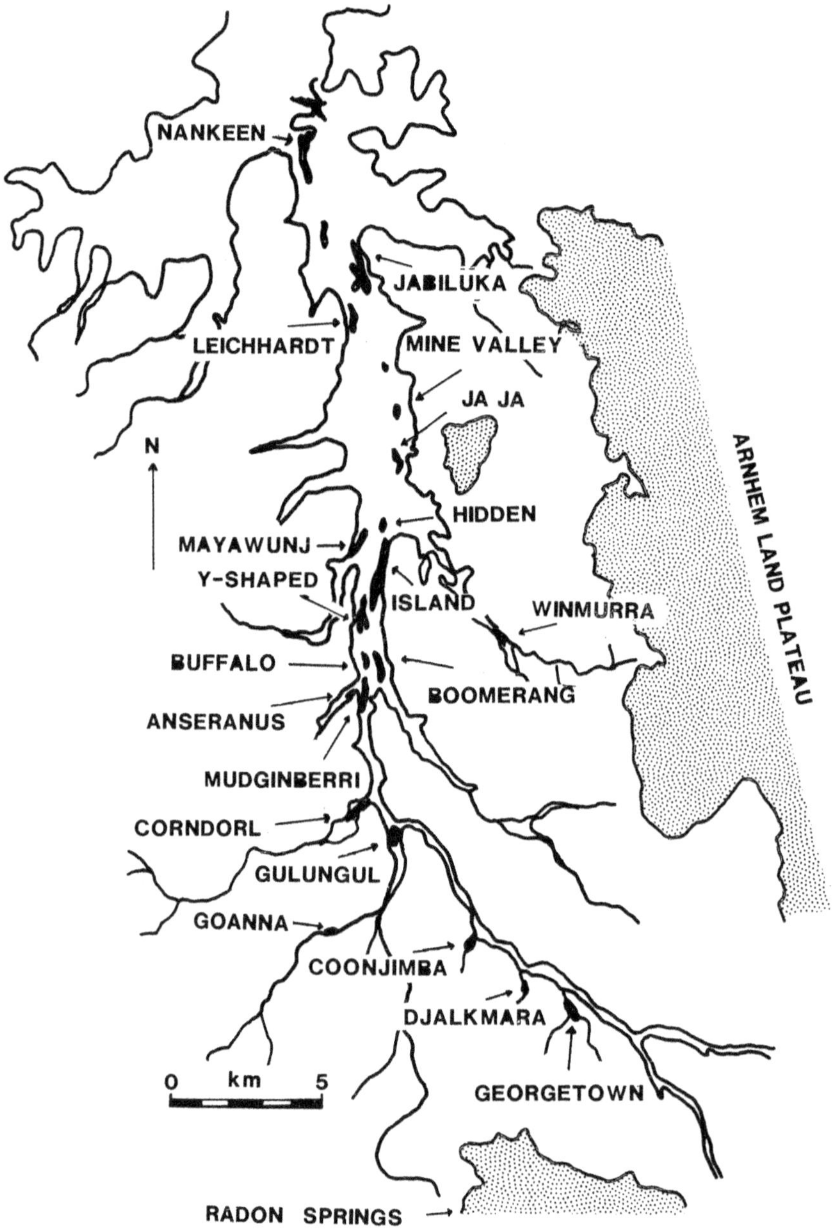

Fig. 13. Map of Magela Creek and floodplain showing the location of the major billabongs (after Finlayson *et al.* 1990a).

natural levee banks and is flooded by freshwater for at least 3–4 months of the year. Upstream of the tidal limit the river is not contained in discrete channels, but rather in multiple channels which disgorge into a broad seasonal basin. In places this merges with the swamps of the blacksoil plain. Flooding of the alluvial plain is entirely by freshwater.

4.2. Vegetation description

4.2.1. General vegetation patterns

A number of vegetation studies have included the freshwater wetlands of the Kakadu region. Lists of macrophytic plant species have been compiled e.g. by Adams *et al.* (1973), Lazarides & Craven (1980), Cowie & Finlayson (1986) and Brennan

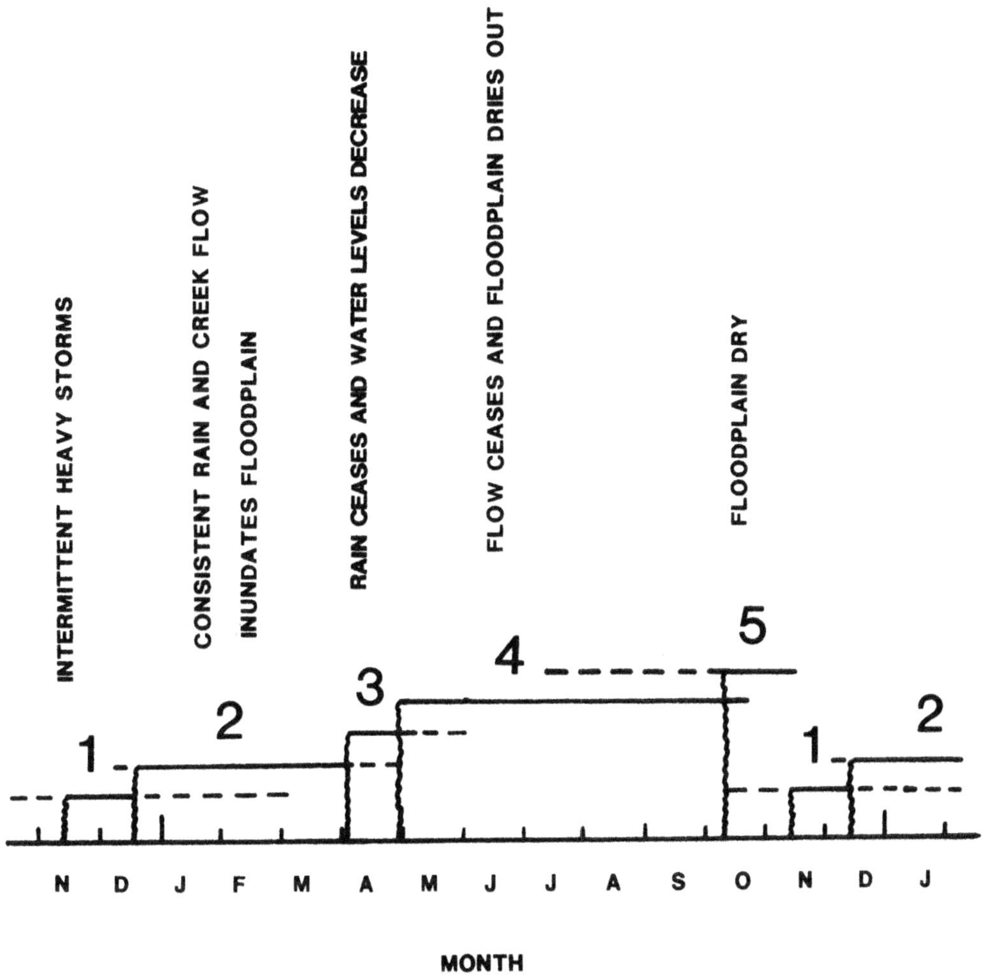

Fig. 14. Seasonal cycle on the Magela Creek and flood plain based on observed events from 1979–85 (adapted from Sanderson *et al.* 1983).

(1990). There are many discrepancies between these lists, partly as a result of the manner in which they were compiled. Except for Brennan (1990) the published lists do not specifically identify the wetland plant species. General vegetation studies of the wetlands in the region are briefly considered in this section, as an introduction to the more specific studies (Table II), considered later.

In a generalised structural and floristic description of freshwater wetlands of Australia, Briggs (1981) lists a number of wetland types that occur in the region. All of these are associated with the seasonally inundated flood plains (including the fringing woodland and forests) and/or billabongs and are briefly described below. Where the taxonomic nomenclature used by Briggs (1981) has been superceded, the more recent name is given in parentheses. In general, the nomenclature follows that in Dunlop (1987).

i) Paperbark swamp forests: dominated by tree species including *Melaleuca viridiflora*, *Melaleuca cajaputi*, and *Melaleuca leucadendron* (*M. leucadendra*), and to a lesser extent *Barringtonia gracilis* (*B. acutangula*) and *Pandanus* spp. These forests are inundated by up to a metre of water during the wet season and are dry at other times. Sedges and floating-leaved and submerged aquatic plants form an understorey. The relationship between these species and the hydrological regime is shown in Fig. 15, adapted from Briggs (1981).

ii) Eleocharis sedgelands: dominated by *Eleocharis dulcis* and *Eleocharis sphacelata* (and

Table II. Comparison of vegetation classifications of the Magela flood plain (from Finlayson *et al.* 1989).

Story (1976) 1:500 000	Williams (1979) 1:350 000	Sanderson *et al.* (1983) 1:100 000	Finlayson *et al.* (1989) 1:100 00
Paperbark forest (*Melaleuca leucadendra, Melaleuca argentea*)	**Forest** (*Melaleuca viridiflora, Melaleuca nervosa, Melaleuca leucadendra, Melaleuca cajaputi*)	**Tree communities not mapped** The following communities occur amongst woodland: – fringing herbland (*Najas tenuifolia, Blyxa* spp.) – fast-flow herbland (*Blyxa* spp., *Caldesia oligococca*) – fringing closed grassland (*Hygrochloa aquatica, Nymphoides* spp., *Pseudoraphis spinescens*) – grassland (*Pseudoraphis spinescens*)	**Melaleuca open forest and woodland** (*Melaleuca viridiflora, Melaleuca cajaputi, Melaleuca leucadendra*) **Melaleuca woodland** (*Melaleuca leucadendra*) **Mixed grass/sedgeland** occurs amongst woodland (*Oryza meridionalis, Hymenachne acutigluma, Pseudoraphis spinescens, Eleocharis* spp.)
	Mixed herbfield (*Eleocharis* spp., *Fimbristylis littoralis, Oryza meridionalis*)	**Grassland** (*Oryza meridionalis, Hygrochloa aquatica*) **Closed grassland** (*Hygrochloa aquatica*)	**Grassland** (*Oryza meridionalis*)
Herbaceous swamp (*Caldesia oligococca, Hymenachne acutigluma, Commelina lanceolata, Nymphoides* spp., *Nymphaea* spp.)	**Grassland** (*Pseudoraphis spinescens*) **Undulating annual swamp and grassland** (*Pseudoraphis spinescens, Panicum paludosum, Ludwigia adscendens, Polygonum attenuatum, Nymphoides indica*)	**Herbland** (*Najas tenuifolia, Nymphaea* spp., *Pseudoraphis spinescens*) (*Pseudoraphis spinescens*) **Grassland** (*Pseudoraphis spinescens*) **Sedgeland** (*Eleocharis* spp., *Pseudoraphis spinescens*)	**Grassland** (*Pseudoraphis spinescens*)
		Not present on area mapped	**Grassland** (*Hymenachne acutigluma*)
	Annual swamp (*Eleocharis* spp., *Nymphoides indica*) **Perennial swamp** (*Chara* sp., *Melaleuca leucadendra, Nelumbo nucifera, Hymenochaeta grossa, Hymenachne acutigluma*)	Not present on area mapped	**Herbfield** (*Nelumbo nucifera, Nymphoides indica, Nymphaea* spp., *Hymenachne acutigluma, Leersia hexandra, Ludwigia adscendens*)
		Deep water community (*Najas tenuifolia, Nymphaea* spp., *Leersia hexandra, Hymenachne acutigluma*)	**Open water community** (*Nymphaea* spp., *Najas tenuifolia, Hydrilla verticillata, Ceratophyllun demersum, Salvinia molesta*)
		Mixed herbland (*Hygrochloa aquatica, Limnophila* spp., *Nymphoides* spp.)	
Sedgeland (*Cyperus* spp., *Eleocharis* spp., Fuirena sp., *Scirpus* sp.)	**Mixed herbfield** (*Eleocharis* spp., *Fimbristylis littoralis, Orzya meridionalis*)	Not present on area mapped	**Sedgeland** (*Eleocharis* spp.) **Sedgeland-grassland** (*Eleocharis* spp., *Hymenachne acutigluma*)

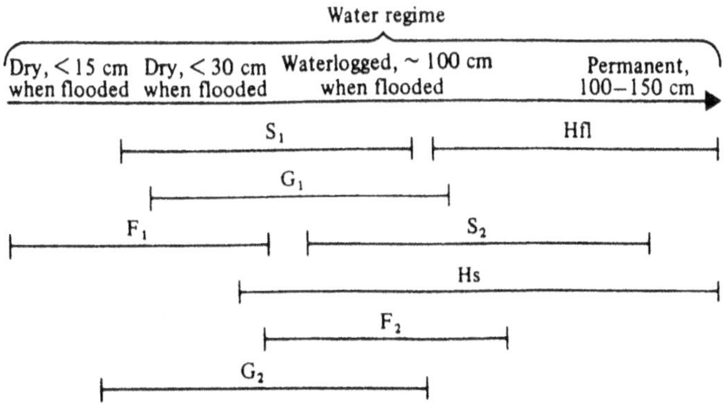

Fig. 15. Generalised zonation in wetlands on the Darwin sub-coastal plain, Northern Territory (F1 *Melaleuca nervosa/Melaleuca viridiflora* forest, F2 *Melaleuca leucadendra* forest, G1 *Oryza meridionalis* grassland, G2 *Pseudoraphis spinescens* grassland, Hfl floating-leaved herbland, Hs submerged herbland, S1 *Eleocharis dulcis* sedgeland, S2 *Eleocharis sphacelata* grassland) (adapted from Briggs 1981).

Eleocharis brassii) in association with other sedge and grass species, most notably *Oryza fatua* (*O. meridionalis*).

iii) Wet grasslands: extensive areas comprised of a host of species. This includes *Pseudoraphis spinescens, Leersia hexandra, Echinochloa* spp., and sedges, mainly *Cyperus, Fimbristylis* and *Fuirena* species.

iv) Floating and floating-leaved herblands: occurring on the seasonally inundated plains and in amongst the paperbark forests: typical species include *Nelumbo nucifera, Nymphaea capensis* and *Nymphaea gigantea* (most probably a mixture of *N. hastifolia, N. macrosperma, N. nouchali, N. pubescens, N. violacea*), and *Nymphoides* spp.

v) Submerged and emergent herblands: occur over a range of sites and may be associated with floating and floating-leaved communities. Common species are *Triglochin procera, Caldesia oligococca, Limnophila indica, Ludwigia adscendens, Utricularia* spp., *Eleocharis* spp. and *Vallisneria spiralis* (*V. gigantea*).

Detailed vegetation maps that included the freshwater wetlands of the Kakadu region were prepared by Story (1969, 1976) as part of the CSIRO land use surveys. Wetland vegetation communities were not as widespread as the terrestrial communities and included paperbark (*Melaleuca* spp.) forests, sedgelands (*Cyperus, Eleocharis, Fuirena* and *Scirpus* spp.), grasslands (*Hymenachne* and *Commelina* spp.) and herbaceous swamps. A similar classification was prepared by Burgmann & Thompson (1982) in the vicinity of Jabiluka Billabong on the Magela flood plain. These authors concluded that, floristically and structurally, the Jabiluka area could be taken as a good sample of much of the vegetation of the region. Wetland plant communities in this sample area included the flood plain macrophyte communities, the flood plain fringe community, and the riparian community.

4.2.2. South alligator flood plain

The vegetation description of this flood plain was done as part of an investigation undertaken by the CSIRO to determine the impact of feral buffalo on the flood plain ecosystem. A vegetation analysis commenced in 1981 with systematic grid sampling at 70 locations in two adjacent areas on the floodplain. One area contained buffaloes, at an estimated density of 7 km^{-2} on the open plains and 15 km^{-2} along the margins of rivers, creeks and waterholes (Ridpath *et al.* 1983), whereas the other area was effectively free of buffaloes. The intensive sampling stopped in 1983 and was replaced by a biannual analysis of the same sites. Whilst the results have not been published a few preliminary analyses have identified very distinctive differences between the wet and dry season floras and have demonstrated the extreme year-to-year variability that exists.

Taylor & Friend (1984) investigated the effect of buffaloes on ground surface features and vegetation lifeform and structure on the South Alligator flood plain and adjacent areas. They found that vegetation structure and plant lifeform attributes were associated with the ground surface features on a seasonal basis. Unfortunately, they do not adequately describe the vegetation of the area to enable a thorough assessment of the botanical features of the flood plain to be presented. Importantly, however, they do point out that the flood plain vegetation can not be considered independently of the surrounding terrestrial habitats when assessing the impact of the mobile buffaloes on the vegetation.

Braithwaite & Werner (1987) briefly reported on some of the changes that have occurred on the South Alligator flood plain following removal of buffaloes in 1982. The perennial grass *Hymenachne acutigluma*, a species favoured as food by the buffaloes (Letts *et al.* 1979) was not common before this time, but now forms extensive and thick stands. Similarly, *Phragmites karka* is re-establishing. It is not known if these changes are detrimental or beneficial to other flood plain plant and animal species, or importantly, whether they represent a temporally stable change.

4.2.3. Magela creek flood plain

Species habitats

A total of 222 plant species were recorded by Finlayson *et al.* (1989) on the flood plain. These were listed in four broad habitat categories – seasonally inundated plain, seasonally inundated fringe zone, billabong and permanent swamp. The fringe zone category covered the edges of the flood plain and therefore included the fringing *Melaleuca* forests/woodlands. The seasonally inundated plain category covered the remainder of the flood plain, except for the permanently wet areas. The seasonally inundated plain and the fringe zone contain 41% and 71% respectively, of the 222 species, compared with 20% in the billabongs and 10% in the permanent swamps (Table III).

Table III. Plant species found in four broad habitat areas on the Magela flood plain (from Finlayson *et al.* 1989).

Habitat	Total species	Annuals	Perennials	Geophytic perennials
Permanent Billabongs	46	19	21	6
Seasonally Inundated Plain	94	57	29	8
Fringe Zone	158	100	50	8
Permanent Swamps	21	5	11	5

Overall, there were 139 annual species with 102 classified as terrestrial and 37 as aquatic species, with one, *Rotala ocultiflora*, difficult to classify. Eighty-nine of the terrestrial species occur in the fringe zone; only 27 are found on the plain which is seasonally inundated for a longer period than the fringe zone. The terrestrial annuals are a diverse group of species with 60 of them classified as herbs, 18 as sedges and 17 as grasses. Twenty-seven of the aquatic annuals are herbs with *Ceratophyllum demersum* and *Nymphoides hydrocharoides* being amongst the more common. The aquatic grasses *Oryza meridionalis* and *Hymenachne acutigluma* are the most common species in the seasonally inundated areas.

There are 68 perennial species, 50 of which occur in the fringe zone (Table III). Thirty-four of the perennials are terrestrial, 26 aquatic, with 8 others difficult to classify. There are 12 terrestrial trees including *Eucalyptus* spp., *Pandanus spiralis*, *Lophostemon lactifluus* and *Syzygium suborbiculare*. The aquatic species are dominated by 12 herbs, including *Hydrilla verticillata*, *Ludwigia adscendens*, *Nelumbo nucifera* and *Nymphoides indica*, and by 5 grasses, including the widespread *Hymenachne acutigluma* and *Pseudoraphis spinescens*.

Geophytic species comprise 14 species. The more widespread include the *Nymphaea* and *Eleocharis* species.

Contained in the total number of plant species were 21 alien species, 13 annuals and 8 perennials (Cowie *et al.* 1988). The perennials include the important flood plain weed species *Salvinia molesta* and *Mimosa pigra* and the terrestrial shrubs *Sida* spp., the vine *Passiflora foetida* and the herb *Stylosanthes hamata*.

Plant communities

Williams (1979) identified six broad vegetation types that were regarded as direct indicators of water depth on the 12 000 ha of the flood plain considered. These are described below with an estimate of the area (ha) and the proportion (as a percentage) of the floodplain that each community, except for the mixed herbfield which extended onto the East Alligator flodplain, occupied at the time of the survey.

i Mixed herbfield: distinct herbaceous communities dominated by *Eleocharis dulcis, Fimbristylis littoralis* or *Oryza fatua* (*Oryza meridionalis*). They occur on heavy clay soils;

ii) Grassland: 2640 ha, 22%: open sward grasslands dominated by *Pseudoraphis spinescens*;

iii) Undulating annual swamp and grassland: 1800 ha, 15%: similar to the above grassland community, but intermixed with annual swamp plant species over an undulating land surface. Major species are *Pseudoraphis spinescens, Panicum paludosum, Ludwigia adscendens, Polygonum attenuatum* and *Nymphoides indica*;

iv) Forest: 3720 ha, 31%: dense cover of trees of the genus *Melaleuca* with *M. nervosa* and *M. viridiflora* in seasonally inundated areas and *M. cajaputi* and *M. leucadendron* (*M. leucadendra*) in areas subjected to prolonged flooding.

v) Annual swamp: 1320 ha, 11%: dominated by a mixture of vascular plants including *Eleocharis dulcis, Nymphoides indica, Ceratophyllum demersum, Utricularia spp., Polygonum attenuatum, Eleocharis sphacelata, Vallisneria spiralis* (*V. gigantea*) and *Hydrilla verticillata*;

vi) Perennial swamp: 2280 ha, 19%: dominated by *Chara* sp. and filamentous algae with *Nelumbo nucifera, Scirpus grossus* (*Hymenochaeta grossa*) and *Hymenachne amplexicaulis* (*H. acutigluma*).

Williams (1979) regarded the Magela flood plain as a freshwater lagoon, cut off at its outlet by the flood plain of the East Alligator River. Within the lagoon, the vegetation units were interpreted as being indicators of water depth. Therefore, at the end of the wet season the mixed herbfield would be expected to dry out first, followed by the grassland, parts of the forests and then the annual swamps. The extent of perennial swamp would vary from year to year as a consequence of the rainfall pattern.

Morley (1981) compiled a list of plant species found on the southern-most part of the floodplain between Nankeen and Mudginberri Billabongs (see Fig. 13 for locations) during 1980 and produced a map of the plant communities. Of the 61 plant species in the list, two were macrophytic algae, *Chara* and *Nitella* species. Overall, 36 plant communities were recognised and described in detail. However, this analysis was not reproducible in subsequent years, being far too detailed. Therefore, Sanderson *et al.* (1983) produced a far less detailed description of the same region with nine plant communities; these are described below.

i) Najas-Nymphaea-Pseudoraphis herbland: 803 ha, 21 %: dominated by *Pseudoraphis spinescens, Najas tenuifolia* and *Nymphaea violacea* in undulating sections of the floodplain and in amongst *Melaleuca* trees;

ii) Pseudoraphis grassland: 1551 ha, 42 %: dominated by *Pseudoraphis spinescens* and *Nymphaea* species in deep water in open areas of the floodplain;

iii) Eleocharis-Pseudoraphis sedge-grassland: 257 ha, 7 %: dominated by *Eleocharis* sp. sedges and *Pseudoraphis spinescens* in shallow water in open areas of the floodplain;

iv) Orzya-Hygrochloa grassland: 94 ha, 3 %: dominated by *Oryza meridionalis* and *Hygrochloa aquatica* growing in open areas that are only inundated for around 4 months of the year;

v) Fast-flow Blyxa herbland: 163 ha, 4 %: *Blyxa* species and *Caldesia oligococca* in fast flowing channels;

vi) Fringing closed Hygrochloa-Nymphoides: 197 ha, 5 %: a mixture of *Hygrochloa aquatica, Nymphoides* species and some *Pseudoraphis spinescens* along the eastern fringe of the floodplain;

vii) Mixed herbland: 260 ha, 7 %: a mix of species dominated by *Hygrochloa aquatica, Pseudoraphis spinescens* and *Nymphoides* species in shallow water near the eastern edge of the floodplain;

viii) Hygrochloa grassland: 172 ha, 5 %: shallow areas at the northern end of the floodplain almost dominated entirely by *Hygrochloa aquatica*;

ix) Deep water community: 203 ha, 5 %: mixture of aquatic plant species with *Nymphaea macrosperma, Najas tenuifolia, Leersia hexandra* and the grass *Hymenachne acutigluma* being plentiful.

The above classification was considered by Sanderson *et al.* (1983) to be more reproducible than the detailed scheme of Morley (1981). How-

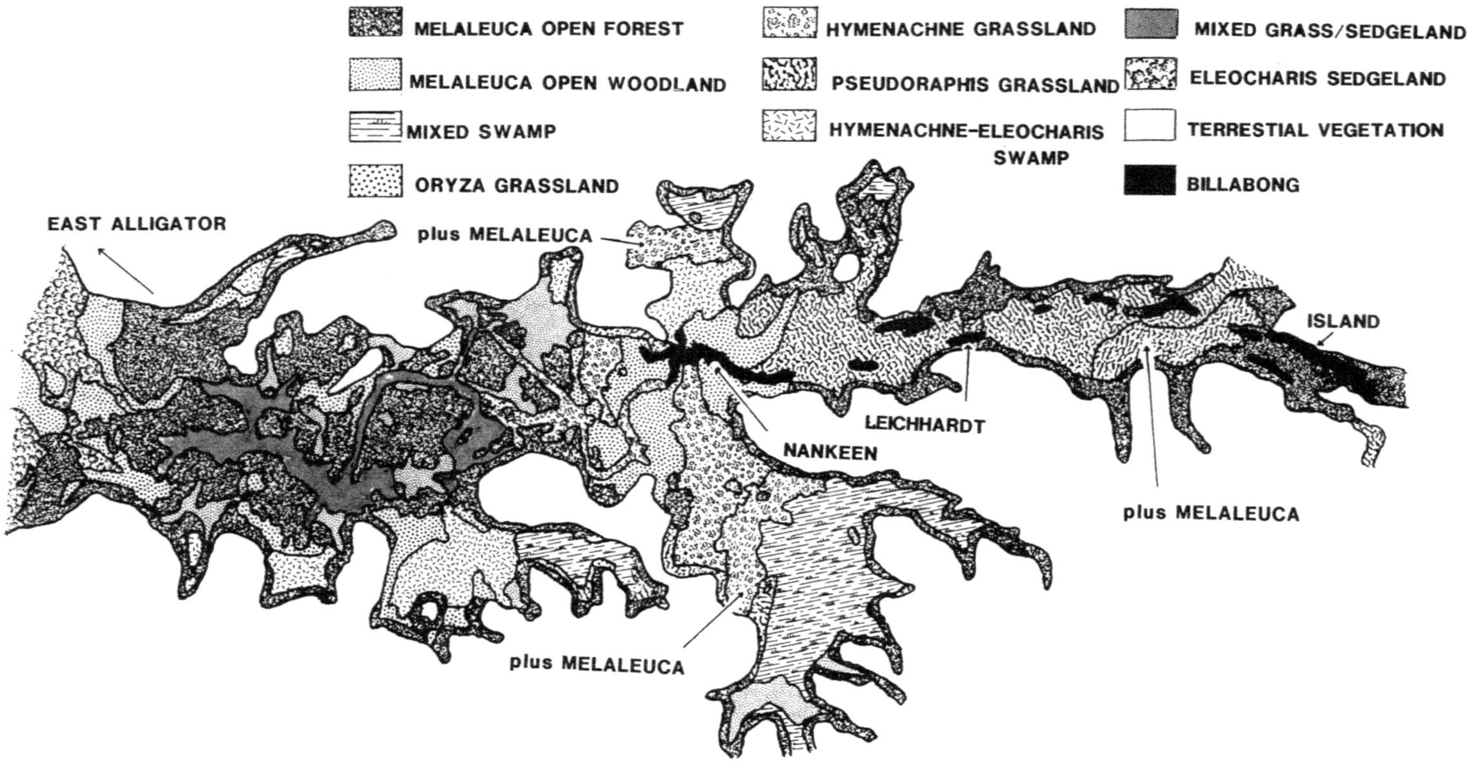

Fig. 16. Vegetation map of Magela Creek flood plain (1:100 000 scale) (adapted from Finlayson *et al.* 1990a).

ever, the classification also did not extend beyond Nankeen Billabong (see Fig. 13 for location).

Finlayson *et al.* (1989) prepared a generalised vegetation map of the Magela flood plain (Fig. 16) based on wet season vegetation maps and descriptions. During the wet season, many of the aquatic plants reach their peak biomass and are easier to differentiate and map. An estimate of the area (ha) of each community is given, as well as the proportion (as a percentage) of the 22 000 ha of the flood plain that each community was estimated as occupying at the time of the surveys. This description is more detailed than that of Williams (1979) and included the fringing seasonally inundated areas of the flood plain. Overall, ten communities were identified. These are briefly described below.

i) Melaleuca open forest and woodland: tree canopy cover of 10–70%, 7390 ha, 34%: comprises areas dominated by one or more *Melaleuca* species – *M. viridiflora* and *M. cajaputi* around the edges and at the northern end of the flood plain, and *M. leucadendra* in back-swamps inundated for 6–8 months. The understorey varies, apparently in response to water depth and degree of shading;
ii) Melaleuca open woodland: tree canopy cover < 10%, 1290 ha, 6%: consists of *Melaleuca leucadendra* in areas inundated for over 6 months. Understorey species are usually the same as those in adjacent areas of the flood plain;
iii) Nelumbo-Nymphoides herbland: 2090 ha, 9%: a mixed community seemingly dominated by the water lilies *Nelumbo nucifera* and *Nymphoides indica* that occur in permanently and semi-permanently wet areas;
iv) Orzya grassland: 2730 ha, 12%: dominated by *Oryza meridionalis* towards the end of the wet season. In the dry season it consists of bare ground and dead *Oryza meridionalis* stems with persistent *Phyla nodiflora* and *Ludwigia adscendens* as xerophytic forms, and *Pseudoraphis spinescens*;
v) Hymenachne grassland: 1930 ha, 9%: dominated by *Hymenachne acutigluma* throughout the year;
vi) Pseudoraphis grassland: 3050 ha, 14%: dominated by the perennial emergent grass *Pseudoraphis spinescens* which has a turf-like habit during the dry season and grows up through the water during the wet season;
vii) Hymenachne-Eleocharis grass-sedgeland: 1290 ha, 6%: consists of swampy areas that dry out seasonally and are dominated by *Hymenachne acutigluma* or *Eleocharis* spp., which are slower to establish;
viii) Mixed grass-sedge-herbland: 1120 ha, 5%: contains a variety of species, the dominant species depending on the topographic situation. *Oryza meridionalis* occurs on the drier sites with *Pseudoraphis spinescens* in slightly wetter places, while *Eleocharis* spp. and *Hymenachne acutigluma* occur in the deeper sites. On sites that remain flooded for 10–11 months *Nymphoides indica* and *Nymphaea macrosperma* may be present;
ix) Eleocharis sedgeland: 960 ha, 4%: *Eleocharis* spp. dominate during the wet season, but are replaced by annual herbs during the dry season;
x) Open-water community: 160 ha , 1%: permanent billabongs, flow channels, shallow waterholes contain *Nymphaea macrosperma* and *Nymphaea pubescens* and a number of submerged plant species. Floating grass mats comprised of *Leersia hexandra, Hymenachne acutigluma* and *Urochloa mutica* along with the herb *Ludwigia adscendens*, occur along the banks of the billabongs.

4.2.4. Seasonally inundated billabongs

Billabongs are strictly defined as oxbow lakes (Bayly & Williams 1977). In the Kakadu region, however, this definition is not followed and all waterholes and lagoons are generically referred to as billabongs (Finlayson *et al.* 1989). Walker & Tyler (1985) have categorised the billabongs of Magela Creek and flood plain (Table IV) into channel (depressions in flow channels), backflow (located on small tributaries and initially filled by water from the main creek) and flood plain (generally remnants of deep channels on the flood plain). Many of these contain water all year round, although the backflow billabongs can dry out during an extended dry period. Additionally, the different categories of billabongs have differing physico-chemical characteristics that are most pronounced late in the dry season (Table IV).

Table IV. Classification of Magela Creek billabongs based on late dry season physico-chemical characteristics (adapted from Walker & Tyler 1984, Finlayson 1988, Finlayson *et al.* 1990a).

Type of billabong	Examples	General characteristics
Channel	Mudginberri Island Buffalo	Maintain ionic character of wet season with Na/Mg HCO_3 dominance and relatively low total phosphorus, turbidity and conductivity.
Backflow	Georgetown Corndorl Coonjimba	NaCl ions become more dominant in the dry season with higher turbidity and total phosphorus than channel billabongs.
Flood plain	Jabiluka Leichhardt Nankeen	SO_4 dominance with trend to NaCl during dry season with higher total phosphorus and turbidity than channel billabongs and higher conductivity than backflow billabongs.

The vegetation is also affected by the morphology of the billabongs. The flood plain and channel billabongs are, on the whole, deeper and have steeper sides than the backflow billabongs (Morley 1981). Thus, in the former, the vegetation, with the exception of the floating aquatic plants, was restricted to a narrow belt around the edge of the billabongs, whereas the backflow billabongs were at times almost completely covered with emergent, submerged and floating-leaved plants (Finlayson *et al.* 1993). Similarly, the vegetation of the flood plain billabongs was greatly influenced by the adjacent plant communities on the seasonally inundated flood plain (e.g. grass mats extending across the flood plain and into the billabongs), whereas the backflow billabongs often had terrestrial vegetation in close proximity and abutting the fringing *Melaleuca* woodlands.

Finlayson *et al.* (1994) reported on semi-quantitative analyses of vegetation dominance in three backflow billabongs (Georgetown, Coonjimba and Djalkmara) along Magela Creek and two flood plain billabongs (Leichhardt and Jabiluka) on Magela flood plain (for locations see Fig. 13).

The three backflow billabongs had a generalised vegetation zonation consisting of:

i) fringing *Melaleuca* spp. woodland in seasonally inundated areas, particularly on the northern and southern ends of Djalkmara and Coonjimba billabongs;
ii) a mix of grass and sedge species in seasonally inundated areas shaded by a woodland canopy;
iii) a belt of *Eleocharis* spp. sedges in water that is usually < 1.5 m deep during the wet season;
iv) a small area of open water usually between 1.5–2.0 m deep in the wet season;
v) patches of waterlilies and submerged plants along the boundary between the sedges and the open water.

Overall, the dominant plant species, in terms of biomass and 'ground' cover, in these three billabongs are the *Melaleuca* spp. trees and the geophytic, perennial *Eleocharis* spp. sedges. The annual species do not contribute greatly to the plant biomass or 'ground' cover in these billabongs.

The two flood plain billabongs had a generalised vegetation zonation comprising:

i) fringing *Melaleuca* spp., *Pandanus aquaticus*, and *Barringtonia acutangula* trees along the western bank;
ii) mixture of grass and sedge species and a few trees interfacing with the flood plain grass communities along the eastern bank;
iii) mix of grass, herb and sedge species overlaying a floating mat of *Salvinia molesta* plants extending from the banks towards the middle of the billabong;
iv) a discontinuous fringe of submerged plants and waterlilies along the edge of the floating mat.

The most obvious change in the vegetation of the backflow billabongs over the last decade is an increase in abundance of *Eleocharis* spp. This has been attributed to the virtual elimination of buffaloes from the vicinity of the billabongs. Grazing, pugging and wallowing by buffaloes previously prevented these plants from establishing the dense stands that now occur and, in some cases, almost choke the billabongs. The development of these dense stands of sedges could lead to further changes in the vegetation as a result of increased siltation. In addition, as drainage outlets that were formerly kept open by buffaloes infill, more water could be retained in the billabongs. Increased siltation and longer water retention could have a major influence on the vegetation structure.

The flood plain billabongs are much deeper than the backflow billabongs and have not been as drastically affected by buffaloes. Jabiluka contains the only known stand of reed (*Phragmites karka*) in the Magela: Braithwaite & Werner (1988) have reported the re-appearance of this species on the South Alligator flood plain following removal of buffaloes. The mats of the floating *Salvinia molesta* that occur in these billabongs can completely cover the water surface and support secondary growths of grasses, sedges and even *Melaleuca* seedlings. Consequently, the floating-leaved waterlilies are now less common.

4.3. *Productivity*

The only information collected on the productivity of the vegetation of the freshwater wetlands is that from the Magela flood plain (Finlayson 1988, 1991a). This records seasonal changes in dry weight of three aquatic grass species over an 18 month period and litterfall from *Melaleuca* trees over 32 months.

4.3.1. *Grasslands*

Changes in above-ground biomass (dry weight/unit area) for the dominant aquatic grasses *Pseudoraphis spinescens, Hymenachne acutigluma* and *Oryza meridionalis* were determined on samples collected during 1983–84 (Finlayson 1991a). The dry weight biomass of each plant appears to be influenced by water depth which in itself is a function of the rainfall and surface water flow in Magela Creek. Whilst there is an association between biomass and water depth the "correlation" appears broadly positive for two species, but negative for the third, *Hymenachne acutigluma* (Fig. 17).

At the start of the wet season *Pseudoraphis spinescens* undergoes a change from a turf-like habit on a nearly dry plain to elongated culms that extend up through the water as depth increases. The maximum biomass recorded was 1.67 ± 0.21 kg m^{-2} during May 1984 when the water level was falling. Over the next 8 weeks the culms senesced and the plants reverted to a turf-like appearance. *Hymenachne acutigluma*, unlike *Pseudoraphis spinescens*, was growing in a perennial swamp and had a semi-erect/creeping habit with short internodes, and horizontal culms anchored to the substrate by roots at the nodes. Biomass increased after the first floods, but then decreased from 0.78 $\pm$ 0.10 to 0.23 ± 0.03 kg m^{-2} following a large increase in water level (Fig. 17). A maximum dry weight of 1.41 ± 0.10 kg m^{-2} was recorded at the end of the wet season when the water level was falling. *Oryza meridionalis* is an annual species that germinates after the first rains of the wet season and continues to grow as the plain fills with water. The maximum biomass, 0.51 ± 0.10 kg m^{-2}, occurred in February.

Productivity of these species was determined using the minimum-maximum method of estimation. *Oryza meridionalis* had a maximum dry weight of 0.51 ± 0.10 kg m^{-2} which represents the annual productivity. *Pseudoraphis spinescens* had two growth periods with dry matter production of 1.06 ± 0.23 kg m^{-2} and 0.85 ± 0.03 kg m^{-2}. *Hymenachne acutigluma* similarly had two growth periods with 0.96 ± 0.26 kg m^{-2} and 1.19 $\pm$ 0.12 kg m^{-2}. As these occurred within the one year the annual productivity for these two species was 1.91 kg m^{-2} and 2.09 kg m^{-2} respectively.

Productivity of these grasses is not as high as those reported for some of the large emergent aquatic plant species growing elsewhere. However, *Pseudoraphis spinescens* and *Hymenachne acutigluma*, in particular, are relatively productive when compared to the range for wetland species reported by Bradbury & Grace (1983). As these three grasses are widespread and dominate 35% of the flood plain area (Fig. 20) and the flood plain supports about another 220 plant species the Magela flood plain can be described as highly productive.

4.3.2. *Melaleuca woodland*

Productivity data for the widespread *Melaleuca* woodlands and forests on the Magela flood plain

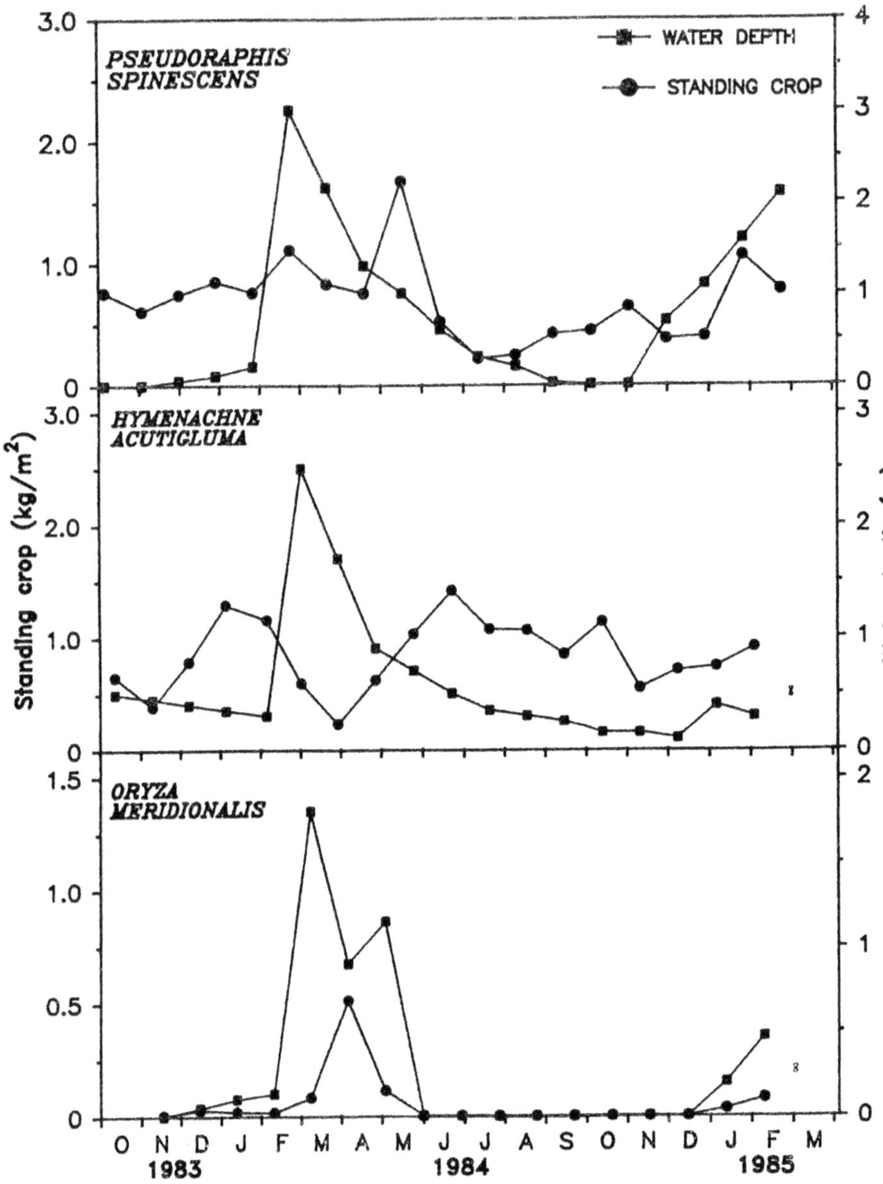

Fig. 17. Above-ground biomass (dry weight) and water depths at sampling sites on the Magela flood plain. Error bars are least significant differences (P = 0.05) for the biomass (adapted from Finlayson 1991a).

are available indirectly through an analysis of litterfall data (Finlayson 1988, 1991b, Finlayson *et al.* 1993) which has been shown to be linearly related to tree biomass and is an indirect measure of tree productivity (Spain 1984).

In an intensively sampled *Melaleuca* forest on the Magela flood plain the total litterfall was around 0.7 kg m^{-2} y^{-1}, which places the forest at the lower end of the litterfall range for Australian tropical forests (Finlayson *et al.* 1993). At another site on the flood plain, less intensively investigated, a value of around 1.5 kg m^{-2} y^{-1} was recorded (Finlayson 1988); a relatively high value compared to other forests (Spain 1984, Lonsdale 1988). Comparative data for *Melaleuca* forests is limited to two studies; *Melaleuca cuticularis* in temperate south-western Australia with 0.43 kg m^{-2} y^{-1} (Congdon 1979) and to *Melaleuca* sp. at a mixed *Melaleuca*-mangrove site in tropical north-eastern Australia where *Melaleuca* litter comprised 0.39 kg m^{-2} y^{-1} of the total litterfall of 0.94 kg m^{-2} y^{-1} (Duke 1982).

The weight of litterfall, in particular leaf material which comprised 70% of the litterfall weight in the *Melaleuca* forest (Finlayson *et al.* 1993), is regarded as an indication of the primary productivity of a forest (Bray & Gorham 1964). Thus, tropical forests with high rates of litterfall are regarded as highly productive. Based on an analysis of the relationship between total litterfall and latitude done by Lonsdale (1988) the value of 0.7 kg m^{-2} y^{-1} is within the range recorded for other forests at the same latitude. The higher 1.5 kg m^{-2} y^{-1} reported by Finlayson (1988), is at the upper limit in the same analysis. The difference between sites reported by Finlayson (1988) should be considered when making an assessment of the productivity of these forests/woodlands. Not all sites have the same level of litterfall, varying from 0.8 kg m^{-2} y^{-1} to 1.5 kg m^{-2} y^{-1}. This reflects, at least in part, differences in density of trees on the different sites. In summary, the litterfall in the *Melaleuca* forests on the Magela flood plain is either equivalent to or higher than that in many forests elsewhere in Australia, indicating that it is highly productive.

4.4. Vegetation succession

Based on an empirical model of plant succession in wetlands, developed by Van der Valk (1981), changes in vegetation that occur as a result of changes in the annual hydrological pattern (i.e. water depth and period of inundation) have been broadly described (Finlayson 1991b). Information for this approach came from general vegetation studies, described above, and an analysis of the seedbank in grasslands on the Magela flood plain (Finlayson *et al.* 1990a,b).

In undertaking this description it was necessary to recognise that not only does the model have limitations (e.g. it does not take into account interactions between the plants), but that the input information (i.e. ecological information on the many plant species involved) was limited. Despite these limitations it is possible to use the model to provide a framework around which to predict changes in the vegetation patterns, i.e. compare the predicted to the actual situation. If the model output is rejected it means that other factors, in addition to those used in the model, are involved. These could then be investigated.

For the purposes of validating the model the 1983–84 hydrological cycle was considered. Seasonal changes in species dominance in three widespread grass communities under these hydrological conditions is shown in Fig. 18. The species "successional states" used in the model are explained in Fig. 19 (see Van der Valk 1981). The importance of water depth and period of inundation in determining these patterns was borne out by application of the model to the data from 1983–84. For example, the *Oryza*-community underwent pronounced changes in species occurrence and biomass dominance over the wet-dry cycle. The sparsely distributed dry season annuals (AS-I) were replaced by wet season annuals (AS-II), dominated by *Oryza meridionalis* and the vegetatively reproducing *Eleocharis* spp. (VS-(II)). In terms of successional changes the introduced weeds *Mimosa pigra* (PS-I) and *Salvinia molesta* (VD-(II)) have the potential to alter both the seasonal vegetation changes and succession between years.

Mimosa pigra can survive throughout both the dry and wet season almost regardless of the growing conditions (see summary in Cowie *et al.* 1988) and can rapidly spread by seed once it is introduced. In contrast, the vegetatively reproducing *Salvinia molesta* is widespread during the wet season but restricted to billabongs during the dry season; as it does not have propagules in the sediment seedbank it can only survive in areas of permanent water. However, it does have the growth potential (Finlayson 1984) to completely cover open water areas and change the vegetation structure and composition (e.g. the loss of wet season annuals). The "successional states" of the plants can be used to determine which plants are likely to survive changes in the seasonal flooding pattern and also the likely survival of introduced species between seasons and between years from propagules stored within the wetland.

The occurrence of feral animals on the flood plains has also been responsible for changes in the vegetation pattern (Finlayson *et al.* 1994). Of particular importance has been the impact of the water buffalo. Within Kakadu National Park

Community	DRY SEASON		WET SEASON	
Pseudoraphis community	DRY SEASON no standing water AS-I *Cyperus* spp. *Fimbristylis* *Glinus* *Heliotropium* VS-I *Polygonum* *Pseudoraphis* ** PS-I *Mimosa*	⇒ ⇒ ⇒ flooding ⇒ ⇒ ⇒	WET SEASON standing water AS-II *Blyxa* * *Hygrochloa* * *Najas* * *Nymphoides* spp. * *Utricularia* spp. VS-I *Polygonum* *Pseudoraphis* ** VS-II *Eleocharis* spp.* *Nymphaea* * VD-II *Salvinia* * PS-I *Mimosa*	⇒ ⇒ ⇒ drawdown ⇒ ⇒ ⇒
Hymenachne community	DRY SEASON shallow standing water some exposed and moist areas VS-I *Ludwigia* *Pseudoraphis* VS-II *Azolla* *Eleocharis* spp. *Hymenachne* ** *Lemna* *Nelumbo* * *Nymphaea* *Urochloa* VD-II *Salvinia*	flooding ⇒ ⇒ ⇒	WET SEASON deeper standing water than during dry AS-II *Aeschynomene* spp. * *Oryza* VS-I *Ludwigia* *Pseudoraphis* VS-II *Azolla* *Eleocharis* spp. * *Hymenachne* * *Lemna* *Nelumbo* *Nymphaea* *Urochloa* VD-II *Salvinia*	drawdown ⇒ ⇒ ⇒
Oryza community	DRY SEASON exposed dry areas AS-I *Coldenia* * *Commelina* *Digitara* *Heliotropium* *Phyla* * VS-I *Ludwigia* *Pseudoraphis* PS-I *Mimosa* PS-II *Isoetes*	flooding ⇒ ⇒ ⇒	WET SEASON standing water AS-II *Aeschynomene* spp. *Blyxa* spp. *Hygrochloa* *Ipomoea* *Maidenia* *Nymphoides* spp. *Oryza* ** *Utricularia* spp. VS-I *Ludwigia* *Pseudoraphis* VS-II *Eleocharis* spp. * *Nymphaea* VD-II *Salvinia* PS-I *Mimosa* PS-II *Isoetes*	drawdown ⇒ ⇒ ⇒

Fig. 18. Predicted species succesion due to water level fluctuation on the Magela flood plain during the 1983–84 wet-dry cycle. The dominant species in each community is indicated by ** and the next by * (A = annual, V = vegetative, P = perennial; S = short-lived propagules, D = long-lived propagules; I = propagules established in areas devoid of standing water, II = propagules established in standing water) (from Finlayson 1991b).

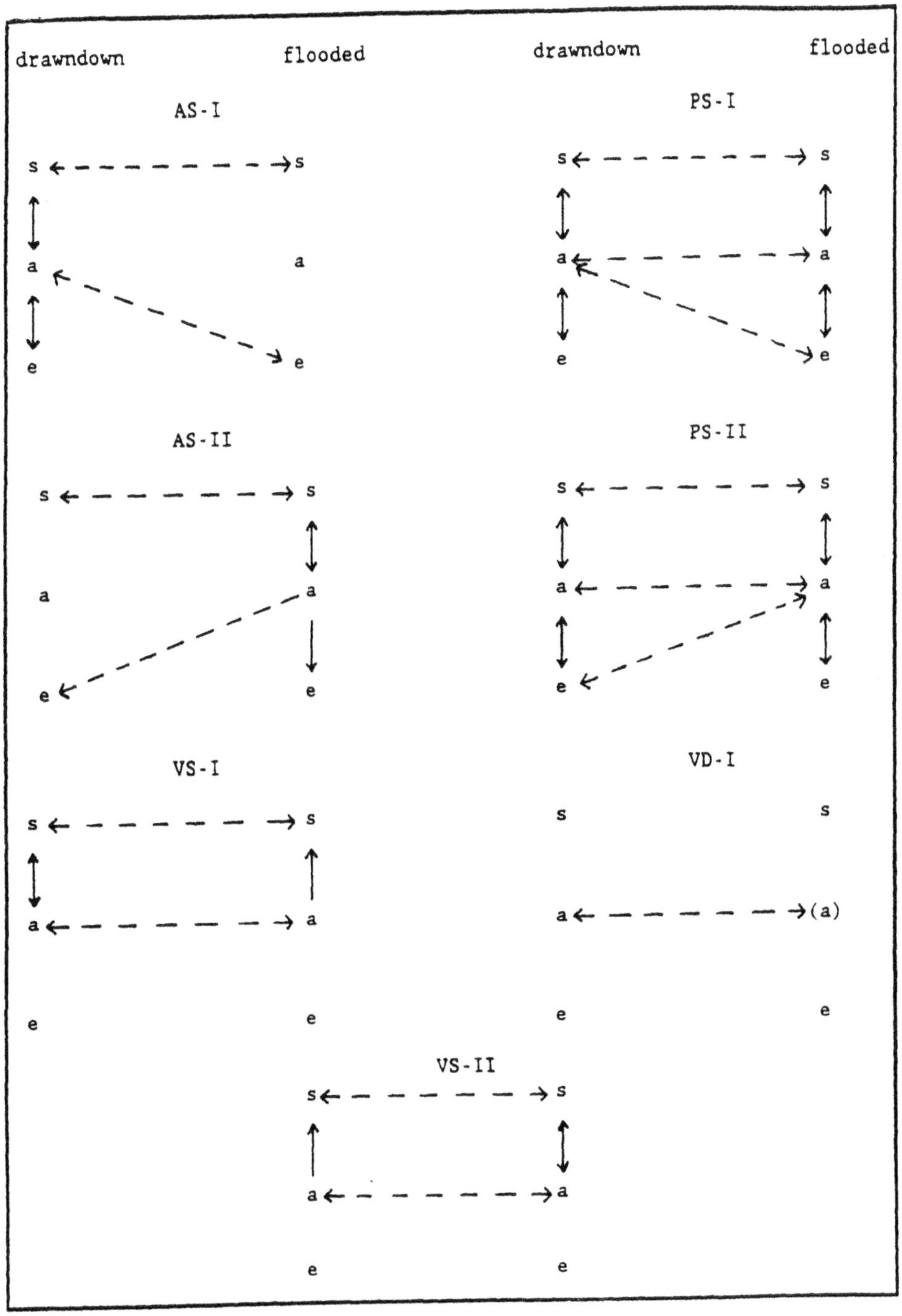

Fig. 19. Potential species transitions between two 'environmental situations' – increasing (i.e. flooding) and decreasing (i.e. drawdown) periods in a wetland. Solid lines represent potential transitions within an environmental situation and dashed lines represent transitions between environmental situations. The species states are: s – present as long-lived propagules in a persistent seed-bank; a – mature adults; and e – locally extinct. If establishment is dependent on dispersal from another site adult populations are indicated in parentheses (from Finlayson 1991b).

ANCA has a policy to remove most of these animals. Thus, whilst the presence of buffalo previously caused widespread changes to the flood plain vegetation, it is the changes that occur as a result of the removal of the buffalo that are now of interest. Such changes could be predicted by using the empirical vegetation model; plants that have the capability to spread can be identified by their "successional states". For example, plants such as *Salvinia molesta* can only spread if vegetative propagules are introduced, whereas there is every expectation that seeds of *Phragmites karka* could already be present in the sediment seedbank. Once a species has become established (or re-established) in the wetland its most likely means of expanding and surviving the seasonal changes can also be assessed. Furthermore, by assessing the "successional states" of weedy species their potential to establish and spread in the various habitats can be predicted.

The seasonal cycles in the main plant communities on the flood plains can be partly shown in the model diagrams (Fig. 18). Any substantial changes to this diagram can therefore be noted and decisions made about their management importance. Further information on successional directions following major disturbance (e.g. the introduction of weeds or feral animals) is needed if management of this dynamically variable flora (and hence the habitat) is to be prescriptive, rather than reactive.

5. References

Adams, L.G., Byrnes, N. & Lazarides, M. 1973. Floristics of the Alligator Rivers area. In: Alligator Rivers region environmental fact-finding study, part XI, physical features and vegetation (vol. II). AGPS, Canberra. Unpaginated.

ANPWS 1980. Nomination of Kakadu National Park for inclusion in the World Heritage List. AGPS, Canberra. 32 pp.

ANPWS 1987. Nomination of Kakadu National Park Stage II for inclusion in the World Heritage List. AGPS, Canberra. 26 pp.

Ball, M.C. 1988. Organisation of mangrove forests along natural salinity gradients in the Northern Territory: an ecophysiological perspective. In: Wade-Marshall, D. & Loveday, P. (eds) North Australia: Progress and Prospects, vol. 2 Flood plain Research. pp. 84–100. ANU Press, Darwin.

Ball, M. C. & Pidsley, S.M. 1988. Establishment of mangrove seedlings in relation to salinity. In: Larson, H.K., Michie, M.G. & Hanley, J.R. (eds) Proceedings of a Workshop on Research and Management of Darwin Harbour. pp. 123–134. North Australian Research Unit, ANU Press, Darwin.

Bayly, I.A.E. & Williams, W.D. 1977. Inland Waters and Their Ecology. Longmans, Melbourne. 314 pp.

Bradbury, J.K. & Grace, J. 1983. Primary production in wetlands. In: Gore, A.J.P. (ed.) Ecosystems of the World, 4A Mires: Swamp, Bog, Fen and Moor. pp. 285–310. Elsevier, Amsterdam.

Braithwaite, R.W. & Werner, P.A. 1987. The biological value of Kakadu National Park. Search 18: 296–301.

Bray, J.R. & Gorham, E. 1964. Litter production in forests of the world. Adv. Ecol. Res. 2: 101–157.

Brennan, K. 1990. Checklist of vascular plants of the Alligator Rivers Region, Northern Territory. Supervising Scientist for the Alligator Rivers Region Open File Record 62. 44 pp.

Briggs, S.V. 1981. Freshwater wetlands. In: Groves, R.H. (ed.) Australian Vegetation. pp. 335–360. Cambridge University Press, Cambridge.

Bunt, J.S., Williams, W.T. & Clay, H.J. 1982. River water salinity and the distribution of mangrove species along several rivers in north Queensland. Aust. J. Bot. 30: 401–412.

Bunt, J.S., Williams, W.T. & Bunt, E.D. 1985. Mangrove species distribution in relation to tide at the seafront and up rivers. Aust. J. Mar. freshwater Res. 36: 481–492.

Burgman, M.A. & Thompson, E.J. 1982. Cluster analysis, ordination and dominance-structural classification applied to diverse tropical vegetation at Jabiluka. Aust. J. Ecol. 7: 375–387.

Chappell, J. & Grindrod, J. 1985. Pollen analysis: key to past mangrove communities and successional changes in north Australian coastal environments. In: Bardsley, K.N., Davie, J.D.S. & Woodroffe, C.D. (eds) Coastal and Tidal Wetlands of the Australian Monsoon Region. pp. 225–236. North Australian Research Unit, ANU Press, Darwin.

Chappell, J. & Woodroffe. C.D. 1985. Morphodynamics of Northern Territory tidal rivers and floodplains. In: Bardsley, K.N., Davie, J.D.S. & Woodroffe, C.D. (eds) Coastal and Tidal Wetlands of the Australian Monsoon Region. pp. 85–96. North Australian Research Unit. ANU Press, Darwin.

Clark, R.L. & Guppy, J.C. 1988. A transition from mangrove forest to freshwater wetland in the monsoon tropics of Australia. J. Biogeogr. 15: 665–684.

Congdon, R.A. 1979. Litter fall of the paperbark tree (*Melaleuca cuticularis*) in the marshes of the Blackwood River estuary, western Australia. Aust. J. Ecol. 4: 411–417.

Cowie, I.D. & Finlayson, C.M. 1986. Plants of the Alligator Rivers Region, Northern Territory. Supervising Scientist for the Alligator Rivers Region Technical Memorandum 17. AGPS, Canberra. 50 pp.

Cowie, I.D., Finlayson, C.M. & Bailey, B.J. 1988. Alien plants in the Alligator Rivers Region, Northern Territory, Australia. Supervising Scientist for the Alligator Rivers Region Technical Memorandum 23. AGPS, Canberra. 26 pp.

Dames & Moore 1984. Mangrove resource delineation, Northern Territory coastline, Stage 1. Report to the Conservation Commission of the Northern Territory, Darwin. 17 pp.

Davie, J.D.S. 1985. The mangrove vegetation of the South

Alligator River, northern Australia. In: Bardsley, K.N., Davie, J.D.S. & Woodroffe, C.D. (eds) Coastal and Tidal Wetlands of the Australian Monsoon Region. pp. 133–152. North Australian Research Unit, ANU Press, Darwin.

Davis, J.H. 1940. The ecology and geologic role of mangroves in Florida. Pap. Tortugas Lab. 32: 303–412.

Duke, N.C. 1982. Mangrove litter fall data from Hinchinbrook Island, north-eastern Australia. Australian Institute of Marine Science Data Report Coastal studies Series. 284 pp.

Dunlop, C.R. 1987. Checklist of vascular plants of the Northern Territory. Conservation Commission of the Northern Territory, Darwin, Australia. 87 pp.

Finlayson, C.M. 1984. Growth rates of Salvinia molesta in Lake Moondarra, Mount Isa, Australia. Aquat. Bot. 18: 257–262.

Finlayson, C.M. 1988. Productivity and nutrient dynamics of seasonally inundated flood plains in the Northern Territory. In: Wade-Marshall, D. & Loveday, P. (eds) North Australia: Progress and Prospects, vol. 2 Flood plain Research, pp. 58–83. ANU Press, Darwin.

Finlayson, C.M. 1991a. Primary production and major nutrients in three grass species on a tropical flood plain in northern Australia. Aquat. Bot. 41: 263–280.

Finlayson, C.M. 1991b. Plant ecology and management of an internationally important wetland in monsoonal Australia. In: Kusler, J. & Day, S. (eds) Proceedings of an international Symposium on Wetlands and River Corridor Management. pp. 90–98. Association of State Wetland Managers, New York.

Finlayson, C.M. & van der Valk, A.G. 1994. Wetland classification and inventory: a summary. Vegetatio 118: 185–192.

Finlayson, C.M. & Von Oertzen, I. 1993. Wetlands of Northern Australia. In: Whigham, D.J., Dykyova, H. & Heijny, S. (eds) Wetlands of the World. pp. 195–304. Kluwer Academic Publishers, Dordrecht.

Finlayson, C.M., Bailey, B.J. & Cowie, I.D. 1989. Macrophytic vegetation of the Magela flood plain, northern Australia. Office of the Supervising Scientist Research Report No 5. 38 pp.

Finlayson, C.M., Bailey, B.J. & Cowie, I.D. 1990a. Characteristics of a seasonally flooded freshwater system in monsoonal Australia. In: Whigham, D.J., Goode, D.F. & Kvet, J. (eds) Wetland Ecology and Management – Case Studies. pp. 141–162. Kluwer Academic Publishers, Dordrecht.

Finlayson, C. M., Cowie, I. D. and Bailey, B. J. 1990b. Sediment seedbanks in grassland on the Magela Creek flood plain, northern Australia. Aquat. Bot. 38: 163–176.

Finlayson, C.M., Cowie, I.D. & Bailey, B.J. 1993. Litterfall in a Melaleuca forest on a seasonally inundated flood plain in tropical northern Australia. Wetl. Ecol. & Manage. 2: 177–188.

Finlayson, C.M., Bailey, B.J., Freeland, W.J. & Fleming, M. 1988. Wetlands of the Northern Territory. In: McComb, A.J. & Lake, P.S. (eds) The Conservation of Australian Wetlands. pp. 103–106. Surrey Beatty & Sons, Sydney.

Finlayson, C.M., Thompson, K., von Oertzen I.A.M.L. & Cowie, I.D. 1994. Vegetation of five Magela Creek billabongs, Alligator Rivers Region, northern Australia. Supervising Scientist for the Alligator Rivers Region Technical Memorandum. AGPS, Canberra. (in press)

Galloway, R.W. 1982. Distribution and physiographic patterns of Australian mangroves. In: Clough, B.F. (ed.) Mangrove Ecosystems in Australia: Structure, Function and Management. pp. 31–54. ANU Press, Canberra.

Grindrod, J. 1988. The palynology of Holocene mangrove and saltmarsh sediments, particularly in northern Australia. Rev. Palaeobot. Polynol. 55: 229–245.

Hegerl, E.J., Davie, P.J.F., Claridge, G.F. & Elliot, A.G. 1979. The Kakadu Park mangrove forests and tidal marshes, Vol. 4: A review of the literature and results of a field reconnaissance. Unpublished report to ANPWS, Canberra. 32 pp.

Hope, G., Hughes, P.J. & Russell-Smith, J. 1985. Geomorphological field work and the evolution of the landscape of Kakadu National Park. In: Jones, R. (ed.) Archaelogical Research in Kakadu National Park. pp. 229–240. ANPWS Special Publication 13. AGPS, Canberra.

Jones, R. (ed.) 1985. Archaelogical Research in Kakadu National Park. ANPWS Special Publication 13. AGPS, Canberra. 317 pp.

Knighton, A.D., Mills, K. & Woodroffe, C.D. 1991. Tidal-creek extension and saltwater intrusion in northern Australia. Geol. 19: 831–834.

Lazarides, M. & Craven, L.A. 1980. Report on the Kakadu National Park Flora Project – Checklist of the flora of Kakadu National Park. Unpublished report by CSIRO Division of Plant Industry, Canberra.

Leach, G.J. & Burgin, S. 1985. Litter production and seasonality of mangroves in Papua New Guinea. Aquat. Bot. 23: 215–224.

Letts, G.A., Bassingthwaighte, A. & De Vos, W.E.L. 1979. Feral Animals in the Northern Territory. N.T. Government Printer, Darwin. 234 pp.

Lonsdale, W.M. 1988. An analysis of litterfall in forests of the world. Ann. Bot. 61: 319–324.

Lugo, A.E. 1980. Mangrove ecosystems: successional or steady state? Biotropica (supplement) 12: 65–72.

Lyster, S. 1985. International Wildlife Law. Grotius Ltd, Llandysul, U.K. 470 pp.

McComb, A.J. & Lake, P.S. 1988. Conservation of Australian Wetlands. Surrey Beatty, Sydney. 196 pp.

Messel, H., Wells, A.G. & Green, W.J. 1979. The Alligator region rivers system. Monograph 4. Surveys of tidal river systems in the Northern Territory of Australia and their crocodile populations. Pergamon Press, Oxford. 70 pp.

Morley, A. (ed.) 1981. A review of Jabiluka environmental studies. Pancontinental Mining Limited, Sydney. (six volumes, unpaginated)

Phillips, B. 1993. Introduction. In: Usbank, S. & James, R. (compilers) A Directory of Important Wetlands in Australia. pp. 1.1–1.10. Australian nature Conservation Agency, Canberra.

Ramsar Convention Bureau 1990. Directory of Wetlands of International Importance. IUCN, Gland, Switzerland. 796 pp.

Ridpath, M.G., Begg, R.J., Dudzinski, M.L., Forbes, M.A. & Graham, A. 1983. Counting the same populations of large tropical animals from the ground and from the air. Aust. Wildl. Res. 10: 487–498.

Russell-Smith, J. 1985. A record of change: studies of holocene vegetation history in the South Alligator River region, Northern Territory. Proc. Ecol. Soc. Aust. 13: 191–202.

Sanderson, N.T., Koontz, D.V. & Morley, A.W. 1983. The ecology of the vegetation of the Magela Creek flood plain: upper section from Oenpelli road crossing to Nankeen Billabong. Unpublished report in Scientific Workshop, Environmental Protection in the Alligator Rivers Region, Jabiru. pp. 33.1–33.9.

Sasekumar, A. & Loi, J.T. 1983. Litter production in three mangrove forest zones in the Malay Peninsula. Aquat. Bot. 17: 283–290.

Semeniuk, V. 1980. Mangrove zonation along an eroding coastline in King Sound, north western Australia. J. Ecol. 68: 789–812.

Spain, A.V. 1984. Litterfall and the standing crop of litter in three tropical Australian rainforests. J. Ecol. 72: 947–961.

Stocker, G.C. 1970. The effects of water buffalo on paperbark forests in the Northern Territory. Aust. For. Res. 5: 29–34.

Story, R. 1969. Vegetation of the Adelaide-Alligator. In: Story, R., Williams, M.A.J., Hooper, A.D.L., O'Ferrall, R.E. & McAlpine, J.R. (eds) Lands of the Adelaide-Alligator area, Northern Territory. pp. 114–130. CSIRO Land Research Series 25. CSIRO, Melbourne.

Story, R. 1976. Vegetation of the Alligator Rivers area. In: Story, R., Galloway, R.W., McAlpine, J.R., Aldrick, J.M. & Williams, M.A.J. (eds) Lands of the Alligator Rivers area, Northern Territory. pp. 89–111. CSIRO Land Research Series 38. CSIRO, Melbourne.

Taylor, J.A. & Dunlop, C.R. 1985. Plant communities of the wet-dry tropics of Australia: the Alligator Rivers Region. Proc. Ecol. Soc. Aust. 13: 83–128.

Taylor, J.A. & Friend, G.R. 1984. Ground surface features attributable to feral buffalo, Bubalus bubalis, I. Their distribution relative to vegetation structure and plant lifeform. Aust. Wildl. Res. 11: 303–309.

Thom, B.G. 1967. Mangrove ecology and deltaic geomorphology, Tabasco, Mexico. J. Ecol. 55: 301–343.

van der Valk, A.G. 1981. Succession in wetlands: a Gleasonian approach. Ecol. 62: 688–696.

Walker, T.D. & Tyler, P.A. 1985. Tropical Australia, a dynamic limnological environment. Verh. Inter. Verein. Limnol. 22: 1727–1734.

Wells, A.G. 1984. Mangrove Vegetation in Northern Australia. Ph.D. Thesis, University of Sydney, Australia. 236 pp.

Wells, A.G. 1985. Grouping of tidal systems in the Northern Territory-Kimberley region of Western Australia on presence/absence of mangrove species. In: Bardsley, K.N., Davie, J.D.S. & Woodroffe, C.D. (eds) Coastal and Tidal Wetlands of the Australian Monsoon Region. pp. 119–132. North Australian Research Unit, ANU Press, Darwin.

Wightman, G.M. 1989. Mangroves of the Northern Territory. Northern Territory Bot. Bull. 7, Conservation Commission of the Northern Territory, Darwin. 130 pp.

Williams, A.R. 1979. Vegetation and stream pattern as indicators of water movement on the Magela flood plain, Northern Territory. Aust. J. Ecol. 4: 239–247.

Williams, A.R. 1984. Changes in Melaleuca forest density on the Magela flood plain, Northern Territory, between 1950 and 1975. Aust. J. Ecol. 9: 199–202.

Williams, C.K. & Ridpath, M.G. 1982. Rates of herbage ingestion and turnover of water in feral swamp buffalo Bubalis bubalis in relation to primary production in a cyperaceous swamp in monsoonal northern Australia. Aust. Wildl. Res. 9: 397–408.

Wolanski, E., Chappell, J., Ridd, P. & Vertessey, R. 1988. Fluidization of mud in estuaries. J. Geophys. Res. 93: 2351–2361.

Woodroffe, C.D. 1988. Changing mangrove and wetland habitats over the last 8000 years, northern Australia and south-east Asia. In: Wade-Marshall, D. & Loveday, P. (eds) North Australia: Progress and Prospects, vol. 2 Flood plain Research. pp. 1–33. ANU Press, Darwin.

Woodroffe, C.D. & Bardsley, K.N. 1988. The distribution and productivity of mangroves in Creek H, Darwin Harbour. In: Larson, H.K., Michie, M.G. & Hanley, J.R. (eds) Proceedings of a Workshop on Research and Management of Darwin Harbour, pp. 82–121. North Australian Research Unit, ANU Press, Darwin.

Woodroffe, C.D., Thom, B.G. & Chappell, J. 1985a. Development of widespread mangrove swamps in mid-Holocene times in northern Australia. Nat. 317: 711–713.

Woodroffe, C.D., Bardsley, K.N., Ward, P.J. & Hanley, J.R. 1988. Production of mangrove litter in a macrotidal embayment, Darwin Harbour, Northern Territory, Australia. Estuar. Coastal Shelf Sci. 26: 581–598.

Woodroffe, C.D., Chappell, J., Thom, B.G. & Wallensky, E. 1985b. Geomorphology of the South Alligator River and plains, Northern Territory. Proceedings of a conference on Coasts and Tidal Wetlands of the Australian Monsoon Region. NARU Monograph, AGPS, Canberra. pp. 5–16.

Woodroffe, C.D., Chappell, J., Thom, B.G. & Wallensky, E. 1986. Geomorphological Dynamics and Evolution of the South Alligator Tidal River and Plains. North Australian Research Unit, Darwin. 190 pp.

Woodroffe, C.D., Chappell, J., Thom, B.G. & Wallensky, E. 1989. Depositional model of a macrotidal estuary and flood plain, South Alligator River, northern Australia. Sediment. 36: 737–756.

Woodroffe, C.D., Thom, B.G., Chappell, J., Wallensky, E., Grindrod, J. & Head, J. 1987. Relative sea level in the South Alligator River region, north Australia, during the Holocene. Search 18: 198–200.

CHAPTER 6

Weed ecology

IAN D. COWIE

Abstract. Perennial weeds capable of invading relatively undisturbed native plant communities represent one of the greatest challenges for land management authorities in the Kakadu region. With their potential for dispersal and rapid population growth, these invasive species (especially *Mimosa pigra*) can severely modify natural communities in the region over a comparatively short period, especially the relatively vulnerable wetlands. Once established, factors such as persistent soil seed banks coupled with high plant fecundity and poor access to infestations during the wet season can make these species very difficult to remove.

Most weed species in the region are tropical annual herbs. The proportion of weed species in the flora of the region (5.8% or 89 species) is typical of that for adjoining areas and for similar biomes in conservation reserves on other continents. Rates of weed invasion in the region have been relatively low (1.6 species per year since 1948), but the large number of introduced species now in cultivation (>305) represents a potential source of weeds. Of these 119 are weeds in some situations in other countries, and 19 cultivated species have become weeds in nearby areas.

The management of alien plant species is subject to a complex regulatory framework, which provides for control of the growing, spread and introduction of certain declared noxious weeds and other introduced plants (including controls on the importation of some materials likely to contain weeds). Provision has been made to allow the cultivation of some approved introduced plants in addition to species found naturally in the region. Disturbance by either humans or feral animals appears to be a major factor favouring weed invasions. Control of feral animals and the early revegetation (with native species) of areas disturbed by humans are seen as important in minimising opportunities for weed invasion.

1. Introduction

Historically, early European settlement and land use in the Kakadu region probably resulted in relatively few alien species being introduced. Land use then was extensive with buffalo and crocodile shooting, limited cattle grazing and agriculture at Oenpelli (which became a church mission station in 1925) (Bauer 1964). A number of small uranium mines were in operation in the upper parts of the South Alligator River valley during the 1950s and 1960s. A brief period of pasture development occurred at Munmarlary and Mudginberri in the early 1970s with limited clearing of native vegetation and planting of introduced pasture species. Pressure for the establishment of a national park and for development of the region for tourism and mining during the 1970s (Christian & Aldrick 1977), and subsequent development during the 1980s, led to increased concern about the potential invasion of alien plants.

Weeds have the potential to alter the flora and fauna by replacing native plant species and by modifying the resources available to animals. Weed infestations impede recreation and management, change hydrological regimes, alter fire regimes, interfere with revegetation, hybridise with native taxa, or have a visual impact. The occurrence of three important invasive species, all of them introduced, proved some of these concerns. The South American shrub *Mimosa pigra* had rapidly covered areas of flood plain to the west of the Kakadu region, converting herbaceous swamps into shrublands (Miller *et al.* 1981). A well established population of *Salvinia molesta* was discovered on waterways of the Magela flood plain during 1983 and concurrently six plants of *Eichhornia crassipes* were found on drying mud and were successfully removed (Finlayson 1984a). Mats of *Salvinia* can easily cover whole waterbodies and thus drastically alter water quality (D. Mitchell 1978). These events stimulated research on the invasive alien flora of the region (Cowie *et al.* 1988, Cowie & Werner 1987, 1988, 1993).

In order to maintain some consistency when addressing the status of various introduced plants

C.M. Finlayson and I. von Oertzen (eds), Landscape and Vegetation Ecology of the Kakadu Region, Northern Australia, 113-135.

the following definitions have been adopted:

i) A **weed** in the context of the primary land use of the Kakadu region (i.e. conservation) was any invasive **alien** species of plant. Some native species that colonise disturbed ground (e.g. *Brachyachne convergens)*, have burrs (e.g. *Urena lobata*) or both (e.g. *Triumfetta* spp.) are not treated here.

ii) **Alien species** were those originating outside the Kakadu region that had reached it in active or passive association with humans, usually since European colonisation (Kloot 1984).

iii) **Invasive species** were those alien species reproducing by any means and establishing new individuals without deliberate human assistance, and past the first generation outside of cultivation.

iv) **Adventive species** were those alien species reproducing by any means and establishing new individuals without deliberate human assistance, but not past the first generation outside cultivation.

v) **Persistent species** were those alien species planted by humans and persisting after deliberate human assistance had ceased, but not establishing new individuals without deliberate human assistance.

The botanical nomenclature used in these studies generally followed Dunlop (1987).

2. Rates of invasion and proportion of invasive species

The relationship between the cumulative occurrence of invasive alien plant species in the region and time has been examined by Cowie *et al.* (1988) (Fig. 1). Since 1948, alien species had accumulated at a rate of 1.6 species per year. In comparison, the rate of accumulation of invasive alien species in south-eastern Australia averaged 6 species per year for the past 150 years (Specht 1981), with a similar scenario for the Victorian flora (Ross 1976). The discrepancy reflects differences in the history of settlement, and industrial and agricultural development between southern and northern Australia. The Kakadu region was developed much later and less intensively.

Among a range of regions and conservation areas in north-western Australia, the best predictor of the number of alien plant species was the total number of species in the flora ($y = 17.433 + 0.058x$, $r^2 = 0.931$, $F_{(1,7)} = 93.7$, $P << 0.001$) (Fig. 2). A possible explanation for this observation is that the number of plant species (both native and alien) is related to the habitat diversity in an area. However, at the plant community level different processes may be in operation. Fox & Fox (1986) found a strong inverse relationship between the number of introduced species and the number of

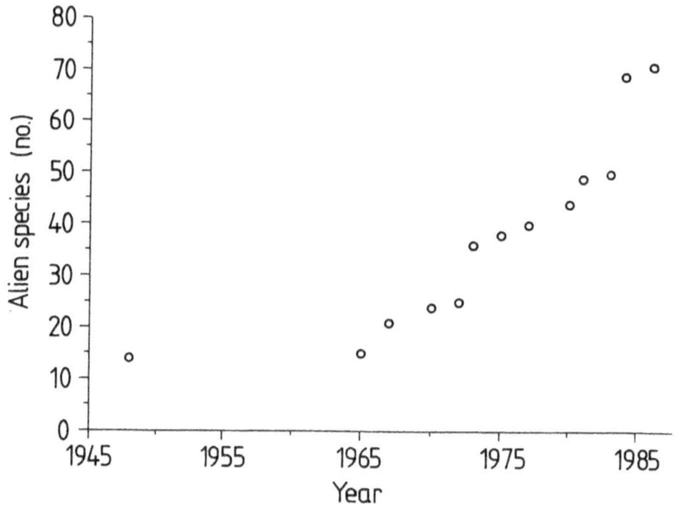

Fig. 1. The cumulative number of alien plant species recorded in the Alligator Rivers Region as a function of time ($y = 0.974e^{0.045x}$, $r^2 = 0.860$, $F_{(11)} = 67.7$, $P << 0.001$) (after Cowie *et al.* 1988). Although an exponential model provided the best fit to the data, it was argued that this shape of curve reflected a recent increase in collecting effort in the Region and that the rate of accumulation was probably linear.

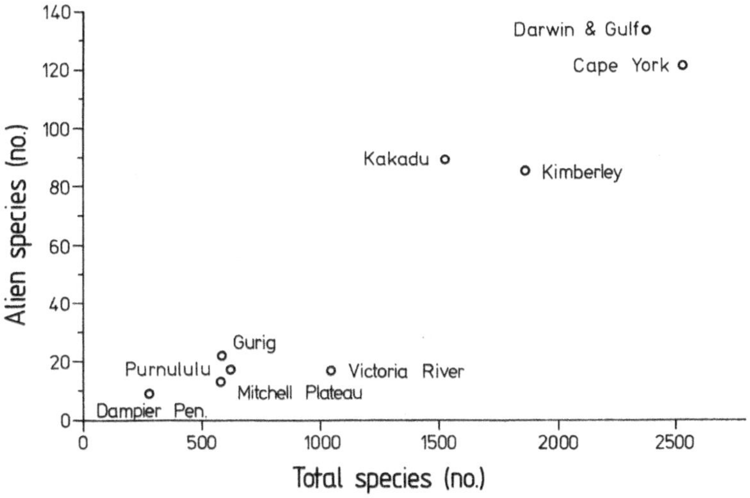

Fig. 2. The number of alien species as a function of the total number of plant species for regions of northern Australia. The data show a strong positive relationship ($y = -17.433 + 0.058x$, $r^2 = 0.931$, $F_{(7)} = 93.7$, $P << 0.001$) (data from Cowie & Werner, 1993). Approximate areas (km^2) of the regions are: Dampier Peninsula Reserves (291); Mitchell Plateau (1 400); Gurig National Park (2 210); Purnululu (Bungle Bungle) National Park and adjacent areas (3 100); Kakadu National Park (19 800); Cape York (130 600); Victoria River Region (233 000); Kimberley Region (302 600); Darwin and Gulf Region (335 000).

native species for temperate coastal shrubland and heathland communities which had been grazed. In the north-western Australian data there was also a tendency for larger areas ($r = 0.60$, $n = 9$, $P < 0.1$) and areas with higher rainfall ($r = 0.59$, $n = 9$, $P < 0.1$) to have more invasive species. Curiously, larger floras had a greater proportion of alien species than did smaller floras ($r = 0.86$, $n = 9$, $P < 0.001$). No explanation is obvious.

A total of 89 invasive alien species were recorded in Kakadu National Park, representing 5.8% of the estimated 1526 vascular plant species in the region (Cowie & Werner 1993). This proportion of invasive species was within the range (2.6–8.8%) recorded for continental reserves dominated by tropical savannas and dry woodlands in other countries (Macdonald & Frame 1988, Usher 1988). It is also consistent with the level of infestation of the Darwin and Gulf regions of the Northern Territory (5.6%) (Dunlop 1987), but greater than those for both the climatically similar Kimberley region of Western Australia (4.6%) (Kenneally 1989) or the Cape York region of north Queensland (4.8%) (Clarkson & Kenneally 1988).

The level of weed infestation in the Kakadu region is low compared with that of many other natural ecosystems (Usher 1988), which, in the case of island nature reserves can reach almost 50% (Brockie *et al.* 1988). The proportion of alien species in the flora is also greater in more developed areas of southern Australia, for example 28% in the Sydney region (Beadle *et al.* 1972) and 22% for the State of Victoria (Willis 1970, 1972). Apart from the above mentioned differences in the history of settlement and development between northern and southern Australia, low soil fertility and the extremes of a monsoonal climate could be contributing factors to the small proportion of alien species in the Kakadu region. Fox & Fox (1986) concluded from a qualitative survey carried out mostly in southern Australia, that communities subject to regular low intensity disturbance (such as the frequently burned open forest and woodland communities which dominate the Kakadu region) were relatively free of invasion compared with communities subject to irregular massive disturbance.

3. Distribution and abundance of invasive species

In their study of the vegetation of the northern part of the Kakadu region, Taylor & Dunlop (1985) recorded the distribution and abundance of 22 species of invasive alien plants occurring in 6 of a total of 11 natural communities studied. However, anthropogenic habitats (those disturbed by humans as described below) were not included in their study. Cowie *et al.* (1988) completed an inventory of invasive alien plant species and collected information on the distribution of 71 species, primarily on anthropogenic habitats. A more comprehensive study of the distribution and abundance of invasive alien plant species on both natural and anthropogenic habitats in Kakadu National Park was undertaken by Cowie & Werner (1993) to provide a basis for the management of these species. The following account is derived largely from that source.

These authors stratified Kakadu National Park into two types of habitat:

i) anthropogenic habitats, i.e. those disturbed during the past 170 years by Europeans and Aboriginal people engaged in non-traditional activities (e.g. commercial shooting of feral water buffalo – *Bubalus bubalis*) or by domestic animals, but not those disturbed by feral animals; and
ii) natural habitats which corresponded generally to vegetation types recognised by Taylor & Dunlop (1985).

Two measures of abundance were used – plant cover and frequency of occurrence, with sample sites located systematically (with a random start) in nearby suitable habitats along roads and tracks. Cover of each alien species was estimated using a modified Braun-Blanquet cover abundance scale (Mueller-Dombois & Ellenberg 1974). The presence of species in particular habitats, but outside the quadrats was also noted.

Cowie & Werner's (1993) results for the success of particular alien species are in substantial agreement with those of Russell-Smith (1984) on lowland monsoon forest and Taylor & Dunlop (1985) on natural habitats. Mean cover levels of alien plants were relatively low, with only two species (*Hyptis suaveolens* and *Sida acuta*) having mean cover values of >5% on any habitat (miscellaneous disturbed sites and settlements). To gain an appreciation of the ability of alien species to occupy space (independent of the spread of a species) the mean cover of alien species was calculated only on the basis of the quadrat where the plant was present. Using this measure, many species made a major contribution to cover at the often limited number of places where they occurred.

Some species were widespread in the park (*Hyptis suaveolens*, *Sida acuta*, *Passiflora foetida*, *Sida cordifolia*, *Euphorbia hirta* and *Alysicarpus vaginalis*), and occurred at many sites in many habitats (Cowie & Werner 1993). Other species were widespread only in particular habitats. They included *Senna obtusifolia* (syn. *Cassia obtusifolia*) in lowland monsoon forest, margin woodland and lowland riparian habitats, *Malachra fasciata* in *Fimbristylis* sedgeland, *Heliotropium indicum* in flood plain margin "lawn" and *Melaleuca* swamp, and *Stylosanthes hamata* in some anthropogenic habitats. These species formed a significant, locally dominant component of the vegetation where they occurred. In contrast, many species were either sparse or rare. Some 31 species were recorded at less than 5 sample sites, whilst a further 27 species present in Kakadu National Park were not recorded on any sample site.

Some of the alien species which are at present comparatively rare in the park, ultimately may be of greater significance than those which are widespread and common. Some species dominated large areas of relatively undisturbed native plant communities whilst most widespread, common species were only dominant over small areas. Potentially significant species may have been rare (low frequency of occurrence) for a variety of reasons. They may have arrived in the region relatively recently, such as *Salvinia molesta* and *Pennisetum polystachion* (Cowie *et al.* 1988). Others may be subject to intensive control programs (e.g. *Mimosa pigra*), or may have poor natural dispersal between catchments (e.g. *Senna alata* (syn. *Cassia alata*), *Salvinia molesta* and *Brachiaria mutica* (syn. *Urochloa mutica*). These species were highly competitive weeds making a major contribution to plant cover where they occurred. Within the region, *Brachiaria mutica* and *Salvinia molesta*, and outside the region *Mimosa pigra* and *Pennisetum polystachion* have dominated substantial areas of relatively undisturbed native plant communities (Letts 1960, Miller *et al.* 1981, Cowie *et al.* 1988, Cowie & Werner 1993).

Disturbance, either by humans or feral animals appeared to be an important factor affecting the success or otherwise of plant invasions in the Kakadu region. Invasive species were more common in anthropogenic habitats than in natural habitats. The former supported more invasive species (60 vs 37) at higher estimated mean cover values than did natural habitats, even though anthropogenic habitats comprised less than 0.1% of the park. Many of the alien species on anthropogenic habitats had only recently been recorded as invasive (Cowie *et al.* 1988) and were present at few sites, suggesting that they were still near their initial point of establishment. Thus, anthropogenic habitats may well provide a focus for naturalisation and a reservoir of alien species to spread into natural habitats.

Several studies have linked feral buffalo with alien plant invasion. Braithwaite *et al.* (1984), working in monsoon forest (which frequently adjoins flood plains), found that the abundance of the invasive annual plants *Senna obtusifolia* and *Senna occidentalis* (syn. *Cassia occidentalis*) was positively correlated with buffalo use. Following the experimental removal of buffalo at Kapalga (in Kakadu) Minchin & Dunlop (1989) found that the abundance of invasive weeds in natural plant communities declined. Both Taylor & Dunlop (1985) and Cowie & Werner (1993) found natural habitats adjacent to flood plains and rivers (i.e. monsoon forest, margin woodland and lowland riparian habitats) to be among the most severely infested natural communities. The plant communities adjoining flood plains are also those most heavily used by feral water buffalo (Tulloch 1978).

Wetland plant communities appear to be most susceptible to domination by invading plant species. Perennial invasive species such as *Brachiaria mutica*, *Senna alata*, *Salvinia molesta*, and *Mimosa pigra* are all able to dominate certain wetland communities while in the drier upland communities *Pennisetum polystachion* is amongst the few perennial species able to dominate. Apart from disturbance by feral buffalo, the disturbance caused by recurrent flooding may be a contributing factor. Consistent with these results, Fox & Fox (1986) in their survey of mostly southern Australian invasions, found that plant formations subject to irregular massive disturbance such as flooding were very susceptible to invasion, compared with less disturbed community types.

In addition to disturbance, higher nutrient levels, lower species diversity of vegetation, and the structural simplicity and dynamic nature of wetland plant communities may be factors contributing to plant invasions of riparian and wetland vegetation. The interaction between these factors in local plant invasions remains unclear.

4. Characteristics of the invasive flora

The invasive alien flora is represented by 29 families with Poaceae having the largest number of species, followed by Fabaceae, Malvaceae, Asteraceae Caesalpiniaceae, and Amaranthaceae (Cowie & Werner 1993). This pattern is similar to that observed for other areas of northern Australia (Cowie & Werner 1993). Poaceae, Fabaceae and Asteraceae also dominate the alien flora of south-eastern Australia (Specht 1981). These three families are, after the Orchidaceae, the worlds's largest plant families (Heywood 1978). Williamson & Brown (1986) for Great Britain and Esler & Astridge (1987) for New Zealand concluded that on a global basis larger families tend to contribute more invasive species than smaller families.

Although the majority of alien species appeared to originate in the New World tropics (49 species), the Old World figure (34 species) may be an underestimate, reflecting the difficulty in distinguishing human introductions from those plants arriving naturally. As A. Mitchell (1978) argues, it is likely that many years of regular Macassan contact with northern Australia (mid 1600s to early 1900s) resulted in plant introductions (other than *Tamarindus indicus*) before the first European botanist visited the area in 1803. Unrecorded plant introductions are also likely to have resulted from importation of stock and supplies from nearby Indonesia during the period of early settlement of northern Australia by Europeans. Interpretation of what is introduced and what is native is further confounded by the possibility that natural dispersal of weedy species into northern Australia may have occurred during periods of lower sea level (Barlow 1981). The alien flora also included three species native to or from other parts of Australia but, alien to Kakadu National Park.

The life-forms of 71 alien species in the region are listed by Cowie *et al.* (1988). The largest group consists of annual forbs (18), with annual shrubs and perennial grasses the next largest groups (11 each). Perennial grasses were more numerous than annual grasses, while annuals were more numerous than perennials among the shrubs, climbers and forbs. Overall, annuals were more numerous than perennials. Trees were poorly represented, although they are strongly represented in the native flora. Pteridophytes and gymnosperms were not present in the alien flora but are also poorly represented in the native flora. Given that most of the vegetation in the region consists of woodland and open forest, these observations are consistent with those of Fox & Fox (1986) who noted that on an Australia-wide basis these plant formations support very few successful tree weeds, most weeds being in the understorey.

A significant proportion of the alien flora seems to have its origins in crop and pasture plants introduced before Kakadu National Park was declared. Six species (*Brachiaria mutica, Cenchrus ciliaris, Macroptilium atropurpureum, Macroptilium lathyroides, Stylosanthes humilis*, and *Stylosanthes guyanensis*) were planted at Mudginberri and Munmarlary for pasture improvement during the late 1960s and early 1970s (I.L. Miller personal communication). It is also suspected that several other pasture species were introduced to the region during this period (*Andropogon gayanus, Calopogonium mucunoides and Stylosanthes viscosa*). *Chloris gayana*, *Cenchrus ciliaris* and *Stylosanthes hamata* (used elsewhere for pasture improvement) were probably introduced to the region for soil stabilisation on mines and along roadsides. A further ten species are fruiting, ornamental or medicinal plants that have escaped from cultivation (*Alternanthera dentata, Anacardium occidentale, Annona squamosa, Cassia fistula, Delonix regia, Gmelina arborea, Jatropha* spp., *Mangifera indica* and *Senna alata*). The fact that several of these species are now important weeds in the region, indicates the potential importance as weeds of other cultivated species which have more recently been introduced to the area.

5. Weed potential of alien plants in cultivation

5.1. Species in cultivation

An inventory of alien species in cultivation now or in the recent past in Kakadu National Park was compiled by surveying house yards and public places in each settlement (Cowie & Werner, 1987). The status of each species at each settlement was classified as: i) present in cultivation only; ii) persisting after abandonment; iii) adventive; or iv) invasive. Species of potted plants and seeds on sale locally were also recorded. The scientific names used generally follow Graf (1981). The names used should be regarded as provisional, as in some cases, reference to the relevant taxonomic literature or to properly determined herbarium specimens was not possible.

In excess of 305 alien plant species were cultivated or had been cultivated in the recent past (Cowie & Werner 1987, 1988). Most species were of tropical origin with 65 shrubs, 94 herbs, 62 trees, 25 arborescent monocotyledons (mostly palms) and the remainder vines or epiphytes. Eleven species were native to the northern part of the Northern Territory (but not the Kakadu region) while 23 species were native to other parts of Australia. Of the species originating overseas, 81 were from Asia, 85 from the Americas, and 22 from Africa. A wide variety of families were represented in the cultivated flora with the Arecaceae (22 species), Araceae (19), Leguminosae (18), Euphorbiaceae (17), Liliaceae (14), Verbenaceae and Poaceae (11 each), and Acanthaceae, Asteraceae, and Myrtaceae (9 each) most strongly represented. Thirty three species were persistent at sites where they had been abandoned, while 19 species were adventive and 17 species had become invasive. A total of 81 species of alien plant were on sale in the region, either as potted plants or as seed. Twelve of these were also in cultivation. During the survey, 55 cultivated species were found to be invasive, adventive or persistent, although few were important weeds.

Jabiru, the largest settlement, with 256 species and Jabiru East (the former construction camp for the Ranger Mine and Jabiru) with 91 species were the two most important settlements in terms of the number of alien species in cultivation. Jabiru East has now been depopulated, dismantled and most cultivated plants have been removed. Other settlements in decreasing order of number of species were Nourlangie Camp (59 species), Mudginberri (51 species), Cooinda (50 species), and East Alligator (47 species).

5.2. *Cultivated species as weeds*

The large number of alien species in cultivation could provide a substantial reservoir of species which have the potential to become naturalised. In Great Britian, Williamson & Brown (1986) found that approximately 66% of introduced vascular plant species (excluding those in artificial environments) became established outside cultivation, while 8% of the introduced species became pests. Thus, the pest species represented 12% of the established species. This figure is close to the approximately 14% of invasive plant species regarded as important weeds in Kakadu. If it was assumed that other parts of this relationship applied for the species currently in cultivation in Kakadu, then approximately 25 would become important weeds. Thus, an attempt was made using information on the climatic origin and weed status elsewhere in the world, to determine which alien species in cultivation would be most likely to become invasive.

The weed status in other tropical countries of alien plants in cultivation, recent past cultivation or available as seed in Kakadu was established from the literature (principally Morton 1976, Holm *et al.* 1979 and various floras including Hutchinson & Dalziel 1954–72 and Backer & van den Brink 1963–68) as an indicator of the plant's potential to become invasive. Since climatic similarity of source and reception areas is known to be important in biological invasions (Swincer 1986) the analysis was restricted as far as possible to tropical regions. Countries with wet-dry tropical climates similar to that of the region were identified using world climatic maps (Troll 1966). Countries with large areas experiencing temperate climates and small tropical areas were excluded to avoid biasing the data. The responses of alien species in relevant countries were categorised into four groups based on whether plants were known as: i) agricultural; or ii) non-agricultural weeds in tropical countries with (a) or without (b) areas experiencing wet-dry climates. A plant could be in more than one group.

Of the plant species in cultivation within Kakadu National Park, a total of 119 (including 19 species available as seed) were known to be invasive in other tropical countries. However, relatively few (20) species cultivated in the region were recorded as being non-agricultural weeds elsewhere (Table I). Twelve of this group were already known to be adventive or invasive in the northern part of the Northern Territory. The largest group (Table II) comprised of 76 species reported as non-agricultural weeds in countries experiencing rainy tropical or humid summer tropical climates. Of this group, 17 species were already invasive locally, 4 species were adventive and 11 persisted after abandonment. Six species from the group were native to the northern parts of the Northern Territory, but not to Kakadu National Park. Another group of 32 species (of which nine were already invasive locally) were agricultural weeds in countries with comparable climates elsewhere. A forth group of 26 species were, at most, agricultural weeds in countries with rainy tropical or humid summer tropical climates.

Table I. Species cultivated in Kakadu which have been recorded as weeds in tropical countries with areas experiencing wet-dry tropical climates ("Weeds" were those species which were invasive, or were environmental or ruderal weeds in other countries) (after Cowie & Werner 1988).

Invasive	Adventive	Persistent	Cultivated only
Anacardium occidentale	*Canna indica*	*Caladium bicolor*	*Ipomoea batatas*
Catharanthus roseus		*Kalanchoe pinnata*	*Luffa aegyptiaca*
Celosia argentia		*Quisqualis indica*	*Ocimum basilicum*
Euphorbia heterophylla			*Passiflora edulis*
Ipomoea quamoclit			*Portulaca grandiflora*
Lantana camara			
Leucaena leucocephala			
Melia azedarach			
Senna alata			
Tamarindus indicus			
Wedelia trilobata			

Table II. Species cultivated in Kakadu which have been recorded as weeds in tropical countries *lacking* areas with wet-dry tropical climates ("Weeds" were those species which were invasive, or were environmental or ruderal weeds in other countries.) Codes for the status of each species in these other tropical countries are "a" major weeds, "b" medium weeds and "c" minor weeds. Species native to the N.T. but not to Kakadu are prefixed by an asterisk. (after Cowie & Werner 1988)

Invasive	Adventive	Persistent	Cultivated only
Anacardium occidentale	Antigonon leptopus[c]	Bambusia vulgaris	*Adenanthera pavonina[b]
Axonopus compressus	Caesalpinia pulcherrima[b]	Capsicum annum	*Casuarina equisetifolia a
Cassia fistula	Canna indica[c]	Citrus spp.[c]	*Cucumis melo
Catharanthus roseus[c]	Cryptostegia madagascariensis[c]	Cocos nucifera[c]	*Schefflera actinophylla
Celosia argentea	Jatropha multifida[b]	Hibiscus rosa-sinensis[c]	*Terminalia catappa[c]
Delonix regia[c]		Kalanchoe pinnata[b]	*Vitex trifolia[c]
Euphorbia heterophylla		Manihot esculenta[c]	Ageratum houstonianum
Ipomoea quamoclit		Psidium guajava[b]	Allamanda cathartica[c]
Jatropha curcas[b]		Quisqalis indica	Alstonia scholaris[b]
Jatropha gossypifolia[b]		Sanseveria trifasciata[b]	Argyreia nervosa[c]
Lantana camara		Tecoma stans[c]	Arundo donax
Leucaena leucocephala[b]		Thevetia peruviana[b]	Barleria cristata[b]
Mangifera indica[b]			Bauhinia variegata[b]
Melia azedarach[b]			Calliandra surinamensis[b]
Senna alata			Carica papaya[b]
Stachytarpheta cayennensis			Citrullus lanatus
Tamarindus indicus[c]			Cucumis sativus
Wedelia trilobata[b]			Euphorbia lactea[c]
			Euphorbia tirucalli
			Ficus elastica[c]
			Hedychium coronarium[c]
			Hemigraphis colorata
			Impatiens balsamina
			Ipomoea batatas
			Jacaranda mimosaefolia[c]
			Malpigia coccigera
			Malpigia glabra[b]
			Murraya exotica[b]
			Ocimum basilicum
			Pedilanthus tithymaloides[c]
			Persea americana[b]
			Phyllanthus acidulis[c]
			Psidium littorale
			Rhoeo spathacea[c]
			Russelia equisetiformis[c]
			Saccharum officinarium
			Salvia splendens
			Sesbania grandiflora[c]
			Setcreasea purpurea[c]
			Synonium podophyllum[c]
			Thunbergia grandiflora
			Thunbergia laurifolia

Of this last group only six species have been recorded as invasive in the northern parts of the Northern Territory. There was some overlap between the four groups.

Many cultivated species (185) were not recorded as invasive or as weeds either locally or elsewhere. Of the 55 species in cultivation that were already invasive, adventive or persistent in the northern part of the Northern Territory, 17 species were not recorded as being weeds elsewhere.

The proportions of the four groups of weeds that are already persistent, adventive or invasive in the Kakadu region suggest that those species that are invasive or regarded as non-agricultural weeds in countries with wet-dry tropical climates (Table I) would be most likely to become invasive in the region. However, for a variety of reasons, the above analysis can only be used as an indication of the potential of plants to become weeds in the Kakadu region. Some of the species listed as weeds elsewhere have been present in cultivation in the region for many years, but have not become invasive (e.g. *Psidium guajava*, *Passiflora edulis*) (Holtze 1887, Spillett 1972), although *Psidium guajava* is highly invasive in wet tropical areas (Humphries *et al.* 1991).

Some care needs to be exercised in extrapolating weed problems from agricultural to conservation situations. Agricultural weeds may often include native species that are difficult to remove during land development (Swarbrick 1983), but are not highly invasive, or include the persistent remnants of previous crops (e.g. *Ipomea batatas* [sweet potato], *Manihot esculenta* [cassava], *Solanum melongena* [egg plant] and *Saccharum officinarum* [sugar cane]). Similarly, some species which may be important agricultural weeds such as *Echinochloa colonum* may be of little or no consequence in Kakadu.

In this analysis the coarseness of the climatic information regarding the origin of cultivated plants and lack of habitat information remain limitations. Better information and the use of bioclimatic models (e.g. BIOCLIM -Busby 1986a,b) may improve the predictive potential of such analysis).

6. Management of alien species

6.1. Legislation

Management of alien plants in the Kakadu region falls under an extensive and complex regulatory framework of Commonwealth and Territory laws. laws of the Northern Territory apparently apply on Commonwealth land in the Northern Territory, except where Commonwealth legislation is to the contrary. Under the Northern Territory Noxious Weeds Act plants may be declared as one of several categories of noxious weeds, depending on the extent of control measures to be taken against them. Many of the alien species in Kakadu National Park have been declared under this Act. The Act requires that landholders (including ANCA and the mining companies) i) eradicate; ii) control the growing and spread; or iii) control the introduction of the declared plants, depending on the category in which they have been declared.

The import and use of alien plant species in the Kakadu region is tightly controlled under the Northern Territory Uranium Mining (Environmental Control) Act 1979. Under subsections 6(a) and (b) of the Environmental Requirements for the Ranger Project listed in Schedule 1 of this act, it would appear that the Ranger joint venture partners are legally required to prevent the introduction of alien flora by their employees or contractors to the Ranger Project Area and to Kakadu National Park. Exceptions may be granted by the Director of the Australian National Conservation Agency (ANCA), under the Kakadu National Park Plan of Management (ANPWS 1991) or pursuant to regulations made under the National Parks and Wildlife Conservation Act 1975. Similar provisions apply to the Nabarlek Uranium Project Area, except that introductions are at the discretion of the supervising authority (in this case the Northern Territory Department of Mines and Energy).

The town of Jabiru lies on land leased from Kakadu National Park and was set up under the Jabiru Town Development Act 1978 (Northern Territory). This Act regulates the introduction, propagation and cultivation of plants in Jabiru. The definition of plant includes dead plant material such as hay and seeds. Lists of alien plant species regarded as noxious and of species approved for use in the town have been made under the Jabiru Town Development (Plants) Bylaws of this Act. At the present time, the approved list includes some 116 species of alien plants, and all species native to the Park. This list has recently been reviewed (Brock & Cowie 1992). It would appear that the Jabiru Town Council has powers to apply penalties, to enter premises and to destroy plants not approved for use in Kakadu National Park. However, on the recommendation of the Director of ANCA, permits may be issued for the introduction or cultivation of non-approved plants.

The National Parks and Wildlife Conservation Act 1975 (Commonwealth) allows the Director of

ANCA to take measures considered necessary for the control of animals and plants that are not 'wildlife' in Kakadu. Under this Act the Director is required to draw up a Plan of Management for Kakadu. This plan is apparently legally binding (although in some cases the regulations may not exist to enforce the management prescriptions). This Act, together with the Plan of Management appear to prevent the importation by any person of non-approved alien plants and plant materials to Kakadu National Park, to leased parts of Kakadu (e.g. South Alligator Holiday Village) and to prevent these categories of plants being transported through Kakadu to excisions such as Cooinda. Again, exceptions may be granted by the Director of ANCA.

6.2. Invasive species

6.2.1. Control

The problem of weed control in national parks is in many cases fundamentally different from weed control in agricultural situations. In the parks, the problem is often to kill one or a few species of plants whilst leaving most species unharmed, effectively the opposite of conventional weed control practices in agriculture (Swarbrick 1987). Thus, control practices often need to be discriminating, with a minimum of undesirable side effects. Biological control, selective use of herbicides and discriminating mechanical control can fulfil this need. Where invasive species dominate the native flora, control programs combining fire, biological control and active revegetation may be ecologically more appropriate and acceptable to traditional owners, park managers and the public than programs relying entirely on chemical methods (Groves 1990). In an area such as the Kakadu region where there are a variety of weeds of various population sizes and with differing priorities for control, one or more control techniques may be necessary. However, as Stafford (1990) notes, it is easy to fall into the trap of believing that if a weed species is controlled or removed then the weed problem is solved. Often the removal of one weed simply creates a gap for other weeds to occupy. The most effective way to fill such gaps in conservation areas is by revegetation with native species.

Many species have only recently been recorded as invasive in the Park and exist as small populations at a few locations. Control (or even elimination) of these species is more likely to be successful if implemented early (Usher 1988). Although the potential of these species to become weeds in the region is generally not known, it is likely that a few could become much more abundant. Other invasive species were considered too well established to be eliminated from the Park (e.g. *Brachiaria mutica, Senna obtusifolia, Hyptis suaveolens* and *Salvinia molesta*). For these species, biological control, habitat manipulation (especially the control of disturbance) with selective mechanical and chemical control were considered appropriate.

Factors such as fire which may affect population processes could, potentially, be used to manipulate the abundance of some species. In many instances it will be necessary to obtain a greater understanding of how these factors operate before they can be used to manipulate weed populations. Further, year to year variations in abundance due to natural causes need to be taken into account. Research programs on biological control for weeds primarily of conservation areas, such as *Brachiaria mutica,* are likely to be beyond the current resources of park managers. However, where alien plants are also serious weeds of other land uses in northern Australia, biological control programs have been implemented (e.g. *Sida* spp., *Mimosa pigra*, *Salvinia molesta*) (Wilson & Lonsdale 1987).

Accidental introductions as a result of increased tourism and development are likely to continue despite current management policies (ANPWS 1991). Lonsdale & Lane (1990) have found that tourists' vehicles can carry considerable numbers of seeds, some of them from plants known to be tropical weeds, into the Park. The importance of this method of transport is likely to depend on the rate at which seed from cars reaches sites suitable for establishment. Transportation of hay has also been shown to be an important means of dispersal for weeds (Thomas *et al.* 1984). Within Kakadu National Park, the importance of plant materials (including hay), and construction materials in the spread of weeds has long been recognised and their importation has been regulated. Regular monitoring of areas likely to receive 'new' alien species (e.g. camp grounds, settlements and tracks) and the continuation of current restrictions on the importation of plants, plant material and construction materials are important steps in controlling weed species.

6.3. Alien species in cultivation

Of the 374 species in cultivation, recent past cultivation or available as seed, some 288 (including species known as tropical weeds) were not at the time of survey approved for use in the region (Cowie & Werner 1987). Among the steps recommended by these authors to reduce the risk of cultivated species becoming weeds were: i) to enforce existing regulations as far as possible by removing undesirable species; ii) to control the supply of plants for sale within the Park; iii) to educate newly arrived residents about the regulations and hazards to the environment of introduced weeds; iv) to review the list of approved species; and v) to routinely monitor high risk areas (e.g. vacant land where garden clippings are dumped) for newly escaped species.

Settlements within Kakadu National Park have recently (April 1992) been resurveyed for alien species in cultivation or escaping from cultivation and the lists of approved and prohibited plant species reviewed and expanded (Brock & Cowie 1992). Prohibited plant species are to be removed either immediately or as houses become vacant. As most housing accommodation in Kakadu National Park is rented either from the Northern Territory or Commonwealth Governments or mining companies, the process of removing plants from vacant houses is simplified. Many of the older European settlements (some abandoned) located near spring-fed creeks and flood plain margins present a particular problem. At most of these locations, a number of alien species have already escaped from cultivation (e.g. *Senna alata*, *Delonix regia*, *Cassia fistula*, *Jatropha* spp.), sometimes forming extensive stands. To prevent further spread of invasive species from these sites, both Cowie & Werner (1987) and Brock & Cowie (1992) recommended that removal of invasive species at these locations have a high priority and that all alien species persisting or adventive at abandoned settlements be removed. A list of native species recommended for cultivation has been drawn up to encourage the horticultural use of Kakadu's unique flora, although research into propagation techniques is required before some species can be brought into cultivation. It was also recommended that alien species in cultivation be resurveyed every five years and that this survey be brought into synchrony with the Plan of Management for the Park.

7. Major invasive species

7.1. Mimosa pigra

Mimosa pigra, a spiny leguminous shrub of South American origin usually grows 3 to 4 m tall (Backer & van den Brink 1963–68, Lonsdale *et al.* 1985). It is the major potential weed of the flood plains (Fig. 3), posing an enormous conservation problem in the region. It forms monospecific, tall shrublands over many square kilometres, with plant densities typically of 1 m^{-2} (Lonsdale & Segura 1987). The effect of mimosa on flood plains outside of Kakadu has been to convert aquatic herblands, grasslands and sedgelands to shrublands with a consequent loss of native plants and animals (Miller *et al.* 1981, Lonsdale *et al.* 1985, Braithwaite *et al.* 1989). The species can also invade the understorey of tree dominated communities, with the probable prevention of regeneration of the native tree species (W.M. Lonsdale & R.W. Braithwaite unpublished).

The fruit of *Mimosa pigra* is a pod which breaks into dehiscent, one seeded articles that float, or can be dispersed in mud adhering to vehicles or animals (Backer & van den Brink 1963–68, Lonsdale *et al.* 1985). Stands of mimosa in northern Australia have a seed production rate of approximately 9100 seeds m^{-2} y^{-1} with seed set occurring between the middle of the wet season and the middle of the dry season (Lonsdale *et al.* 1985, Lonsdale 1988). Although most seeds are short-lived in the soil, the high seed output and early reproduction (within the first growing season) result in large soil seed banks (c. 12000 m^{-2}) (Lonsdale *et al.* 1988).

Substantial infestations of *Mimosa pigra* occur on most of the flood plains to the west of Kakadu National Park and probably provide a source of seed for new infestations to establish in the region. The only large infestation within the region occurs

Fig. 3. Primary invasion of the shrub *Mimosa pigra* along flood plain channels, Finnis River, NT in the late wet season. Similar infestations on the East Alligator River flood plain have been brought under control. (Photo I. Cowie)

near Oenpelli on the east side of the East Alligator River. This infestation was treated chemically in 1983, but was not treated again until 1988, and again in 1989 (I. Miller personal communication). It is now the subject of a federally funded campaign aimed at eradication.

Within Kakadu National Park many small infestations have been found and brought under control by up to six people employed full time (Skeat *et al.* 1987). Biological, chemical and manual control techniques are used. The strategy adopted has been to eradicate small infestations and, more importantly, to prevent further spread. Thus, more than 1000 km^2 of wetlands are extensively surveyed each year to monitor existing infestations and search for new infestations. Eradication methods include the manual removal of plants, foliar spraying (dicamba) and soil sterilisation (hexazinone). In addition, soil seed banks are destroyed using intense fire. Infestations are fenced to prevent the spread of seed by feral animals and vehicles are washed down. The fact that 25% of infestations found in the park are in the immediate vicinity of road creek crossings or buffalo catching camps highlights the role that human activities play in weed dispersal (Skeat *et al.* 1987).

A joint biological control program involving the Northern Territory Department of Primary Industries and Fisheries and CSIRO Division of Entomology has been under way since the early 1980s, and a research program into the management of *Mimosa pigra* in Thailand and Australia was commenced in 1984 with financial support from the Australian Centre for International Agricultural Research (Wilson & Lonsdale 1987). Many organisms have been screened with several released. Among the first were the bruchid beetles *Acanthoscelides puniceus* and *Acanthoscelides quadridentatus* whose larvae develop in the seeds, thus destroying them. A stem feeding chrysomelid beetle (*Chlamisus* sp.) was also released in 1984 (Wilson & Lonsdale 1987). However, these organisms have so far had no obvious effect on the abundance of *Mimosa.*

Two moths with complimentary feeding strategies, *Carmenta mimosa* and *Neurostrota gunniella* were released in the Northern Territory (including the Kakadu region) in 1989 (Wilson & Flanagan 1990, C.G. Wilson personal communication). The larvae of *Carmenta* are extremely destructive to *Mimosa* plants, tunnelling into the older stems and rapidly killing the branches and eventually the whole plant. *Carmenta* is becoming established on the Adelaide River flood plain and is also expected to have become established in the Kakadu region (C.G. Wilson personal communication). Young larvae of *Neurostrota* mine leaf pinnules while older larvae bore into the young stems, causing death of plant tissues. At an experimental site on the Adelaide River flood plain, this moth spread extremely rapidly with records of the species up to 7.5 km from the initial point of release after one year (Wilson & Flanagan 1990). More than 60% of growing tips within one kilometre of the initial release site were found to be infested, while in close proximity to the release site up to 98% of growing tips were infested. Although initial results appear promising, it is too early to know the ultimate effect of these insects on the density of *Mimosa*.

A study of the abundance and diversity of native insects on *Mimosa* plants at one site on the Adelaide River flood plain revealed a relative lack of leaf and flower feeding insects compared with the situation in the native range of the plant, indicating the potential usefulness of these types of biocontrol agents (Flanagan *et al.* 1990); a number of flower feeding (and seed feeding) weevils have been screened (Lonsdale *et al.* 1989), with a bud feeder, *Apion aculeatum* released (Forno 1992). Two fungi, *Phloeosporella* sp. and the rust *Diabole cubensis*, are reported to severely debilitate *Mimosa pigra* in Mexico and are being investigated as potential biological control agents (Lonsdale *et al.* 1989).

Fires occur regularly on the seasonally inundated wetlands of the region and have been used by traditional Aboriginal owners to control regeneration of native leguminous tree species such as *Cathormium umbellatum* (D. Lindner personal communication). Biological control in conjunction with fire may provide more effective control than could either in isolation. If biological control agents were to reduce the density of *Mimosa* sufficiently to allow the more flammable grasses and sedges to invade, then fire could also suppress or kill *Mimosa* plants. However, fire can promote the germination of plants by breaking seed dormancy (Miller & Lonsdale 1992).

7.2. Salvinia molesta

Salvinia molesta is a floating aquatic fern that has the potential to form dense floating mats on water bodies (Fig. 4), making it one of the most significant weeds in the region. It is able to rapidly colonise open water, doubling its cover every 36

Fig. 4. A floating mat of *Salvinia molesta* on Leichhardt Billabong, Magela Creek, mid dry season. (Photo I. Cowie)

hours under good growing conditions (Finlayson 1984b). The biology of the species has been reviewed by Harley & Mitchell (1981) and Room (1990). Control techniques are reviewed by Finlayson & Mitchell (1982) and for the Northern Territory by Miller & Wilson (1989).

Salvinia was first found in the region on the southern end of the Magela Creek flood plain during September 1983 (Finlayson 1984a). Other infestations were discovered on a tributary of the East Alligator River during 1987, on a back swamp of Nourlangie Creek during the dry season of 1990 and at Nanambu Creek. Due to the size of the initial infestation when it was first discovered and its inaccessibility, chemical control was regarded as too difficult and expensive, and in the long-term, ineffective. Thus, populations of the weevil *Cyrtobagus salviniae*, a biological control agent, were established. However, in contrast to elsewhere in the tropics (Room *et al.* 1981, Room 1990) and some other places in the Northern Territory (Miller & Wilson 1989), the weevil has not been very effective in controlling the weed (Skeat 1990). The reason for this are under investigation. Skeat (1990) found an inverse relationship between *Cyrtobagus* abundance and water temperature 90 days previously, although the mechanism for this relationship is not clear. Nitrogen levels in the plant could also not be completely ruled out as a factor causing the lack of success with *Cyrtobagus*. CSIRO Division of Entomology is carrying out further investigations into the control of *Salvinia* in Kakadu National Park.

Other control measures implemented in the region include limited removal by hand, mechanical harvesting and chemical control (ANPWS 1991). Mechanical harvesting of an infestation on Island Billabong (Magela Creek system) was trialed in 1990. Although the large, stable mats of *Salvinia* were broken up and removed, significant amounts of the plant remained. The small infestation on Nourlangie Creek is being brought under control using hexazinone (I.L. Miller personal communication). In an effort to prevent the spread of the weed, affected areas have been largely closed to the public and all vehicles passing through infested waterways during the wet season have been inspected (ANPWS 1991).

On the Magela Creek, *Salvinia* covers one or more billabongs each year, usually between September and December (late dry season to early wet season) (I.D. Cowie & C.M. Finlayson unpublished). Once wet season flooding occurs the floating mats of *Salvinia* are usually washed out of the deeper billabongs onto the shallower flood plain. Most plants apparently desiccate as the flood plain dries out during the following dry season. Floating mats may also reform on the permanent-water billabongs at this time. On top of this seasonal cycle there has been a general increase in the extent of mats over time (Skeat 1990). In years with below average rainfall during the wet season (e.g. 1989–90) mats are not flushed out and become more stable.

It is likely that floating mats which completely cover billabongs have detrimental effects on some aquatic species, especially during the late dry season when the aquatic biota may already be under stress. At this time of year dissolved oxygen levels are usually low and water temperatures are high (Walker & Tyler 1984). Natural 'fish kills' sometimes occur (Bishop 1980, Brown *et al.* 1983). Elsewhere, water beneath unbroken mats of *Salvinia* has lowered oxygen concentrations, light levels and pH, and raised temperatures compared with open water (D. Mitchell 1978). An unbroken floating mat is also likely to reduce the amount of habitat available for other aquatic plants, pelagic fish and birds feeding in open water.

Salvinia seems to have increased the stable size of the floating mat communities that fringe the western edges of some billabongs by providing a substrate for colonisation by the grass *Leersia hexandra* and the stoloniferous herb *Ludwigia adscendens* (Finlayson *et al.* 1995). This change may benefit some species. Naturally occurring floating mats based on the similar floating aquatic herb *Pistia stratiotes* are an important nesting habitat for the estuarine crocodile (*Crocodylus porosus*) on the Finnis River to the west of the region (Hill & Webb 1982).

7.3. Brachiaria mutica (Urochloa mutica)

Paragrass is a stoloniferous perennial grass of African origin and is naturalised in most tropical regions of the world (Hitchcock 1950, Bor 1960, Holm *et al.* 1977, Lazarides 1980). It was introduced to the northern parts of the Northern Territory before 1910 as a pasture species and introduced to the Kakadu region at Oenpelli soon after (Wesley-Smith 1973). In 1922 'several acres' were planted at Oenpelli and had spread to cover an estimated area of 3100 ha by 1960 (Letts 1960). The grass was apparently introduced by humans to the Magela Creek flood plain and several back swamps of the South Alligator River flood plain at various times during the 1970s (I.L. Miller unpublished, D. Lindner personal communication, S. Jacobs personal communication).

Brachiaria mutica appears to rarely set seed under local conditions, and to spread mainly by vegetative means (Holm *et al.* 1977, Calder 1982). Letts (1960) also noted that buffalo tended to pull up stolons from the wet ground whilst feeding, an action that may have served to spread the grass (I.L. Miller personal communication). During the wet season, any stolons dropped would have a high chance of survival. Indeed, a standard technique for establishing the species in pastoral situations is to drop pieces of stolon into swamps during the wet season (Miller 1970). The marginal rate of spread of one outlying population was estimated at c.10 m over one year (I.D. Cowie unpublished). Experimental plantings of *Brachiaria mutica* made by Cowie & Werner (1988) were able to establish in native communities dominated by *Pseudoraphis spinescens*, *Eleocharis sphacelata*, *Hymenachne acutigluma* and *Oryza rufipogon*. In a nearby area, paragrass had previously established a near monoculture over the range of topographic positions supporting these same communities. This strongly suggests that it can displace the above native plant communities, although grazing by feral buffalo may also have played a role. The displacement of *Oryza* and *Eleocharis* spp. which form a major component of the diet of Magpie Geese (*Anseranas semipalmata*) is of concern for the long-term conservation of this species.

Anecdotal evidence suggests that *Brachiaria mutica* has responded more rapidly to the reduction in buffalo grazing pressure than have native grasses such as *Phragmites karka* (D. Lindner personal communication). It also appears to be better at recolonising ground after dry season fire than does the native *Hymenachne acutigluma*. Paragrass establishes long stolons over the bare ground while *Hymenachne* grows as isolated tussocks. However, *Hymenachne* seems to be tolerant of wetter conditions than does *Brachiaria mutica* (Calder 1982).

Although most populations of para grass in the region are too large to be readily removed (Fig. 5), small infestations should be removed by a combination of chemical and mechanical means. Prospects for the biological control of grasses is generally poor and for this species they are also limited by its importance as a fodder plant in pastoral areas. Further research into the ecology of the species is necessary and may reveal aspects which could be exploited in management.

7.4. Pennisetum polystachion

Pennisetum polystachion is a tall, vigorous perennial grass (Fig. 6) of African origin, but now of a pantropical distribution (Brunken 1979a,b). Although not yet widespread in the region it appears to have the potential to become much more common. Near Darwin, it is a weed of disturbed places and is one of the few weeds able to invade relatively undisturbed *Eucalyptus* open forest and woodland. For this reason it deserves special further study.

It has been postulated that the apparent ability of *Pennisetum polystachion* to produce a more substantial and later-curing fuel load than do the native grasses would result in hotter, more extensive fires in the late dry season, if this species were to spread through Kakadu National Park

Fig. 5. A dense, almost monospecific sward of the perennial grass *Brachiaria mutica* on a back swamp at Munmarlary in Kakadu National Park. The fence encloses the site of a former *Mimosa* infestation. (Photo I. Cowie)

Fig. 6. The tall perennial grass *Pennisetum polystachion* is one of the few species capable of invading relatively undisturbed savanna and remains green until well into the dry season. (Photo I. Cowie)

(Braithwaite 1985). The altered fire regime which may be promoted by this species has detrimental consequences for the native vegetation, on re-vegetated areas and around settlements (Hoare *et al.* 1980, Braithwaite & Estbergs 1985).

Overseas, this species has been found to have a high seed production, with estimates ranging from 8000 to 350000 seeds per plant (Fernandez 1980, Noda *et al.* 1985, Suwanarak 1987). Seeds are surrounded by plumose bristles which appear to aid in wind dispersal. The diaspores also lodge easily in machinery and in construction materials such as gravel and timber. The result of seed germination experiments elsewhere are somewhat contradictory, with one study reporting little dormancy whilst others report that dormancy breaks over a period of several months (Fernandez 1980, Pemadasa & Amaransinghe 1982a, Noda *et al.* 1985). The plant can produce a comparatively large amount of organic matter at relatively low soil nutrient concentrations (Amaransinghe & Pemadasa 1982, Pemadasa & Amaransinghe 1982b).

7.5. *Andropogon gayanus*

A tall perennial grass of African origin, this species appears to have become more abundant to the west of the Kakadu region in recent years. Extensive areas have been planted for pasture improvement and it has invaded road sides, especially on the Adelaide River flood plain, and some areas of open forest. *Andropogon gayanus* produces a large above-ground biomass and may cause management problems similar to those of *Pennisetum polystachion*. Existing populations in the region (near Mudginberri) are viewed with concern and their removal is being attempted (J. Russell-Smith personal communication). The grass was introduced to the Northern Territory for pasture improvement and during field trails was found to have 'excellent growth at all sites sown' (a variety of woodland and flood plain fringe sites) and to show strong recovery after burning (Cameron *et al.* 1984). It persisted well and spread slowly. In Africa, it occurs throughout the continent and commonly grows in a zone above permanent swamps (Thompson 1985); typically on heavy clay soils in the upper parts of flood plains.

7.6. *Hyptis suaveolens*

Hyptis suaveolens is an aromatic annual shrub (Fig. 7) of tropical American origin and is wide-spread in the Kakadu region where it has probably been present for nearly 150 years (A. Mitchell 1978, Cowie *et al.* 1988). Little is known of

Fig. 7. An infestation of the annual herb *Hyptis suaveolens* along an ephemeral creek. Heavy use of such areas by feral animals has promoted weed invasion. (Photo I. Cowie)

its biology. In South America, seed from one population exhibits great variation in individual weight which is correlated with different germination requirements (Wulff 1973). Such germination polymorphism may allow the species to exploit a greater range of sites for establishment than species of uniform germination requirements (Harper 1965). The pericarp of the nutlets swells to sticky, gelatinous mass when soaked in water (Keng 1978, van der Pijl 1982). This adaptation probably facilitates the dispersal of the seed on animals and vehicles.

Hyptis occurs commonly in disturbed situations and flood plain margin communities (Taylor & Dunlop 1985, Cowie & Werner 1993). In the less common monsoon forest community, Braithwaite *et al.* (1984) found that the presence of *Hyptis* was related to recent fire, rather than being associated primarily with disturbance caused by feral buffalo. (Although the habitat changes which allow monsoon forest to burn may be caused by buffalo!) Stocker & Mott (1981) observed that invasion of plant communities by *Hyptis suaveolens* resulted in an increase in the woody perennial component of the vegetation, purportedly because of the less intense fires this species supports.

A biological control program for *Hyptis suaveolens* was started but has now stopped due to lack of funding (Wilson & Lonsdale 1987, C.G. Wilson personal communication). The initial work of selecting biological control agents produced a promising rust fungus, but revealed a paucity of insects which feed on the species in its area of origin (Wilson & Lonsdale 1987, C.G. Wilson personal communication). A closely related species, *Hyptis spicigera*, has known insecticidal properties (Lambert *et al.* 1985). The fetid aromatic leaves of *Hyptis suaveolens* are reported to have been used for control of bed bugs in Asia (Uphof 1968, Keng 1978) and together with the lack of insects the plant supports, suggests the possibility that this species also has insecticidal properties. *Hyptis suaveolens* is a vigorous weed of disturbed places even in its native range (Wulff & Medina 1971), where a full complement of predators and pathogens presumably exists. Thus, although biological control may ultimately reduce the density of *Hyptis suaveolens*, the weed is likely to remain a problem.

7.7. Sida species

Sida acuta, *Sida cordifolia* and *Sida rhombifolia* are pantropical, perennial shrubs growing to 2 m tall. *Sida acuta* and *Sida cordifolia* are common weeds of disturbed areas, but *Sida rhombifolia* is much less common (Cowie & Werner 1993).

Little is known of the biology of the two major weed species. *Sida acuta* has two mechanisms controlling seed dormancy. The embryo has an after ripening requirement, and the seed has an impermeable coat resulting in approximately 30% of the seeds remaining dormant after one wet season (Mott 1980). This persistent seed bank makes the species more difficult to control. Dispersal of seed is probably aided by the presence of one (*Sida rhombifolia*) or two sharply pointed awns at the apex of the mericarp. In the case of *Sida cordifolia*, the awns are barbed.

A program for the biological control of these species in the Northern Territory has been commenced (Wilson & Lonsdale 1987). A leaf feeding chrysomelid beetle (*Calligrapha pantherina*) for the control of *Sida acuta* and *Sida rhombifolia* was released at several hundred places across the northern part of the Northern Territory, including the Kakadu region during the 1989–90 wet season (C.G. Wilson personal communication). The slow rate at which these insects spread means that spread by humans is likely to be important in establishing them as an effective biological control agent. The beetles are reported to graze the host plants heavily, resulting in a reduction of up to 94% in the amount of seed produced (C.G. Wilson personal communication). However, the ultimate effect of this insect on the density of *Sida* species in the region is not yet known.

7.8. Senna and Cassia species

The two most common species of *Senna* in the region are *Senna obtusifolia* (syn. *Cassia obtusifolia*) and *Senna occidentalis* (syn. *Cassia occidentalis*). Both are annual shrubs, probably of tropical American origin, but now of almost circumtropical distribution (Irwin & Barneby 1982). In the Kakadu region, these plants are weeds of disturbed places and especially flood plain margin communities. The perennial shrub, *Senna alata* (syn. *Cassia alata*) is an important (although not yet widespread) weed of the comparatively rare spring-fed creeks in the region. The tree *Cassia fistula* is invading riparian vegetation at one location in the region, and persists at several others.

Research elsewhere indicates that *Senna obtusifolia* produces seed with a hard seed coat (Creel *et al.* 1968, Eastin 1981). The former authors found that only 10% of unscarified seed germinated within the first 30 days and germination had reached only 15% after 12 months. In comparison, Singh (1968) reported 40% germination of seed (at 30°C) after storage for one year. A hard seed coat would serve to prevent germination too early in the dry season. Creel *et al.* (1968) report that crops infested with the weed can produce up to 900 kg ha^{-1} of weed seed. Seed production at this level together with the longevity of the seed could lead to a substantial build up of seed in the soil, making the plant difficult to remove. Locally, *Senna obtusifolia* has longer lived seed than does *Senna occidentalis* (M.W. Lonsdale unpublished).

The optimum temperature for germination of seed of overseas populations of *Senna obtusifolia* is reported as between 20 and 30 °C, with maximum seedling growth between 30 and 36 °C (Creel *et al.* 1968, Eastin 1981). The species is reported to prefer acid soils, growing well between pH 3.2 and 7.9 with optimum growth between pH 5.5 and 6 (Creel *et al.* 1968).

A program for the biological control of *Senna obtusifolia* in Queensland is apparently under way (C.G. Wilson personal communication) and, if successful, may ultimately lead to the release of effective biocontrol agents in the Kakadu region and elsewhere in the Northern Territory.

7.9. Pennisetum pedicellatum

Pennisetum pedicellatum is a vigorous annual grass growing to 1.2 m tall. It is of West African origin, but now occurs patchily throughout the tropics of the Old World (Parham 1955, Bor 1960, Holm *et al.* 1977, Noda *et al.* 1985). It has been present in the northern part of the Northern Territory since 1960 (Cameron *et al.* 1984) and has probably not yet reached its ultimate distribution. In the Kakadu region, this species is primarily a weed of disturbed habitats, including flood plain margin communities.

Mott (1980) studied the germination and establishment of the species at Katherine in the Northern Territory. He found that seed was dormant at seed fall and required high and alternating temperature for three months to remove an after ripening requirement. Almost all seed was nondormant by the end of the dry season with nearly 80% of the total wet season germination occurring after the first two rainy periods of more than 20 mm rainfall. Germination of buried seed was delayed. At the end of the wet season, no viable seed of *Pennisetum pedicellatum* remained in the soil. Germination varied little with temperatures between 19 and 37 °C.

7.10. Passiflora foetida

Passiflora foetida is a perennial vine of South American origin and was introduced at Darwin around 1882 (Holtze 1892), but is now widespread. It is common along sandy stream banks, flood plain fringes, margins of monsoon forest and in disturbed places. Little is known of the ecology of the species, except that the sweet pulp of the round orange coloured fruit is readily eaten by birds and humans, which may disperse the seed (Holtze 1892, Kleinschmidt & Johnson 1977).

7.12. Other species

Pistia stratiotes, once regarded as introduced in the northern part of the Northern Territory (D. Mitchell 1978) is now widely regarded as native (Aston 1973, Cowie & Finlayson 1986, Dunlop 1987, Gillett *et al.* 1989). Evidence advanced by the latter authors included that the earliest records of the species in the Northern Territory date back to 1888 (not long after the first permanent European settlement) and then in areas far remote from established settlement. They also found that the species supported a complex of native organisms which kept it at sub-economic levels, and that the biogeography of *Pistia stratiotes* and associated plant species, together with recent geological history were consistent with native status. However, the species is still a declared noxious weed under Northern Territory legislation. A biological control agent (the beetle *Neohydronomus pulchellus*) has been released (Wilson & Lonsdale 1987), but has apparently failed to establish (C.G. Wilson personal communication).

In the context of Kakadu National Park, *Pistia* is not regarded as a weed (Cowie & Werner 1987), although under the Northern Territory Noxious Weeds Act landholders are required to prevent further spread and eradicate small infestations. The species can be locally abundant and occurs on a few widely scattered back swamps in the region, but is not usually common. Populations undergo occasional irruptions during which the plant may form dense floating mats, but at other times plants are present at low densities (D. Lindner personal communication, G. Miles personal communication).

The tall perennial shrub *Leucaena leucocephala* has been recorded at Mudginberri and in some gardens at Jabiru (Cowie & Werner 1987, 1988). Around Darwin and other towns in the northern part of the Northern Territory, this species is a troublesome weed of wetter disturbed sites, occurring as thick stands on vacant allotments and on some road sides. It has also invaded and dominated some areas of coastal vine thicket disturbed by cyclonic winds. However, the long term effects of this species on succession in this habitat are not known. Although it has not yet become well established outside urban areas in the Northern Territory it is of some concern to the Kakadu National Park authority (ANPWS 1991).

Some alien species (*Acanthospermum hispidum, Alternanthera pungens, Cenchrus echinatus* and *Tribulus* spp.) have trample burrs (a group not present in the native flora) which make them important weeds of campgrounds and settlements. *Tribulus* spp. are well known for their thorny burrs which cause injury to feet and skin as well as puncturing tyres of bicycles and (reputedly) motor vehicles (Holm *et al.* 1977). The other species have spiny burrs which can puncture skin and adhere to clothing causing discomfort and inconvenience to visitors. Although these species were found at only a few localities around the region, recent increases in human activities have the potential to spread them more widely.

8. Acknowledgements

I wish to thank Prof. P. Werner and Dr. J. Russell-Smith for their constructive comments on earlier drafts of this chapter. Much of the work reported here was supported by funds from the Australian National Parks and Wildlife Service to the CSIRO Division of Wildlife and Ecology and forms parts of reports to that organisation (e.g.

Cowie & Werner, 1987, 1988). The opinions expressed in this paper reflect those of the author, not ANCA.

9. References

Amarasinghe, L. & Pemadasa, M.A. 1982. The ecology of a montane grassland in Sri-Lanka. 4. Vegetative growth of four major grasses. J. Ecol. 70: 721–730.

Aston, H.I. 1973. Aquatic Plants of Australia. Melbourne University Press, Melbourne. 194 pp.

ANPWS 1991. Kakadu National Park Plan of Management. Commonwealth of Australia, Canberra. 186 pp

Backer, C.A. & van den Brink, R.C.B. 1963–68. Flora of Java vI-vIII. Wolters-Noordhoff N.V., Groningen. vI 648 pp, vII 641 pp, vIII 761 pp.

Barlow, B.A. 1981. The Australian flora: its origin and evolution. Flora of Australia 1: 25–75.

Bauer, F.W. 1964. Historical geography of white settlement in part of northern Australia. Part II. the Katherine-Darwin region. CSIRO, Australia. Division of Land Research Regional survey Divisional Report No. 64/1. 284 pp.

Beadle, N.C.W., Evans, O.D., & Carolin, R.C. 1972. Flora of the Sydney Region. Reed, Sydney. 724 pp.

Bishop, K.A. 1980. Fish kills in relation to physical and chemical changes in Magela Creek (East Alligator River System, Northern Territory) at the beginning of the tropical wet season. Aust. Zool. 20: 485–500.

Bor, N.L. 1960. The grasses of Burma, Ceylon, India and Pakistan. Pergamon Press, London. 284 pp.

Braithwaite, R.W. 1985. The implications for management. In: R.W. Braithwaite, R.W. (ed.) Kakadu Fauna Survey Final Report, Vol. 4. pp. 877–899. Report to Australian National Parks and Wildlife Service, Canberra.

Braithwaite, R.W. & Estbergs, J.A. 1985. Fire patterns and woody vegetation trends in the Alligator Rivers Region of Northern Australia. In: Tothill, J.C. & Mott, J.J. (eds) Ecology and Management of the Worlds Savannas. pp. 359–364. Australian Academy of Science, Canberra.

Braithwaite, R.W. Lonsdale, W.M. & Estbergs, J.A. 1989. Alien vegetation and native biota: the impact of *Mimosa pigra*. Biol. Conserv. 48: 189–210.

Braithwaite, R.W., Dudzinski, M.L. Ridpath, M.G. & Parker, B.S. 1984. The impact of water buffalo on the monsoon forest ecosystem in Kakadu National Park. Aust. J. Ecol. 9: 309–22.

Brock, J. & Cowie, I.D. 1992. Kakadu National Park: Review of Approved Plant List and Survey of Alien Plants at Settlements within the Park. Unpublished report to Australian National Parks and Wildlife Service. 41 pp.

Brockie, R.E., Loope, L.L., Usher, M.B. & Hamann, O. 1988. Biological invasions of island nature reserves. Biol. Conserv. 44: 9–36.

Brown, T.E., Morley, A.W., Sanderson, N.T. & Tait, R.D. 1983. Report of a large fish kill resulting from natural acid water conditions in Australia. J. Fish. Biol. 22: 335–350.

Brunken, J.N. 1979a. Cytotaxonomy and evolution in *Pennisetum* section *Brevivalvula* Gramineae in Tropical Africa. Bot. J. Linn. Soc. 79: 37–50.

Brunken, J.N. 1979b. Morphometric variation and the classification of *Pennisetum* section *Brevivalvula* Gramineae in tropical Africa. Bot. J. Linn. Soc. 79: 51–64.

Busby, J.R. 1986a. A bioclimatic analysis of *Nothofagus cunninghamii* (Hook.) Oerst. in southeastern Australia. Aust. J. Ecol. 11: 1–7.

Busby, J.R. 1986b. Bioclimate Prediction System (BIOCLIM) User's Manual. Version 2.0. Bureau of Flora and Fauna, Canberra.

Calder, G.J. 1982. *Hymenachne acutigluma* in the Northern Territory. Technical Bulletin No. 46, Department of Primary Production, Darwin. 58 pp.

Cameron, A.G., Miller, I.L., Harrison, P.G. & Fritz, R.J. 1984. A review of pasture plant introduction in the 600–1500 mm rainfall zone of the Northern Territory. Technical Bulletin No. 71. Department of Primary Production, Darwin. 81 pp.

Christian, C.S. & Aldrick, J.M. 1977. Alligator Rivers Study: a review report of the Alligator Rivers Region environmental fact-finding study. Australian Government Publishing Service, Canberra. 174 pp.

Clarkson, J.R. & Kenneally, K.F. 1988. The floras of Cape York and the Kimberley: a preliminary comparitive analysis. Proc. Ecol. Soc. Aust. 15: 259–266.

Cowie, I.D. & Finlayson, C.M. 1986. Plants of the Alligator Rivers Region, Northern Territory. Supervising Scientist for the Alligator Rivers Region Technical Memorandum 17. AGPS, Canberra. 50 pp.

Cowie, I.D. & Werner P.A. 1987. Weeds in Kakadu National Park – A Survey of Alien Flora. Report to Australian National Parks and Wildlife Service, Canberra. 127 pp.

Cowie, I.D. & Werner P.A. 1988. Weeds in Kakadu National Park: A Survey of Alien Plants – Phase II. report to Australian National Parks and Wildlife Service, Canberra. 42 pp.

Cowie, I.D. & Werner, P.A. 1993. Alien plant species invasive in Kakadu National Park, tropical northern Australia. Biol. Conserv. 63: 127–135.

Cowie, I.D., Finlayson, C.M. & Bailey, B.J. 1988. Alien plants in the Alligator Rivers Region, Northern Territory, Australia. Supervising Scientist for the Alligator Rivers Region Technical Memorandum 23. AGPS, Canberra. 26 pp.

Creel, J.M., Hoveland, C.S. & Buchanan, G.A. 1968. Germination, growth and ecology of Sicklepod. Weed Sci. 16: 396–400.

Dunlop, C.R. (ed.) 1987. Checklist of vascular plants of the Northern Territory. Technical Report No. 26, Conservation Commission of the Northern Territory, Darwin. 87 pp.

Eastin, E. F. 1981. Sicklepod germination as influenced by seed treatment and temperature. Miscellaneous Publication No. MP-1481, Texas Agricultural Experiment Station. 7 pp.

Esler, A.E. & Astridge, S.J. 1987. The naturalisation of plants in urban Auckland, New Zealand 2. Records of introduction and naturalisation. New Zeal. J. Bot. 25: 523–538.

Fernandez, D.B. 1980. Some aspects in the biology of *Pennisetum polystachyon* (L.) Schult. Philippine J. Weed Sci. 7: 1–10.

Finlayson, C.M. 1984a. Salvinia, water hyacinth and mimosa in the Alligator Rivers Region, Northern Territory. Aust. Weeds 3: 83.

Finlayson, C.M. 1984b. Growth rates of *Salvinia molesta* in Lake Moondarra, Mount Isa, Australia. Aquat. Bot. 18: 257–262.

Finlayson, C.M. & Mitchell, D.S. 1982. Management of salvinia (*Salvinia molesta*) in Australia. Aust. Weeds 2: 71–76.

Finlayson, C.M., Thompson, K., von Oertzen, I.A.M-L. & Cowie, I.D. 1995. Macrophytes of five Magela Creek billabongs, Alligator Rivers Region, northern Australia. Supervising Scientist for the Alligator Rivers region Technical Memorandum (in press). AGPS, Canberra.

Flanagan, G.J., Wilson, C.G. & Gillett, J.G. 1990. The abundance of native insects on the introduced weed *Mimosa pigra* in Northern Australia. J. Trop. Ecol. 6: 219–230.

Forno, I.W. 1992. Biological control of *Mimosa pigra*: research undertaken and prospects for effective control. In: Harley, K.L.S. (ed.) A Guide to the Management of *Mimosa pigra*. pp. 38–42. CSIRO, Canberra.

Fox, M.D. & Fox, B.J. 1986. The susceptibility of natural communities to invasion. In: Groves, R.H. & Burdon, J.J. (eds) Ecology of Biological Invasions. pp. 57–66. Australian Academy of Science, Canberra.

Gillett, J.G., Dunlop, C.R. & Miller, I.L. 1989. Occurrence, origin, weed status and control of water lettuce (*Pistia stratiotes* L.) in the Northern Territory. Plant Protect. Quart. 3: 144–148.

Graf, A.B. 1981. Tropica (3rd edn.). Roehrs Co., East Rutherford. 1152 pp.

Groves, R.H. 1990. Weed control in conservation reserves and amenity areas of Australia. In: Heap, J.W. (ed.) Proceedings of the Ninth Australian Weeds Conference. pp. 97–102. Crop Science Society of South Australia, Adelaide.

Harley, K.L.S. & Mitchell, D.S. 1981. The biology of Australian weeds 6. *Salvinia molesta* D.S. Mitchell. J. Aust. Inst. Agric. Sci. 47: 67–81.

Harper, J.L. 1965. Establishment, aggression and cohabitation in weedy species. In: Baker, H.G. & Stebbins, G.L. (eds) The Genetics of Colonizing Species. pp. 243–268. Academic Press, New York.

Heywood, V.H. (ed.) 1978. Flowering Plants of the World. Oxford University Press, Oxford. 335 pp.

Hill, R. & Webb, G. 1982. Floating grass mats of the Northern Territory floodplains – an endangered habitat? Wetlands 2: 45–50.

Hitchcock, A.S. 1950. Manual of the grasses of the United States (2nd edn) United States Department of Agriculture Miscellaneous Publication No. 200. Washington, U.S. Govt. Printing Office. 680 pp.

Hoare, J.R.L., Hooper, R.J., Cheney, N.P. & Jacobsen, K.L.S. 1980. A report on the effects of fire in tall open forest and woodland with particular reference to fire management in Kakadu National Park in the Northern Territory. Report to Australian National Parks and Wildlife Service, Canberra. 59 pp.

Holm, L., Pancho, J.V., Herberger, J.P. & Plucknett, D.L. 1979. A Geographical Atlas of World Weeds. John Wiley and Sons, New York. 609 pp.

Holm, L.G., Plucknett, D.L., Pancho, J.V. & Herberger, J.P. 1977. The World's Worst Weeds: Distribution and Biology. East-West Centre and University Press of Hawaii, Honolulu. 391 pp.

Holtze, M. 1892. Introduced Plants in the Northern Territory. Trans. Royal Soc. South Aust. 15: 1–4.

Holtze, N. 1887. Annual report on the Experimental and Botanical Gardens, Darwin. In: Government Resident's Report on Northern Territory for the year 1887. pp 30–40. South Australian Government, Adelaide.

Humphries, S.E., Groves, R.H. & Mitchell, D.S. 1991. Plant invasions of Australian Ecosystems: the incidence of environmental weeds in Australia. Kowari 2: 1–188.

Hutchinson, J. & Dalziel, J.M. 1954–1972. Flora of West Tropical Africa. Crown Agents, London. vI 828 pp, vII 544 pp, vIII 574 pp.

Irwin, H.S. & Barneby, R.C. 1982. The American *Cassinae*. Mem. the New York Bot. Garden 35: 1–918.

Keng, H. 1978. Labiatae. Flora Malesiana Series I, 8: 301–394.

Kenneally, K.K. 1989. Checklist of vascular plants of the Kimberley, Western Australia. Handbook No 14, Western Australian Naturalist Club, Nedlands, Perth. 108 pp.

Kleinschmidt, H.E. & Johnson, R.W. 1977. Weeds of Queensland. Queensland Department of Primary Industries, Brisbane. 237 pp.

Kloot, P.M. 1984. The introduced elements of the flora of southern Australia. J. Biogeog. 11: 63–78.

Lambert, J.D.H., Gale, J., Arnason, J.T. & Philogene, B.J.R. 1985. Bruchid control with traditionally used insecticidal plants *Hyptis spicigera* and *Cassia nigricans*. Insect Science Applic. 6: 167–170.

Lazarides, M. 1980. The Tropical Grasses of Southeast Asia. J.Cramer, Vaduz. 108 pp.

Letts, G.A. 1960. Cattle Management and pasture improvement in the high rainfall belt of the N.T. – An historical review of the establishment and spread of para grass at the Oenpelli Mission. (unpublished). Department of the Northern Territory, Darwin.

Lonsdale, W.M. 1988. Litterfall in an Australian population of *Mimosa pigra*, an invasive tropical shrub. J. Trop. Ecol. 4: 381–392.

Lonsdale, W.M. & Lane, A.M. 1990. Vehicles as vectors of weed seeds in Kakadu National. In: Heap, J.W. (ed.) Proceedings of the Ninth Australian Weeds Conference. pp. 134–136. Crop Science Society of South Australia, Adelaide.

Lonsdale, W.M. & Segura, R. 1987. A demographic study of native and introduced populations of *Mimosa pigra*. In: Lemerle, D. & Leys, A.R. (eds) Proceedings of the 8th Australian Weeds Conference. pp. 163–166. Council of the Australian Weed Science Societies, Sydney.

Lonsdale, W.M., Harley, K.L.S. & Gillett, J.D. 1988. Seed bank dynamics in *Mimosa pigra*, an invasive tropical shrub. J. Appl. Ecol. 25: 963–976.

Lonsdale, W.M., Miller, I.L. & Forno, I.W. 1989. The biology of Australian Weeds. 20. *Mimosa pigra* L. Plant Protect. Quart. 4: 119–131.

Lonsdale, W.M., Harley, K.L.S. & Miller, I.L. 1985. The biology of *Mimosa pigra* L. In: Proceedings (2) of the tenth Conference of the Asian-Pacific Weed Science Society, Chiang Mai, Thailand.

Macdonald, I.A.W. & Frame, G.W. 1988. The invasion of introduced species into nature reserves in tropical savannas and dry woodlands. Biol. Conserv. 44: 67–93.

Miller, I.L. 1970. Para grass. Turnoff, 2: 2. Northern Territory Administration, Darwin.

Miller, I.L. & Lonsdale, W.M. 1992. Ecological management of *Mimosa pigra*: use of fire and competitive pastures. In: Harley, K.L.S. (ed.) A Guide to the Management of *Mimosa pigra*. pp.104–106. CSIRO.

Miller, I.L. & Wilson, C.G. 1989. Management of *Salvinia* in the Northern Territory. J. Aquat. Plant Manage. 27: 40–46.

Miller, I.L., Nemestothy, L. & Pickering, S.E. 1981. *Mimosa pigra* in the Northern Territory. Technical Bulletin No 51. Department of Primary Production, Darwin. 23 pp.

Minchin, P.R. & Dunlop, C.R. 1989. Impact of feral water buffalo on savanna ecosystems in the Australian monsoon tropics. In: IAVS Symposium on forests of the world, Uppsala, Sweden, August 1989. Unpaginated.

Mitchell, A.S. 1978. An historical overview of exotic and weedy plants in the Northern Territory. Proceedings of the 1st Conference of the Council of Australian Weed Science Societies, pp. 145–153.

Mitchell, D.S. 1978. Aquatic weeds in Australian Inland Waters. Australian Government Publishing Service, Canberra. 189 pp.

Morton, J.F. 1976. Pestiferous spread of many ornamental and fruit species in South Florida. Proceedings, Florida State Horticult. Soc. 89: 348–353.

Mott, J.J. 1980. Germination and establishment of the weeds *Sida acuta* and *Pennisetum pedicillatum* in the Northern Territory. Aust. J. Exper. Agric. Anim. Husb. 20: 463–469.

Mueller-Dombois, D. & Ellenberg, H. 1974. Aims and Methods of Vegetation Ecology. Wiley, New York. 547 pp.

Noda, K., Chaiwiratnukal, L. Kananajirawong, S. & Teerawatsakul, M. 1985. Some biological characteristics of *Pennisetum* spp. in Thailand. Proceedings of the tenth Asian-Pacific Weed Science Society Conference, Chiangmai, Thailand. pp 75–80.

Parham, J.W. 1955. The Grasses of Fiji. Bulletin No. 30, Department of Agriculture, Fiji. 166 pp.

Pemadasa, M.A. & Amarasinghe, L. 1982a. The ecology of a montane grassland in Sri-Lanka. 3. Germination of three major grasses. J. Ecol. 70: 483–490.

Pemadasa, M.A. & Amarasinghe, L. 1982b. The ecology of a montane grassland in Sri-Lanka. 3. Interference in populations of 4 major grasses. Journal of Ecology, 70: 731–744.

Room, P.M. 1990. Ecology of a simple plant-herbivore system: biological control of *Salvinia*. Trends in Ecol. Evol. 5: 74–79.

Room, P.M., Harley, K.L.S., Forno, I.W. & Sands, D.P.A. 1981. Successful biological control of the floating weed *Salvinia*. Nat. 294: 78–80.

Ross, J.H. 1976. An analysis of the flora of Victoria. Muelleria 3: 169–176.

Russell-Smith, J. 1984. The status and condition of monsoon vine-forests in the Kakadu Region: A management report. Report to Australian National Parks and Wildlife Service, Canberra. 37 pp.

Singh, J.S. 1968. Comparison of growth performance and germination behaviour of seeds of *Cassia tora* L. and *C. obtusifolia* L. Trop. Ecol. 9: 64–81.

Skeat, A.J. 1990. Biological control of *Salvinia molesta* in Kakadu National Park, Northern Territory. In: Heap, J.W. (ed.) Proceedings of the Ninth Australian Weeds Conference. pp. 130–133. Crop Science Society of South Australia, Adelaide.

Skeat, A.J., von Oertzen, I. & Maddison, J. 1987. Control of *Mimosa pigra* in Kakadu National Park. In: Lemerle, D. & Leys, A.R. (eds) Proceedings of the eighth Australian Weeds Conference. pp. 464–466. Weeds Society of New South Wales, Sydney.

Specht, R.L. 1981. Major vegetation formations in Australia. In: Keast, A. (ed.) Ecological Biogeography of Australia. pp. 163–297. Dr W. Junk, The Hague.

Spillett, P.G. 1972. Forsaken Settlement. Lansdowne Press, Sydney. 196 pp.

Stafford, J.L. 1990. Revegation with native grasses in conservation/amenity areas. In: Heap, J.W. (ed.) Proceedings of the Ninth Australian Weeds Conference. pp. 140–143. Crop Science Society of South Australia, Adelaide.

Stocker, G.C. & Mott, J.J. 1981. Fire in the tropical forests and woodlands of northern Australia. In: Gill, A.M., Groves, R.H. & Noble, I.R. (eds) Fire and the Australian Biota. pp 423–439. Australian Academy of Science, Canberra.

Suwanarak, K. (ed.) 1987. Conclusions and Recommendations. Workshop on *Pennisetum setosum*, 25–26 November 1987, Wangtai Hotel, Ban Don, Surat Thani, Thailand. pp 111–117.

Swarbrick, J.T. 1983. Working list of weeds in Queensland, the Northern Territory and northern Western Australia. Aust. Weeds 4: 156–164.

Swarbrick, J.T. 1987. Weeds in urban bushland – challenge and opportunity. Plant Protect. Quart. 2: 54.

Swincer, D.E. 1986. Physical characteristics of sites in relation to invasions. In: Groves, R.H. & Burdon, J.J (eds) Ecology of Biological Invasions: an Australian Perspective. pp. 67–76. Australian Academy of Science, Canberra.

Taylor, J.A. & Dunlop, C.R. 1985. Plant communities of the Wet-Dry Tropics of Australia: the Alligator Rivers Region. Proc. Ecol. Soc. Aust. 13: 83–127.

Thomas, A.G., Gill, A.M., Moore, P.H.R. & Forcella, F. 1984. Drought feeding and the dispersal of weeds. J. Aust. Inst. Agric. Sci. 50: 103–107.

Thompson, K. 1985. Emergent plants of permanent and seasonally-flooded wetlands. In: Denny, P. (ed.) The ecology and management of African wetland vegetation. pp. 43–107. Junk, Dordrecht.

Troll, C. 1966. Seasonal climates of the earth. In: Landsberg, H.E., Lippmann, H., Pattern, K.H. & Troll, C. (eds) World maps of climatology 3rd ed. pp 19–28. Springer-Verlag, Berlin.

Tulloch, D.G. 1978. The water buffalo, *Bubalus bubalis*, in Australia : grouping and home range. Aust. Wildl. Res. 5: 327–354.

Uphof, J.C. 1968. Dictionary of Economic Plants (2nd edn.). Cramer, Lehre. 591 pp.

Usher, M.B. 1988. Biological invasions of nature reserves: a search for generalizations. Biol. Conserv. 44: 119–135.

van der Pijl, L. 1982. Principles of Dispersal in Higher Plants. 3nd edn. Springer-Verlag, Berlin. 214 pp.

Walker, T. & Tyler, P. 1984. Tropical Australia, a dynamic

limnological environment. Verhand. Inter. Verein. Theoret. Limnol. 22: 1727–1734.
Wesley-Smith, R.N. 1973. Para grass in the Northern Territory –parentage and propagation. Trop. Grasslands 7: 249–250.
Williamson, M.H. & Brown, K.C. 1986. The analysis and modelling of British Invasions. Phil. Trans. Royal Soc. Lond. B 314: 505–22.
Willis, J.H. 1970. A handbook to plants in Victoria. I. Melbourne University Press, Melbourne. 481 pp.
Willis, J.H. 1972. A handbook to plants in Victoria. II. Melbourne University Press, Melbourne. 830 pp.
Wilson, C.G. & Flanagan, G.J. 1990. Establishment and spread of *Neurostrota gunniella* on *Mimosa pigra* in the Northern Territory. In: Heap, J.W. (ed.) Proceedings of the Ninth Australian Weeds Conference. pp. 505–507. Crop Science Society of South Australia, Adelaide.
Wilson, C.G. & Lonsdale, W.M. 1987. Recent advances in weed biocontrol in Northern Australia. In: Lemerle, D. & Leys, A.R. (eds) Proceedings of the 8th Australian Weeds Conference. pp. 89–92. Weed Society of New South Wales, Sydney.
Wulff, R. 1973. Intrapopulational variation in the germination of seeds in *Hyptis suaveolens*. Ecol. 54: 646–649.
Wulff, R. & Medina, E. 1971. Germination of seeds in *Hyptis suaveolens* Poit. Plant Cell Physiol. 12: 567–579.

CHAPTER 7

Plant-animal interactions

ALAN N. ANDERSEN and RICHARD W. BRAITHWAITE

Abstract. The diversity, complexity and functional importance of interactions between plants and animals is illustrated by focussing on three case topics: nutrient cycling by termites, habitat selection by vertebrates, and seed predation by ants. Termites are a conspicuous component of the soil fauna of tropical savannas, and are the premier decomposer insects. Unlike other decomposers, termites fundamentally alter the distribution of carbon and other nutrients, enriching the soils associated with their nests and mounds, and depleting soils elsewhere. Termites might also play a significant role in the global budgets of the "greenhouse" gases carbon monoxide and methane. The key factors influencing habitat selection by vertebrates are considered to be moisture and nutrient availability, and fire. The responses of mammals, lizards and birds to a moisture gradient associated with seasonal creeklines is outlined. Fire typically occurs every 1–3 years at most sites, and is a powerful modifier of vegetation patterns. Some vertebrates, such as granivorous and scavenging birds, exploit the immediate post-fire habitat, whereas others respond to the longer-term effects of fire on vegetation structure. Ants are major seed-harvesters throughout the Kakadu region, as they are throughout most of Australia. Eighteen harvester species have been recorded from a single savanna site, with nest densities averaging over 1600 ha^{-1}. In general terms, insects are pre-eminent over mammals as primary consumers in the Kakadu region, and this is attributed to low soil fertility. The resulting faunal assemblages contrast sharply with the mammalian-dominated systems characteristic of eutrophic African savannas.

1. Introduction

Interactions between plants and animals embrace many of the processes fundamental to ecosystem structure and function. As a field of ecology the subject has received increasing attention over the last decade (e.g. Thompson 1982, Crawley 1983, Miller & Miller 1986, Abrahamson 1989, Bernays 1989, Huxley & Cutler 1991). However, there have been few relevant studies in the Kakadu region. Despite the tendency of many authors (e.g. Edwards & Wratten 1980, Crawley 1983, Miller & Miller 1986, Bernays 1989, New 1989) to treat the subject as one fundamentally concerned with herbivory, it is an extremely diverse and complex field.

The interactions between plants and animals may be tight, such as in the classic concept of coevolution between plants and their pollinators and insect herbivores (Faegri & van der Pijl 1979, Crawley 1983), but may equally be loose and diffuse, such as in the provision of habitat by plants (Partridge 1978, Cody 1985), or in the acceleration of nutrient cycling by soil invertebrates (Lal 1987, de Bruyn & Conacher 1990). The interactions may be detrimental to one partner, such as in herbivory, or may be mutually beneficial, as in pollination and the many ant-plant mutualisms (Janzen 1966, Buckley 1982, Beattie 1985, Huxley & Cutler 1991).

Given the diversity and complexity of this field combined with the paucity of relevant data, it is not possible to provide a comprehensive overview of the structure and function of plant-animal interactions in the Kakadu region in a single chapter. For example, we know of no studies of grazing or folivory by insects anywhere in the Australian seasonal tropics, despite their pre-eminence as native herbivores throughout the region (Andersen & Lonsdale 1990). Similarly, a wide variety of pollination syndromes is evident in the flora of the region, ranging from generalised pollination in eucalypts (Braithwaite 1990a) to specialist bird pollination in species of *Grevillea* and other Proteaceae, but they are virtually unstudied. Therefore, our approach has been to focus on a few topics where sufficient data are available, and which provide an indication of the breadth and significance of interactions that have undoubtedly moulded the ecological landscape of the Kakadu region. These topics are nutrient cy-

C.M. Finlayson and I. von Oertzen (eds), Landscape and Vegetation Ecology of the Kakadu Region, Northern Australia, 137-154.

cling by termites, habitat selection by vertebrates, and seed predation by ants.

2. Nutrient cycling by termites

As throughout much of northern Australia, the soils in the Kakadu region tend to be highly impoverished in nutrients (Mott *et al.* 1985). This suggests that the cycling and redistribution of nutrients play a particularly important role in ecosystem dynamics (see Scholes 1990, Stafford-Smith & Morton 1990). Soil biota, primarily fungi, bacteria and invertebrates, are largely responsible for converting the nutrients contained in plant litter to an inorganic form available for subsequent plant uptake (Holt & Coventry 1990). Micro-organisms play the overwhelmingly dominant role in this mineralisation process in temperate regions (Peterson & Luxton 1982, Seastedt 1984), but in the semi-arid tropics various detritivorous arthropods can also make an important contribution (Crawford 1981). The premier decomposer insects in these systems are termites (Wood 1976, Collins 1981, Whitford *et al.* 1982, Josens 1983).

2.1. Termite richness and abundance

Termite mounds are a conspicuous and often spectacular feature of the Kakadu landscape (Fig. 1; Andersen & Jacklyn 1993). However, these mounds are just the "tip of the iceberg" of termite biomass: not only do most mound-building species have at least partly subterranean nests, but fewer than half of the local termite species build

Table I. Mound-building species of termites in Kakadu National Park (Stages I and II) in relation to total termite species (data from Braithwaite *et al.* 1988)

Family/Genus	Number of species	
	Total	Mound-building
Mastotermitidae		
Mastotermes	1	0
Kalotermitidae		
Neotermes	2	0
Cryptotermes	1	0
Rhinotermitidae		
Heterotermes	3	0
Coptotermes	1	1
Schedorhinotermes	2	0
Termitidae		
Nasutitermes	6	2
Tumulitermes	4	3
Australitermes	1	0
Occultitermes	1	0
Amitermes	6	1
Ahamitermes	1	0
Drepanotermes	1	0
Invasitermes	1	0
Termes	13	3
Microcerotermes	6	6
TOTAL	50	16

Fig. 1. Termite mounds such as this one built by *Nasutitermes triodiae* are a spectacular feature of Kakadu landscapes (Photo B. Curb).

mounds (these are predominantly the grass-eaters; Andersen & Lonsdale 1990). Braithwaite *et al.* (1988) identified 50 termite species in a survey of savanna woodland and open forest sites in Stages I and II of Kakadu National Park, only 15 of which build mounds (Table I).

The richness of termite species in Kakadu National Park appears comparable to that of similar, but better studied savannas of Africa (Josens 1983) (Table II). Australia lacks fungus-growing termites of the sub-family Macrotermitinae (Sands 1969), which accounts for the poorer representation of harvester species in the Kakadu fauna, and the sub-family Apicotermitinae is also absent from Australia (Table II). The taxonomic differences between the continents are probably due more to geological history than to ecological factors; for example, the Macrotermitinae originated in Africa after the break-up of Gondwana.

Termite abundance is notoriously difficult to quantify. Mound densities in northern Australia can exceed 1000 ha^{-1}, but this varies markedly (Lee & Wood 1971). Moreover, estimates of termite biomass based on mound densities are fraught with errors (Andersen & Lonsdale 1990). Lee & Wood (1971) estimated that population sizes of harvester termites in African savannas range from 1000–10000 m^{-2}, representing a biomass range from 5–50 g m^{-2}. Comparable data are not available from Australia.

2.2. *Nutrient enrichment and depletion*

Termites differ from other decomposer organisms in that carbon and other nutrients cycled by them are fundamentally redistributed in the process (Wood 1976, de Bruyn & Conacher 1990). They also differ from most other arthropods in their ability to decompose cellulose, through a combination of gut symbionts and their own digestive processes (Lee & Wood 1971). The redistribution of materials is an outcome of termite sociality. Organic matter is collected by foraging workers and returned to the nest, where it is used either for food or is incorporated into the nest or mound via faeces or salivary secretions. The nutrients are ultimately released into the soil by erosion or leaching processes, or, indirectly through predation on the termites, and the release of alates.

There have been numerous studies of the nutrient dynamics and general pedological effects of termites elsewhere in the world, particularly in African savannas (reviewed by Lee & Wood 1971, de Bruyn & Conacher 1990, Jones 1990). Most of these focus on nutrient enrichment around termite

Table II. Number of species per trophic and taxonomic category in savanna habitats (woodland and open forest), each summed over nine sites within 12700 km^2 of Kakadu National Park (data from Braithwaite *et al.* 1988), compared with similar savannas in Africa (data from Josens 1983; sampling area and intensity variable).

	Zaria, Nigeria	Mokwa, Nigeria	Lamto, Côte d'Ivoire	Kakadu Woodland	Kakadu Open forest
Mean annual rainfall (mm)	1170	1175	1290	1300–1600	
Total no. species	29	31	36	35	36
Feeding largely on:					
grass/grass litter	11	10	10	5	6
wood/leaf litter	16	18	21	20	17
humus/soil	9	9	13	11	12
Family/sub-family					
Mastotermitidae	0	0	0	1	1
Kalotermitidae	0	0	2	0	1
Rhinotermitidae	0	1	1	5	6
Termitinae	10	9	10	20	20
Apicotermitinae	3	3	4	0	0
Macrotermitinae	11	11	9	0	0
Nasutitermitinae	5	7	10	9	8

nests and mounds (termitaria). Similar studies have been conducted in the Townsville region of Queensland (J.A. Holt, A.V. Spain and colleagues, see below) where the savannas are comparable to many of those in the Kakadu region (Mott *et al.* 1985). At one site it was estimated that just two of the mound-building species present, *Amitermes laurensis* and *Nasutitermes longipennis*, were responsible for the mineralisation of at least 250 kg ha^{-1} y^{-1} of organic carbon (Holt 1987, Holt & Coventry 1990). Soil micro-organisms made a far greater contribution (approximately 2300 kg ha^{-1} yr^{-1}), but microbial activity virtually ceased during the dry season, while termites were continuously active throughout the year. If all termites were considered, including subterranean species, it was estimated that they might be responsible for up to 20% of the total carbon mineralised in the ecosystem.

At another site near Townsville, Holt & Coventry (1982) examined enrichment levels of soils beneath the mounds of *Amitermes vitiosus*. Although these soils occupied less than 1% of the total area, they contained about 3% of the total organic carbon, nitrogen, extractable phosphorus, and exchangeable basic cations present in the top 10 cm of soil. Rates of turnover of enriched soil were estimated to be 400 kg ha^{-1} y^{-1}.

Soils of the termite mounds themselves are also substantially enriched, as are the erosion pediments surrounding them (Okello-Oloya *et al.* 1985). Rates of erosion of mound materials have been measured for *Amitermes vitiosus* (Bonell *et al.* 1986). Up to 10% of total mound mass was found to be lost after two wet seasons, and, assuming linear rates over longer time spans, it was estimated that mounds would be completely eroded after about 30 years of natural rainfall. Based on data of Williams (1968) from Brocks Creek in the Northern Territory, Lee & Wood (1971) estimated that erosion redistribution of mound material results in a net accumulation of surface soil of 0.0125 mm y^{-1}. This figure was estimated to be an order of magnitude higher at a nearby site (Lee & Wood 1971). Similar figures were obtained by Holt *et al.* (1980) for a north Queensland site, where it was suggested that the entire A horizons of both red and yellow earths had been derived largely from the erosion of former termite mounds.

The decay of termitaria, and therefore release of accumulated nutrients, can be considerably hastened by the activities of other fauna, particularly by burrowing vertebrates. In Kakadu National Park, the termites which built the termitarium are replaced by other insects, small lizards, rodents, snakes, native cats and bandicoots in something like that order (Braithwaite 1990b).

The effects on plants of nutrient enrichment by termites has been illustrated by glasshouse growth trials (Spain *et al.* 1983). For example, the biomass of the tropical pasture species *Stylosanthes hamata* and *Digitaria ciliaris* grown on nest materials was several-fold greater than that of controls (Spain & Okello-Oloya 1985). Although the remarkable vegetation patterns associated with African termitaria (Wild 1952, Glover *et al.* 1964, Harris 1966) are not known from Australia, nutrient enrichment by termites still has a significant effect in the field. Rapidly growing annual grasses and herbs are associated with termite mounds and their margins, whereas slower growing perennial grasses and sedges predominate elsewhere (Fig. 2). Above-ground biomass diminishes with increasing distance from the mound margin (Spain & Okello-Oloya 1985, Spain & McIvor 1988). Spain & Okello-Oloya (1985) estimated that the zone of nutrient enrichment surrounding termite mounds represented about 5% of the total soil surface.

The enrichment of soils associated with termite nests and mounds requires the depletion of soils elsewhere. Termites are capable of almost completely removing litter from the soil surface, therefore causing substantial nutrient depletion of soils outside termite nests and mounds (Schaefer & Whitford 1981, Jones 1990). Jones (1990) suggests organic carbon in the soil is not just redistributed by termites, but is permanently depleted because decomposition is so thorough; that is, carbon passes straight from termites to the atmosphere as carbon dioxide and methane, thereby bypassing the soil. This might contribute to the low organic carbon content, and therefore low water- and nutrient-holding capacity, of many tropical savanna soils (Jones 1990).

Termites obviously play a key role in the nutrient dynamics of tropical savannas, particularly in the formation of localised patches of nutrient enriched soil in an otherwise infertile landscape

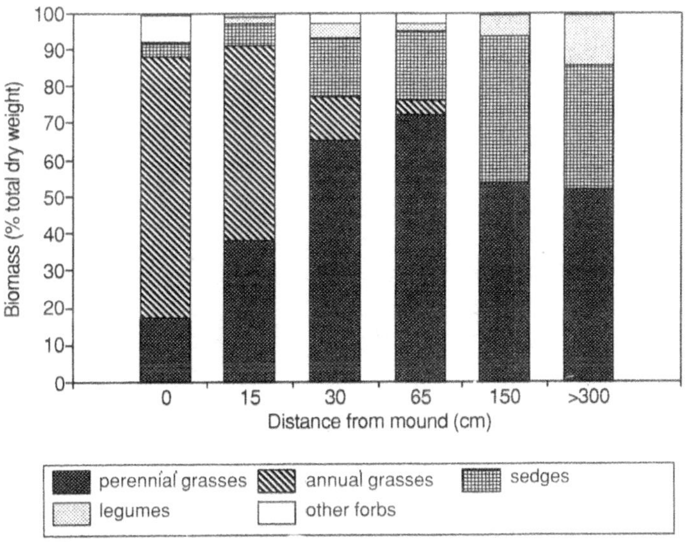

Fig. 2. Changes in herbaceous plant species composition around termite mounds at Antill Plains in northeastern Queensland (data from Spain & McIvor 1988).

(Scholes 1990). However, the overall effects of termites at different spatial and temporal scales are clearly complex and inadequately understood.

In addition to their influence on nutrient cycling, termites can have important effects on soil texture, structure, bulk density, and porosity, all of which are significant for plant growth (Lee & Wood 1971, Lal 1987, de Bruyn & Conacher 1990). Termites might also have an important effect on plants in the future due to their contribution to 'greenhouse'-induced climate change. On a global scale, emissions from decomposition by termites might produce significant proportions of global carbon dioxide and methane released into the atmosphere (Zimmerman *et al.* 1982, Zimmerman & Greensberg 1983, Collins & Wood 1984, Jones 1990).

2.3. *Tree-piping*

Termites might also facilitate nutrient flow in a different way. Janzen (1976) has suggested that the hollow cores of some tropical trees may be an adaptation for trapping nutrients released into them by hollow-nesting vertebrates. In this context, it is interesting to note that in the nutrient-poor savannas of the Kakadu region, an extraordinarily high percentage of trees in open forest and woodland (59% and 40%, respectively of trees > 10 cm DBH) have hollowed trunks (Braithwaite 1985), caused almost entirely by the wood-eating termite *Coptotermes acinaciformis* (Fox & Clark 1974). In the more fertile monsoon forests, this figure is only 7%. The incidence of trunk-hollowing was negatively correlated with total soil phosphorus ($r = -0.55$, $p < 0.001$) and total soil nitrogen ($r = -0.43$, $p < 0.001$) for 20 sites scattered across Kakadu National Park Stages I and II (Braithwaite 1985).

In Kakadu National Park it is the most common trees which are heavily piped: *Eucalyptus miniata*, *Eucalyptus tetrodonta*, *Eucalyptus porrecta* and *Erythrophleum chlorostachys* are the four most common trees in Kakadu Stages I and II, and piping incidences were 79, 70, 66 and 42%, respectively. These hollows are extensively used by vertebrates which feed, nest, defaecate, urinate, die, and generally concentrate nutrients into them. In Kakadu, approximately one-fifth of the bird and amphibian species, half the mammals and a quarter of the reptiles use tree hollows for some purpose (Braithwaite 1985, 1989). This figure is 23% for all vertebrate species in Australia (Ambrose 1980), where soils of low fertility are very common. Further work on rates of nutrient input into tree hollows, and the availability of these nutrients to the hollowed trees, is required before the relationship between *Coptotermes acinaciformis* and some of the eucalypts, mediated by hollow-dwelling vertebrates, can be properly evaluated.

3. Habitat selection by vertebrates

Habitat selection is the differential usage of the area within the geographic range of an animal species (Partridge 1978, Cody 1985). While total biomass and diversity within an ecosystem are generally related to available moisture and nutrients (Rosenzweig 1968, Brown 1981), individual species of plants and animals tend to be maximally adapted to different parts of moisture and nutrient gradients (Whittaker 1975, Austin & Austin 1980). Patterns of animal habitat distribution are also influenced by fire intensity and frequency which interact with the moisture and nutrient gradients. The influence of these factors on habitat selection by vertebrates is considered below.

3.1. Moisture and nutrients

Four general moisture and nutrient gradients occur in the Kakadu region: i) from the coast inland; ii) from the lowlands to the sandstone escarpment; iii) from the major rivers to the eucalypt savanna; and iv) within the savanna from the seasonal creeklines to the savanna ridgetops.

From the coast inland there is a moisture gradient where mean annual rainfall decreases from about 1400 to 800 mm over 300 km. Along the gradient, tropical animals are gradually replaced by arid zone species. For example, the native mice typical of coastal and sub-coastal areas, *Pseudomys nanus* (western chestnut mouse; weight 20–50 g) and *Pseudomys delicatulus* (Delicate Mouse; 6–15 g), are replaced by *Pseudomys calabyi* (Calaby's mouse; 15–25 g) further inland, which is in turn replaced by *Leggadina forresti* (Forest's mouse; 15–25 g) (Woinarski & Braithwaite 1990).

Moving from the lowlands to the sandstone escarpment, up to 100 m in elevation, the rocky country has many seepage areas providing moisture through the long dry season. Freeland *et al.* (1988) found plant diversity and productivity to be greater closer to rocky areas. They argued that this accounts for the unusually large proportion of rock-dwelling mammals in the regional fauna, and for the high degree of escarpment endemism.

From the major rivers to the eucalypt savanna there is a nutrient and moisture gradient generally between 5 and 10 km long and about 30 m in elevation (R.W. Braithwaite unpublished information). The flood plains of the major rivers are fertile, black cracking clays, which contrast with the less fertile lateritic soils of the savannas at slightly higher elevations. The savannas also dry out more quickly following rain. Despite their low productivity, however, the savannas are richest in

Fig. 3. Seasonal creekline within savanna vegetation. Such creeklines provide critical habitat for a range of birds and mammals (Photo D. Braithwaite).

species of plants, insects, lizards, mammals, and granivorous, insectivorous and nectarivorous birds (Braithwaite 1985, 1990a). Snakes, frogs and frugivorous birds are richest in monsoon forest, located between flood plain and savanna.

Within the savanna, from the seasonal creeklines (Fig. 3) to the savanna ridgetops there is primarily a moisture gradient of less than 20 m in elevation and a kilometre in distance. The seasonal creeklines ultimately join the major wetlands, and are joined themselves by higher order drainage lines further along the gradient (R.W. Braithwaite unpublished information). This is the most subtle and perhaps poorly appreciated of the four major gradients, and merits further elaboration here. The vegetation forms distinctive floristic patterns along the gradient. The results of an agglomerative, polythetic, non-hierarchical analysis (ALOC; Belbin 1988) of floristic data from 360 sites on 45 ha of savanna encompassing such a gradient, are presented as ordered in a one dimensional gradient in Fig. 4 (R.W. Braithwaite unpublished information). Eight floristic groups (A-H) are recognised in the analysis. Group A (46% of the study area) represents the very extensive and widespread vegetation type (*Eucalyptus tetrodonta*/*Eucalyptus miniata* savanna) covering over 90% of the region's sub-coastal savannas. The other groups are restricted in extent and usually associated with seasonal drainage lines.

The mean numbers of captures in mammal traps from a 25 month (3 nights per month) survey of the above mentioned study area are shown in Table III. The dasyurid marsupial *Antechinus bellus* (fawn antechinus) was by far the most abundant mammal caught; it was widely distributed, but most common in the creekbed area of floristic type H. The other common dasyurid,

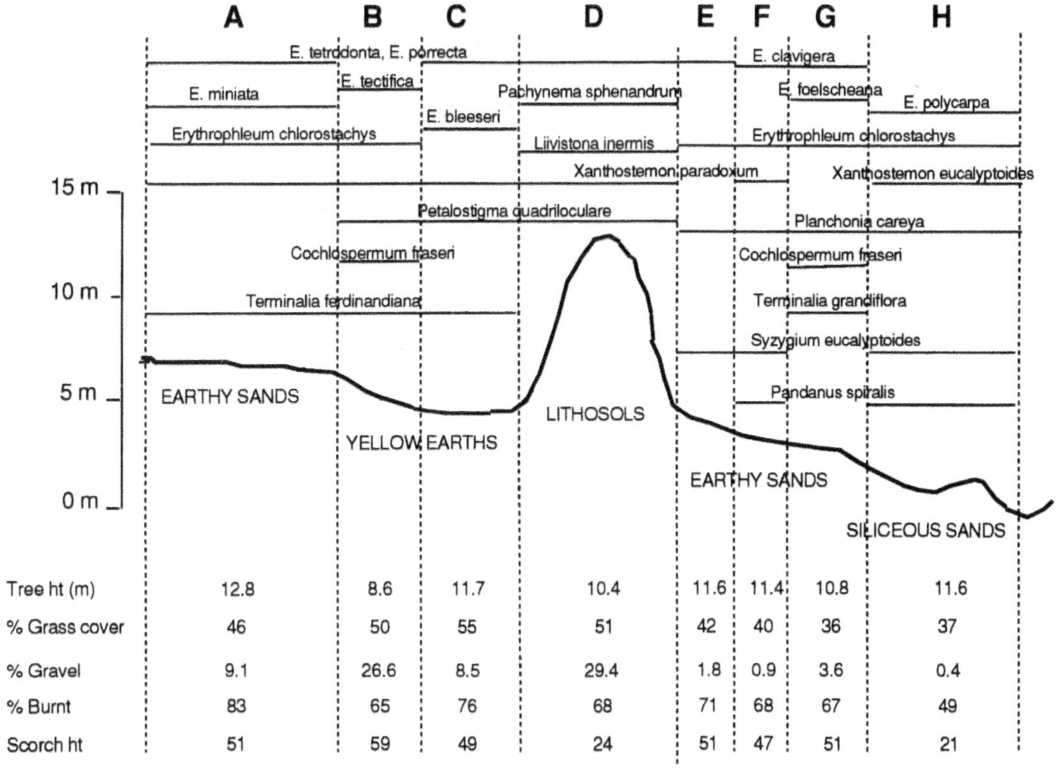

Fig. 4. An altitudinal gradient from ridgetop to seasonal creekline, covering a horizontal distance of approximately 1 km, at the Kapalga Research Station in Stage II of Kakadu National Park. The distributions of the most common woody species are shown (E = *Eucalyptus*). Eight floristic groups (A-H) are recognized, with mean values for some key environmental attributes (maximum tree height; percent cover of perennial grasses; percent gravel in soil; percent area burnt each year; and fire scorch height, expressed as a percentage of maximum tree height) also given. See text for details.

Table III. Habitat preferences of mammals from a 45 ha site at the Kapalga Research Station in Stage II of Kakadu National Park. Data are numbers of captures per 100 trap nights per site, from 360 sites sampled over 25 months. Floristic groups are those shown in Fig. 4.

Species	Floristic Groups A	B	C	D	E	F	G	H	Kruskal-Wallis Statistic	P value
Antechinus bellus	8.70	5.62	7.23	8.24	7.56	8.58	5.57	14.16	33.09	0.000
Dasyurus hallucatus	1.68	2.09	1.47	1.73	1.13	0.76	0.86	0.51	18.39	0.010
Phascogale tapoatafa	0.07	0.04	0.07	0	0.03	0.03	0	0	4.20	0.757
Isoodon macrourus	2.16	2.16	3.30	1.10	3.32	3.39	1.57	1.21	7.96	0.336
Trichosurus arnhemensis	0.66	0.14	0.21	0.24	0.65	0.18	0.08	0.19	0.32	0.946
Melomys burtoni	0	0.07	0	0	0.10	0.06	0.08	0.70	36.19	0.000
Mesembriomys gouldii	0.30	0.29	0.21	0.31	0.62	0.33	0.55	0.89	11.55	0.116
Pseudomys delicatulus	2.39	3.35	1.68	1.26	1.64	0.73	1.18	0.51	44.40	0.000
Rattus colletti	0.10	0.47	0.07	0	0.44	1.67	1.57	1.21	102.00	0.000
Rattus tunneyi	3.44	3.46	3.93	8.55	11.18	19.79	15.29	10.03	129.29	0.000
Total abundance	19.50	17.69	18.18	21.41	26.67	35.52	26.75	29.40	93.73	0.000
Species richness	3.60	3.10	3.37	3.88	3.90	4.09	4.29	4.24	18.00	0.012

Fig. 5. (a) *Isoodon macrourus* (northern brown Bandicoot) (Photo B. Braithwaite); (b) *Rattus colletti* (dusky rat) (Photo E. Slater); c) *Carlia triacantha* (four-fingered skink) (Photo J. Wombey); and d) *Diporiphora bilineata* (two-lined dragon) (Photo J. Wombey).

Dasyurus hallucatus (northern quoll), was also widely distributed, but most common in floristic type G and least in type H. Other species showed different preferences: for example, *Isoodon macrourus* (northern brown bandicoot) (Fig. 5a) and *Rattus colleti* (dusky rat) (Fig. 5b) were all most common in floristic type F, and *Pseudomys delicatulus* (delicate mouse) was most common in type B. In all, a range of vegetation preferences was shown with different species utilising different parts of the drainage gradient (Fig. 6a). An overall gradient in mammal species richness and abundance was also evident (Fig. 6c; see also Braithwaite 1990a). The high variation from month to month adds a second level of complexity to the above patterns. Species abundances changed markedly through the year, and so did their apparent habitat preferences; pooling of monthly data in Table III gives the impression that many species had wide habitat preferences, when at any particular time they did not.

Lizards censused during the trapline checks in the same study were also selective in terms of floristic groups, but showed a different gradient of total richness and abundance to that of mammals (Table IV; Fig. 6c). Small skinks of the genus *Carlia* (Fig. 5c) appeared to be the most selective, with *Carlia gracilis* favouring floristic group H, *Carlia munda* (formerly *C. foliorum*) groups E, F and H, and *Carlia amax* and *Carlia triacantha* both group D (Fig. 6b).

A total of 95 bird nests belonging to 31 species was also recorded during the same study. Total numbers of nests ($X^2 = 4.4$, $p > 0.05$) and numbers of bird species ($X^2 = 5.12$, $p > 0.05$) recorded in each of the eight floristic groups did not differ significantly from random expectation. However, as with mammals and reptiles, different species selectively used different parts of the gradient: for example, the number of nests of finches was greater than expected in the middle parts of the gradient (groups C-F; $X^2 = 14.49$, $p < 0.05$).

Just as small-scale differences in site moisture availability are important in determining the spatial distribution and abundance of the fauna, rainfall appears to have a marked influence on the seasonal distribution and abundance of the fauna at any particular place. In a study of thirty sites in Kakadu National Park Stages I & II that were sampled six times over three years, the mean similarity between samples taken from the same site was calculated as a measure of temporal stability (constancy; Orians 1975). For the non-

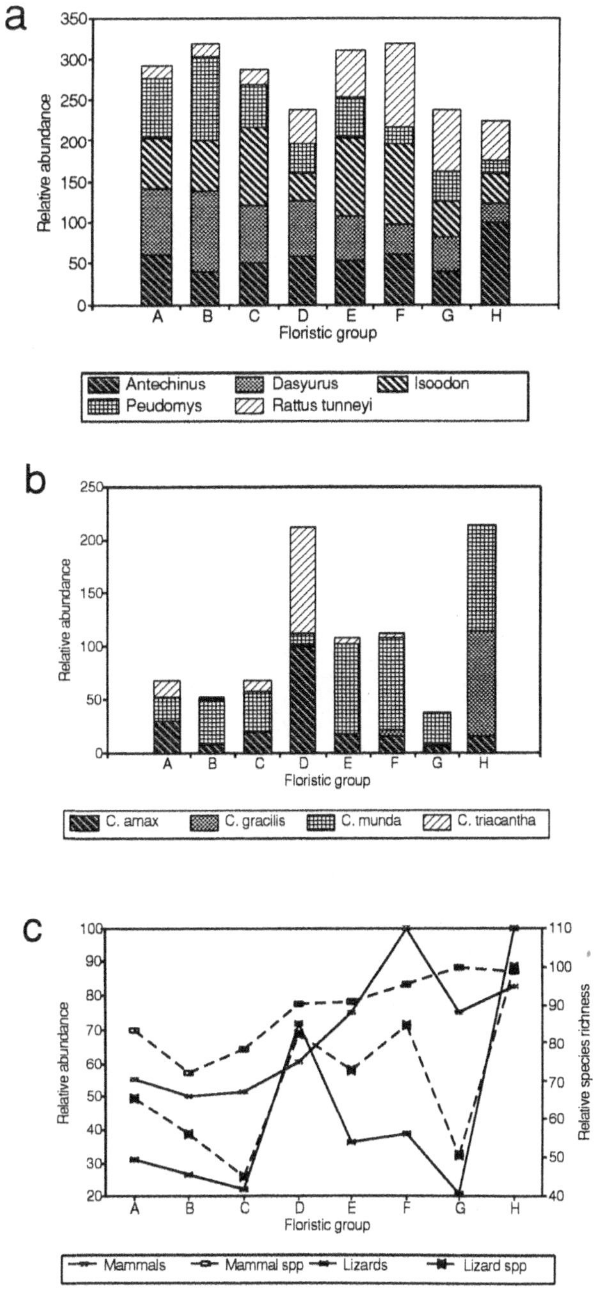

Fig. 6. Distribution of a) mammal species; b) species of the skink genus *Carlia*; and c) total mammal and lizard abundance and richness across the floristic groups shown in Fig. 4. In each case data are scaled with maximum scores given 100% (data from Tables III and IV).

Table IV. Habitat preferences of lizards from a 45 ha site at the Kapalga Research Station. Data are total observations per site between November 1985 and November 1987 (see Table III for further details).

Species	Floristic Groups									
	A	B	C	D	E	F	G	H	Kruskal-Wallis Statistic	P value
Diporiphora bilineata	0.35	0.59	0.0	0.24	0.21	0.25	0.47	0.14	16.40	0.021
Carlia amax	0.99	0.22	0.68	3.35	0.54	0.48	0.18	0.48	53.26	0.000
Carlia gracilis	0.02	0.05	0.0	0.06	0.05	0.29	0.06	3.81	166.83	0.000
Carlia munda	0.26	0.49	0.42	0.12	0.97	0.98	0.35	1.14	36.71	0.000
Carlia triacantha	0.07	0	0.05	0.47	0.03	0.02	0.0	0.0	20.41	0.005
Cryptoblepharus plagiocephalus	0.06	0.05	0.11	0.0	0.13	0.05	0.12	0.19	5.47	0.603
Ctenotus borealis	0.0	0.03	0.0	0.0	0.03	0.02	0.0	0.0	6.08	0.530
Ctenotus essingtoni	0.08	0.0	0.0	0.0	0.08	0.07	0.0	0.10	6.50	0.482
Total abundance	1.87	1.57	1.32	4.29	2.18	2.32	1.24	6.00	37.11	0.000
Species richness	1.22	1.05	0.84	1.53	1.36	1.57	0.94	1.86	14.56	0.042

flying mammals, terrestrial birds and lizards respectively, values of 0.46, 0.66 and 0.83 were obtained (Braithwaite 1985). Using the difference in rainfall from the townsite of Jabiru for successive six-month periods as an inverse measure of climatic similarity, it was possible to estimate the relative impact of climatic variability on these three vertebrate groups. This rainfall dissimilarity measure correlated significantly with the mean similarity value of birds ($r = -0.70$, $p < 0.01$) and native non-flying mammals ($r = -0.58$, $p < 0.05$), but, although the same trend was evident, not with lizards ($r = -0.43$, $p > 0.05$) (Braithwaite 1985).

3.2. Fire

Fire is an important factor affecting the patterning of animal habitats in southern Australia (Catling & Newsome 1981), and is extremely prevalent in the Kakadu region (Andersen & Braithwaite 1992). In savanna woodland, fires are typically annual, and in open forest they occur in two out of three years in any one place (Braithwaite & Estbergs 1985). There is considerable debate over how these fire regimes differ from those traditionally managed by Aborigines, and over the timing, extent and antiquity of traditional aboriginal burning practices (Jones 1969, Horton 1982, Haynes 1985, Braithwaite 1991). The ecological effects of these fires are also poorly understood (Duff & Braithwaite 1990, Andersen & Braithwaite 1992), but they surely must be a potent force shaping the regional fauna.

The ecological effects of fire operate both in the short- and in the long-term. In the short-term, even if relatively few individuals of a species are killed directly by the fires, as often appears to be the case, the subsequent survival and reproduction of individual animals is likely to be substantially affected. Some birds exploit fires for feeding (Braithwaite & Estbergs 1987); for example, *Milvus migrans* (black kite) characteristically hunts along fire fronts, and flocks of seed-eating *Calyptorhynchus magnificus* (red-tailed black cockatoo) often descend upon recently burnt areas. In the longer term, different fire regimes can have an important impact on habitat selection through their effects on vegetation structure (Woinarski 1990, Andersen 1991b).

Studies with lizards have shown that the abundance of these relatively sedentary animals is strongly influenced by fires of different types. In one study (Braithwaite 1987), *Carlia amax* and *Carlia gracilis* appeared to be fire sensitive; *Diporiphora bilineata* (two-lined dragon) (Fig. 5d) and *Carlia triacantha* (Fig. 5c) were apparently favoured by early hot fires; *Cryptoblepharus plagiocephalus* seemed relatively unaffected; *Carlia munda* appeared to be very tolerant of fires, while *Ctenotus* spp. were favoured by low intensity patchy fires with high intensity spots (Braithwaite

1987). Lizards therefore, appear to select habitats created by fires of different intensity and patchiness.

Different types of fires create patchiness of the ground vegetation on different scales, with mild fires creating small patches and intense fires creating huge patches of different species mixed with bare ground. The habitat types of Fig. 4, by virtue of their different moistness, structure and composition, have some propensity to create fires of different types, but prevailing weather conditions are likely to be the dominant factor. The high frequency of fires means that the classical successional paradigm (Fox 1982) of plants being killed by fire and thereby creating an opportunity for a changing parade of new plant and animal species to establish through the years until the next fire, is not possible. Instead, individual fires are more like other short-term modifiers of habitat selection, such as rainfall.

In short, fire and moisture patterns interact to produce constantly changing use of space by many vertebrate species. Floristic groupings using relatively stable woody species composition, however, do show overall patterns in animal species preference (Tables III, IV; Fig. 5). The mammals show the strongest pattern of diversity and abundance with respect to the moisture gradient, and they also show the greatest variability in abundance. These factors are likely to have contributed to the sad record of extinction (Morton 1990) of many of these and related animals from much of inland Australia. In times of drought, populations of many small mammals are forced to retreat to the most favourable sites. The introduction of domestic cattle has led to extensive degradation of these habitats throughout Australia's rangelands, severely compromising their capacity to act as refuges for native fauna (Morton 1990). This experience suggests that the protection of habitats associated with seasonal creeklines is crucial for the maintenance of mammal diversity in savannas of the Kakadu region.

4. Seed-harvesting by ants

Seed-harvesting (= predation) by ants occurs throughout Australia (Andersen 1991a), and is not primarily a desert phenomenon as it is elsewhere in the world (Mares & Rosenzweig 1978, Brown *et al.* 1975, 1979, Abramski 1983). Harvester ants occur in significant numbers in most terrestrial habitats, and are almost universally the most important post-dispersal granivores (Andersen 1991a). Australia has a depauperate fauna of granivorous mammals (Morton 1979), such that mammalian seed predation is often negligible (O'Dowd & Gill 1984, Andersen & Ashton 1985, Morton 1985, Andrew 1986), and avian granivory may only be significant in the arid zone (Morton & Davies 1983). It is possible, however, that some rocky habitats of the Kakadu region and elsewhere in the northern part of the Northern Territory are exceptions. In these environments there is evidence of substantial predation of grass seeds by rodents (species of *Rattus* and *Pseudomys*) and finches (R.W. Braithwaite and J.C.Z. Woinarski unpublished data).

4.1. Harvester ants

The ant fauna of northern Australia is poorly known, but based on collections in the Kakadu region probably supports about 50 harvester species (A.N. Andersen unpublished data). As far as is known, these all belong to one of three myrmicine genera: *Pheidole* (approximately half of the total species), *Monomorium* and *Meranoplus*. It is possible that harvesting species of another myrmicine genus, *Tetramorium*, may also occur, particularly in the more arid southern parts of the region. Almost all harvesting ants outside Australia are also myrmicines (Abramski 1983, Brown *et al.* 1979, Morton & Davidson 1988), but in southern Australia formicine (*Prolasius* and *Melophorus*) and ponerine (*Rhytidoponera*) genera are often the dominant harvesters (Andersen 1991a). The Kakadu region supports many species of *Melophorus* and *Rhytidoponera*, but none are known to eat seeds.

The local diversity of harvester ants in the Kakadu region is high compared with elsewhere in Australia, and exceptional by world standards. Most sites in Australia support about 10 harvester species (Andersen 1991a), but 18 species have been recorded at a single savanna site in Stage II of Kakadu National Park (Table V), the only site in the region for which data are currently available. This is equivalent to almost the entire

Table V. Number of harvester ant nests in two 30 × 30 m plots (separated by 50 m) in savanna woodland in Stage II of Kakadu National Park (data from Andersen & Lonsdale 1990). Numbers in brackets are percentages of total nests.

Harvester ant species	Plot 1	Plot 2
Pheidole (7 spp)	10 (26)	115 (76)
Meranoplus diversus group (3 spp)	25 (64)	27 (18)
Monomorium rothsteini group (2 spp)	3 (8)	4 (3)
other *Meranoplus* and *Monomorium* (6 spp)	1 (3)	5 (3)
Total	39(100)	151 (100)

harvester ant fauna of North American deserts, which comprises about 20 species (Morton & Davidson 1988).

Nest densities of harvester ants in the Kakadu region can also be very high, but are extremely patchy. For example, at the above mentioned savanna site the density of harvester ant nests totalled 0.17 m^{-2} in one plot, but was less than a quarter of this in another plot only 50 m away (Table V). Averaging data from these two plots, nest densities of the major harvesting taxa were as follows: *Pheidole* spp, 669 ha^{-1}; *Meranoplus diversus* group, 289 ha^{-1}; and *Monomorium rothsteini* group, 39 ha^{-1}.

Australian harvester ants range from generalist seed-eating omnivores to specialist granivores (Andersen 1991a), and this holds for the Kakadu region. Most of the species at least partly supplement their diets with insect material, and most eat a variety of seeds (Fig. 7a), apparently selecting them opportunistically on the basis of size, morphology and availability. The major exceptions are species of the *Meranoplus diversus* group which not only are strictly granivorous, but appear to specialise on the seeds of only one or a few plant (primarily grass) species. One of these ants specialises on *Sorghum intrans*, the dominant annual grass in the Kakadu region; its foraging activity is restricted to the time of seed availability, from the end of the wet season, when seed fall occurs, until the first rains at the end of the dry season, when all remaining seeds germinate (Andrew 1986). Its nest middens of discarded *Sorghum* seed husks (Fig. 7b) are a familiar sight in the savannas during the dry season.

Probably the most important harvesters in the region are members of the *Monomorium rothsteini* group. This group is widespread in inland Australia (Briese & Macauley 1981, Davison 1982), and is often extremely abundant in the savannas of the Kakadu region. At the savanna site mentioned in Table V, one of these species was easily the most abundant harvester ant despite its low colony density – this is because colony sizes were extremely large, and workers foraged many metres from their nests (A.N. Andersen unpublished data).

4.2. *Rates of harvesting*

Species of *Pheidole* and *Monomorium* eat the seeds of eucalypts and other woody plants, as well as of grasses and other herbs (Andersen & Ashton 1985). The dominant tree at the savanna site

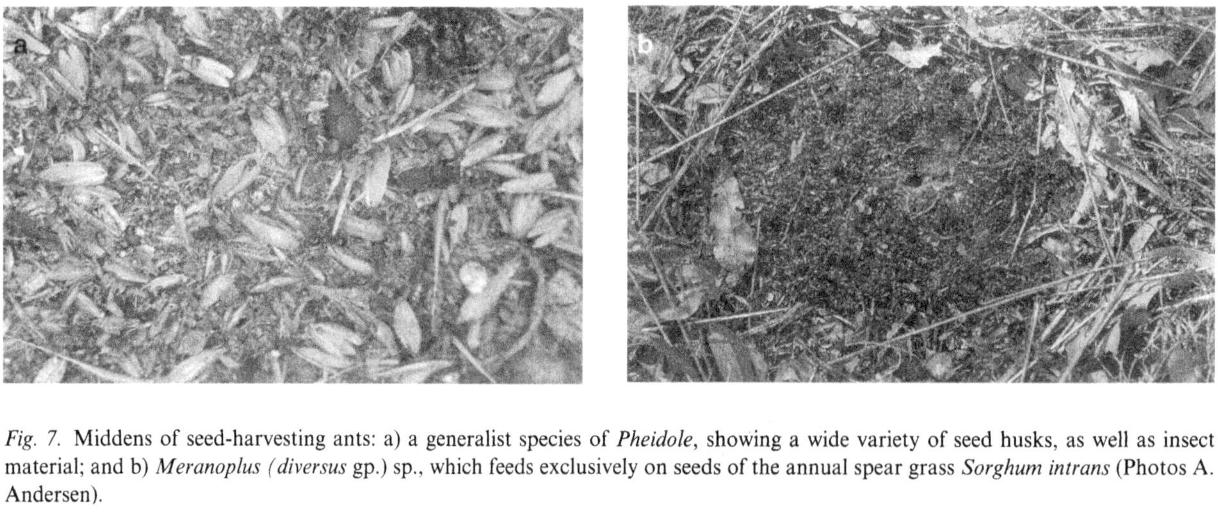

Fig. 7. Middens of seed-harvesting ants: a) a generalist species of *Pheidole*, showing a wide variety of seed husks, as well as insect material; and b) *Meranoplus (diversus* gp.) sp., which feeds exclusively on seeds of the annual spear grass *Sorghum intrans* (Photos A. Andersen).

a

b

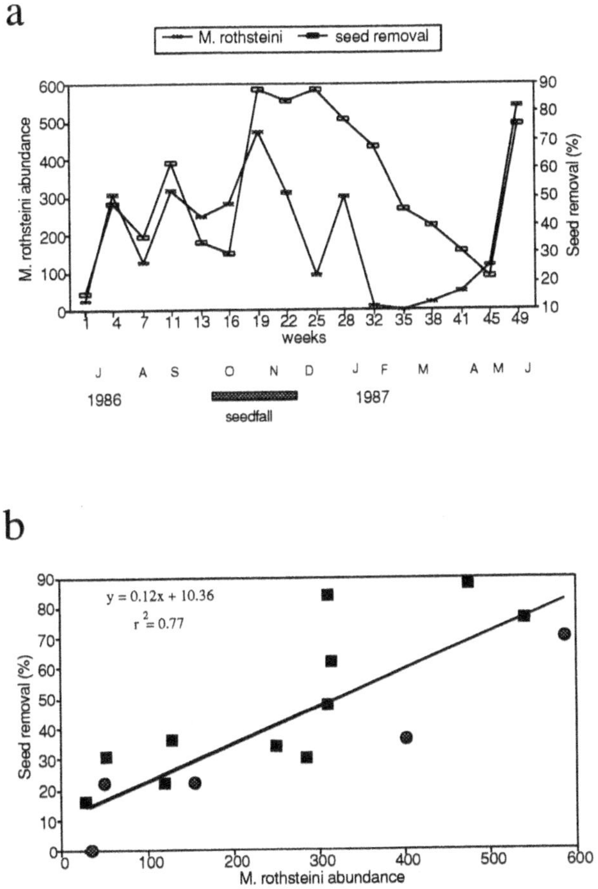

Fig. 8. a) Seasonal changes in rates of removal of *Eucalyptus tetrodonta* seeds from depots, and pitfall captures of seed harvesting *Monomorium 'rothsteini'*. Data were collected from plot 1 in Table V (A.N. Andersen unpublished data). b) Relationship between rates of removal of *E. tetrodonta* seeds and the abundance of *M. 'rothsteini'*, between May and November. Squares represent data from a) (ie. plot 1 in Table V), and circles represent data collected from plot 2.

referred to in Table V was *Eucalyptus tetrodonta*, and ants (primarily a species of the *Monomorium rothsteini* group) can remove up to 90% of its seeds from experimental depots within 48 hrs (Fig. 8a). This figure is similar to those obtained for eucalypts in southern Australia (Ashton 1979, Andersen & Ashton 1985, Wellington & Noble 1985, Andersen 1987). Unlike southern eucalypts, however, *Eucalyptus tetrodonta* does not have persistent fruits, and all seeds are released in the months just prior to the commencement of the wet season. Interestingly, rates of seed removal from experimental depots were highest when seeds were naturally available on the ground, during and immediately after the seedfall period (Fig. 8a). This contrasts with the situation for southern eucalypts, where seeds are stored inside woody fruits and released *en masse* immediately following fire, thereby satiating harvester ants and reducing removal rates (Ashton 1979, O'Dowd & Gill 1984, Wellington & Noble 1985, Andersen 1988).

The high removal rates of *Eucalyptus tetrodonta* seeds during the seedfall period corresponded to the time of highest seasonal abundance of *Monomorium 'rothsteini'* (Fig. 8a). Indeed, throughout the dry season (May to November) removal rates from baits were highly correlated ($r = 0.89$, $p < 0.0001$) with the seasonal abundance of *Monomorium 'rothsteini'* (Fig. 8b). For several months after seedfall, however, removal rates were far higher than would be expected from *Monomorium 'rothsteini'* abundance (Fig. 8a). One explanation is that individual ants forage preferentially for *Eucalyptus tetrodonta* seeds once they become available on the ground. Alternatively, another harvester species might be removing the seeds during this time.

4.3. Impact on plant populations

The impact of harvester ants on plant populations is not straight forward, and depends on patterns of recruitment which in most cases are poorly known. Seed losses will only influence population size if recruitment is limited by seed supply rather than by the availability of 'safe' sites, and in stable populations of long-lived perennials this often does not appear to be the case (Andersen 1989). The role of seedling recruitment in the population dynamics of eucalypts in the Kakadu region is virtually unknown, so it is not possible to assess the consequences of seed losses to harvester ants, particularly given the reliance of most species on vegetative regeneration following fire.

In the case of short-lived plants, however, population sizes are frequently limited by seed supply and therefore directly influenced by losses to seed predators (Borchert & Jain 1978, Reichman 1979, Inouye *et al.* 1980, Risch & Carroll 1986). Although Andrew (1986) initially concluded that *Meranoplus 'diversus'* has no effect on the persistence of *Sorghum intrans* because of compensatory, density-dependent seed production, subsequent modelling of *Sorghum* population dynamics suggests that high densities of *Meranoplus* can in fact have a significant impact on plant density, ultimately leading to local extinction (Watkinson *et al.* 1989).

5. Overview

Plant-animal interactions embody many of the processes fundamental to ecosystem structure and function, and here we have barely scratched the surface of what have undoubtedly been driving forces in the evolution of ecosystems of the Kakadu region. However, one dominant theme emerges: in the Kakadu region insects are pre-eminent amongst animals in their relationships with plants.

The supremacy of insects in Australia's tropical savannas can be linked to low soil fertility. Nutrient-impoverished soils produce grasses low in nutritional value, which are therefore unable to sustain large populations of herbivorous mammals (Andersen & Lonsdale 1990). Insects are the predominant herbivores throughout most of the Kakadu region (Braithwaite *et al.* 1988, Andersen & Lonsdale 1990), with high densities of herbivorous vertebrates being restricted to areas with fertile soils. These are primarily the flood plain sedgelands and grasslands (20% of Stages I & II of Kakadu National Park), where *Rattus colletti* (Redhead 1979, Friend *et al.* 1988) and *Anseranas semipalmata* (Magpie Goose; Frith & Davies 1961) consume considerable quantities of underground storage organs. *Macropus agilis* (agile wallaby) can also be abundant on flood plain margins (Bolton 1974).

It is noteworthy that exotic mammalian herbivores such as water buffalo and pigs have also concentrated their activities on the flood plain margins (e.g. Ridpath & Waithman 1988). Elsewhere, the nutrient status is too poor, and the growing season generally too short and unreliable, to support a large mammalian biomass.

In terms of herbivory, the Kakadu region provides a strong contrast to the famed mammalian-dominated savannas of Africa. It should be noted, however, that within Africa the widespread abundance of mammalian herbivores is restricted to areas of young and rich volcanic soils (Bell 1982, East 1984). Elsewhere in Africa, as in Australia, herbivorous mammals are concentrated in nutrient-rich parts of the landscape (Scholes 1990).

One consequence of the pre-eminence of insects in the Kakadu region is that they have usurped roles normally associated with mammals. The dominance of harvester ants as granivores has already been noted. Another example is in seed dispersal, normally the domain of birds and mammals. Australia has an unusually high incidence of myrmecochory (seed-dispersal by ants), with more than 1500 plant species having their seeds regularly dispersed by ants via a special ant-attracting appendage ('elaiosome'; Berg 1975). The incidence of myrmecochory in the Kakadu regional flora is unknown, but in southern Australia it can be more than 30% (Berg 1975, 1981, Rice & Westoby 1981). Most of the numerous species of *Acacia* (all those with whitish arils) in the Kakadu region, for example, are probably myrmecochorous, but those with large, red arils (e.g. *Acacia auriculiformis*, Noske 1990) are dispersed by birds (Davidson & Morton 1984, O'Dowd & Gill 1986).

A substantial number of plant species in the Kakadu region have fleshy fruits and are therefore presumably also dispersed by birds, but these are concentrated in monsoon forests (Taylor & Dunlop 1985), which support a suite of frugivorous birds restricted to such habitats (Braithwaite 1985). There are few documented cases of seed dispersal by vertebrates in savanna habitats, but one example is the monocotyledonous tree *Pandanus spiralis* (Fig. 9a). It produces pineapple-like aggregates of large (about 6 cm long), orange fruit, with the seeds protected by hard wood at the distal end and a fleshy endoderm at the proximal end. Large parrots, primarily *Cacatua galerita* (sulphur-crested cockatoo), remove the pulpy material with their beaks, leaving behind fibrous parts still attached to the woody distal end containing the kernels (Fig. 9b). These fruits can therefore be readily identified, so that the dispersal shadow created by parrots is easily measured. In one such study (R.W. Braithwaite unpublished data), 201 fruit were counted within five metres of the parent tree, 30 (15%) of which were chewed. No other unchewed fruit were found, but chewed fruit were located at 15, 20, 25, 60, 155 and 180 m from the tree. Although *Pandanus* can form dense clumps, individuals are also very widely dispersed

Fig. 9. a) The monocotyledonous tree *Pandanus spiralis*; b) a A *Pandanus* fruit chewed by cockatoos (left) compared with an unchewed fruit (Photos B. Braitwaite).

through a range of habitats, something which apparently would not occur without parrots.

In searching for an imprint of plant-animal interactions on the Kakadu landscape, the following question can be asked: how has the pre-eminence of insects characteristic of infertile (dystrophic) savannas, both in Australia and in Africa, influenced ecosystem structure and dynamics in a manner not seen in the highly fertile (eutrophic), mammalian dominated savannas of Africa? A higher diversity and biomass of insectivorous vertebrates in dystrophic savannas might be expected, and indeed, there is a positive relationship within the Kakadu region between termite diversity and the diversity of insectivorous reptiles, birds and mammals (Braithwaite *et al.* 1988). The extraordinary diversity of lizards in the Australian arid zone has been linked to an unusually abundant and diverse termite food source (Morton & James 1988).

Given that major qualitative differences in ecosystem structure and function occur between dystrophic and eutrophic savannas, does this mean that there are also important quantitative differences? For example, how do total faunal diversity and biomass compare in the different systems? How does soil fertility affect rates of nutrient cycling and energy flow. What are the implications for the responses of ecosystems to stress and disturbance?

6. References

Abrahamson, W.G. 1989. Plant-Animal Interactions. McGraw-Hill, New York. 480 pp.

Abramski, Z. 1983. Experiments on seed predation by rodents and ants in the Israeli desert. Oecologia 57: 328–332.

Ambrose, G. 1980. Studying tree-holes as habitats: strategies and techniques. Vict. Ornithol. Res. Group Notes 16: 7–14.

Andersen, A.N. 1987. Effects of seed predation by ants on seedling densities at a woodland site in SE Australia. Oikos 48: 171–174.

Andersen, A.N. 1988. Immediate and longer-term effects of fire on seed predation by ants in sclerophyllous vegetation of southeastern Australia. Aust. J. Ecol. 13: 285–293.

Andersen, A.N. 1989. How important is seed predation to recruitment in stable populations of long-lived perennials? Oecologia 81: 310–315.

Andersen, A.N. 1991a. Seed harvesting by ants in Australia. In: Huxley, C.R. & Cutler, D.F. (eds) Ant-Plant Interactions. pp. 493–503. Surrey Beattie & Sons, Sydney.

Andersen, A.N. 1991b. Responses of ground-foraging ant communities to three experimental fire regimes in a savanna forest of tropical Australia. Biotropica 23: 575–585.

Andersen, A.N. & Ashton, D.H. 1985. Rates of seed removal

by ants at heath and woodland sites in southeastern Australia. Aust. J. Ecol. 10: 381–390.

Andersen, A.N. & Braithwaite, R.W. 1992. Burning for the conservation of the Top End's savannas. In: Moffatt, I. & Webb, A. (eds) Conservation and Development Issues in Northern Australia. pp. 117–122. North Australia Research Unit, Darwin.

Andersen, A.N. & Jacklyn, P.M. 1993. Termites of the Top End. CSIRO Press, Melbourne. 30 pp.

Andersen, A.N. & Lonsdale, W.M. 1990. Herbivory by insects in Australian tropical savannas: a review. J. Biogeog. 17: 433–444.

Andrew, M.H. 1986. Granivory of the annual grass *Sorghum intrans* by the harvester ant *Meranoplus* sp. in tropical Australia. Biotropica 18: 344–399.

Ashton, D.H. 1979. Seed harvesting by ants in forests of *Eucalyptus regnans* F. Muell. in central Victoria. Aust. J. Ecol. 4: 265–277.

Austin, M.P. & Austin, B.O. 1980. Behaviour of experimental plant communities along a nutrient gradient. J. Ecol. 68: 891–918.

Beattie, A.J. 1985. The Evolutionary Ecology of Ant-Plant Mutualisms. Cambridge University Press, Cambridge. 182 pp.

Belbin, L. 1988. PATN Pattern Analysis Package Reference Manual. CSIRO Division of Wildlife and Rangelands Research, Canberra. 101 pp.

Bell, R.H.V. 1982. The effect of soil nutrient availability on community structure in African ecosystems. In: Huntley, B.J. & Walker, B.H. (eds.) Ecology of Tropical Savannas. pp. 193–216. Ecological Studies: Springer, New York.

Berg, R.Y. 1975. Myrmecochorous plants in Australia and their dispersal by ants. Aust. J. Bot. 23: 475–508.

Berg, R.Y. 1981. The role of ants in seed dispersal in Australian lowland heathland. In: Specht, R.L. (ed.) Heathlands and Related Shrublands: Analytical Studies. pp. 41–50. Elsevier, Amsterdam.

Bernays, E.A. 1989. Insect-Plant Interactions, volume I. CRC Press, Boca Raton, Florida. 164 pp.

Bolton, B. 1974. An ecological study of the agile wallaby on the coastal plains of the Northern Territory with a comparison between improved and unimproved areas. Unpublished M.V.Sc. Thesis, University of Queensland, Brisbane. 121 pp.

Bonell, M., Coventry, R.J. & Holt, J.A. 1986. Erosion of termite mounds under natural rainfall in semiarid tropical northeastern Australia. Catena 13: 11–28.

Borchert, M.I. & Jain, S.K. 1978. The effect of rodent seed predation on four species of Californian annual grasses. Oecologia 33: 101–113.

Braithwaite, R.W. 1985. The Kakadu Fauna Survey: An Ecological Survey of Kakadu National Park. Report to Australian National Parks and Wildlife Service, Canberra. 1291 pp.

Braithwaite, R.W. 1987. Effects of fire regimes on lizards in the wet-dry tropics of Australia. J. Trop. Ecol. 3: 265–275.

Braithwaite, R.W. 1989. Shelter selection by a small mammal community in the wet-dry tropics. Aust. Mammal. 12: 55–59.

Braithwaite, R.W. 1990a. The implications of Australia's unique biota for ecological processes. J. Biogeog. 17: 347–354.

Braithwaite, R.W. 1990b. More than a home for white ants. Aust. Nat. Hist. 23: 306–313.

Braithwaite, R.W. 1991. Aboriginal fire patterns in monsoonal Australia during the nineteenth century. Search 22: 247–249.

Braithwaite, R.W. & Estbergs, J.A. 1985. Fire patterns and woody vegetation trends in the Alligator Rivers region of northern Australia. In: Tothill, J.C. & Mott, J.J. (eds) Ecology and Management of the World's Savannas. pp. 359–364. Australian Academy of Science, Canberra.

Braithwaite, R.W. & Estbergs, J.A. 1987. Firebirds of the Top End. Aust. Nat. Hist. 22: 298–302.

Braithwaite, R.W., Miller, L. & Wood, J.T. 1988. The structure of termite communities of the Australian tropics. Aust. J. Ecol. 13: 375–391.

Briese, D.T. & Macauley, B.J. 1981. Food collection within an ant community in semi-arid Australia, with particular reference to seed harvesters. Aust. J. Ecol. 6: 1–19.

Brown, J.H. 1981. Two decades of homage to Santa Rosalia: Toward a general theory of diversity. Amer. Zool. 21: 877–888.

Brown, J.H., Grover, J.J., Davidson, D.W. & Lieberman, G.A. 1975. A preliminary study of seed predation in desert and montane habitats. Ecol. 56: 987–992.

Brown, J.H., Reichman, O.J. & Davidson, D.W. 1979. Granivory in desert ecosystems. Ann. Rev. Ecol. System. 10: 201–227.

Buckley, R.C. 1982. Ant-Plant interactions in Australia. Dr W. Junk, The Hague. 162 pp.

Catling, P.C. & Newsome, A.E. 1981. Responses of the Australian vertebrate fauna to fire: an evolutionary approach. In: Gill, A.M., Groves, R.H. & Noble, I.R. (eds) Fire and the Australian Biota. pp. 273–310. Australian Academy of Science, Canberra.

Cody, M.L. (ed.) 1985. Habitat Selection in Birds. Academic Press, Orlando, Florida. 558 pp.

Collins, N.M. 1981. The role of termites in the decomposition of wood and leaf litter in the southern Guinea savanna of Nigeria. Oecologia 51: 389–399.

Collins, N.M. & Wood, T.G. 1984. Termites and atmospheric gas production. Science 224: 84–85.

Crawford, C.S. 1981. Biology of Desert Invertebrates. Springer, Berlin, 314 pp.

Crawley, M.J. 1983. Herbivory: The Dynamics of Animal-Plant Interactions. Blackwell Scientific Publications, Oxford. 437 pp.

Davidson, D.W. & Morton, S.R. 1984. Dispersal adaptation of some *Acacia* species in the Australian arid zone. Ecol. 65: 1038–1051.

Davison, E.A. 1982. Seed utilization by harvester ants. In: Buckley, R.C. (ed.) Ant-Plant Interactions in Australia. pp. 1–6. Dr W. Junk, The Hague.

de Bruyn, L.A.L. & Conacher, A.J. 1990. The role of termites and ants in soil modification: a review. Aust. J. Soil Res. 28: 55–93.

Duff, G.A. & Braithwaite, R.W. 1990. Fire and Top End forests – past, present and future research. In: Roberts, B.R. (ed.) Fire Research in Rural Queensland. pp. 84–97. University of Southern Queensland, Toowoomba.

East, R. 1984. Rainfall, soil nutrient status and biomass of large African savanna mammals. Afric. J. Ecol. 22: 245–270.

Edwards, P.J. & Wratten, S.D. 1980. Ecology of Insect-Plant Interactions. Edward Arnold, London. 60 pp.

Faegri, K. & van der Pijl, L. 1979. Principles of Pollination Ecology. Pergamon, Oxford. 244 pp.

Fox, B.J. 1982. Fire and mammalian secondary succession in an Australian coastal heath. Ecol. 63: 1332–1341.

Fox, R.E. & Clark, N.B. 1974. The incidence of termites in eucalypts of the Darwin area. Aust. Forest. Res. 5: 29–36.

Freeland, W.J., Winter, J.W. & Raskin, S. 1988. Australian rock-mammals: a phenomenon of the seasonally dry tropics. Biotrop. 20: 70–79.

Friend, G.R., Dudzinski, M.L. & Cellier, K.M. 1988. *Rattus colletti* (Rodentia: Muridae) in the Australian wet-dry tropics: seasonal habitat preferences, population dynamics and the effects of buffalo, *Bubalus bubalis*. Aust. J. Ecol. 13: 51–66.

Frith, H.J. & Davies, S.J.J.F. 1961. Ecology of the Magpie Goose, *Anseranas semipalmata*, Latham (Anatidae). CSIRO Wild. Res. 3: 91–141.

Glover, P.E., Trump, E.C. & Wateridge, L.E.D. 1964. Termitaria and vegetation patterns on the Loita plains of Kenya. J. Ecol. 52: 365–377.

Harris, M.V. 1966. The role of termites in tropical forestry. Insectes Sociaux 13: 255–266.

Haynes, C.D. 1985. The pattern and ecology of *munwag*: traditional aboriginal fire regimes in north-central Arnhemland. Proceed. Ecol. Soc. Aust. 13: 203–214.

Holt, J.A. 1987. Carbon mineralization in semi-arid northeastern Australia. J. Trop. Ecol. 3: 255–263.

Holt, J.A. & Coventry, R.J. 1982. The effects of mound-building termites on some chemical properties of soils in northeastern Australia. In: Lee, K.E. (ed.) Proceedings of the 3rd Australasian Conference on Grassland Invertebrate Ecology. pp. 313–319. South Australian Government Printer, Adelaide.

Holt, J.A. & Coventry, R.J. 1990. Nutrient cycling in Australian savannas. J. Biogeog. 17: 427–432.

Holt, J.A., Coventry, R.J. & Sinclair, D.F. 1980. Some aspects of the biology and pedological significance of mound-building termites in a red and yellow earth landscape near Charters Towers, North Queensland. Aust. J. Soil Res. 18: 97–109.

Horton, D.R. 1982. The burning question: Aborigines, fire and Australian ecosystems. Mankind 13: 237–251.

Huxley, C.R. & Cutler, D.F. 1991. Ant-Plant Interactions. Oxford University Press, Oxford. 601 pp.

Inouye, R.S., Byers, G.S. & Brown, J.H. 1980. Effects of predation and competition on survivorship, fecundity,and community structure of desert annuals. Ecol. 61: 1344–1351.

Janzen, D.H. 1966. Coevolution of mutualism between ants and acacias in Central America. Evolution 20: 249–275.

Janzen, D.H. 1976. Why tropical trees have rotten cores. Biotrop. 8: 110.

Jones, J.A. 1990. Termites, soil fertility and carbon cycling in dry tropical Africa: a hypothesis. J. Trop. Ecol. 6: 291–305.

Jones, R. 1969. Fire-stick farming. Aust. Nat. Hist. 16: 224–228.

Josens, G. 1983. The soil fauna of tropical savannas. III. The termites. In: Bouliere, F. (ed.) Ecosystems of the World 13: Tropical Savannas, pp. 505–524. Elsevier, Amsterdam.

Lal, R. 1987. Tropical Ecology and Physical Edaphology. John Wiley & Sons, New York. 732 pp.

Lee, K.E. & Wood, T.G. 1971. Termites and Soils. Academic Press, London. 251 pp.

Mares, M.A. & Rosenzweig, M.L. 1978. Granivory in North and South American deserts: rodents, birds and ants. Ecol. 59: 235–241.

Miller, J.A. & Miller, T.A. (eds) 1986. Insect-Plant Interactions. Springer, New York. 342 pp.

Morton, S.R. 1979. Diversity of desert-dwelling mammals: a comparison of Australia and North America. J. Mammal. 60: 235–264.

Morton, S.R. 1985. Granivory in arid regions: comparison of Australia with North and South America. Ecol. 66: 1859–1866.

Morton, S.R. 1990. The impact of European settlement on the vertebrate animals of arid Australia. Proc. Ecol. Soc. Aust. 16: 201–213.

Morton, S.R. & Davidson, D.W. 1988. Comparative structure of harvester ant communities in arid Australia and North America. Ecolog. Monog. 58: 19–38.

Morton, S.R. & Davies, P.H. 1983. Food of the zebra finch (*Poephila guttata*), and an examination of granivory in birds of the Australian arid zone. Aust. J. Ecol. 8: 235–243.

Morton, S.R. & James, C.D. 1988. The diversity and abundance of lizards in arid Australia: a new hypothesis. Amer. Natural. 132: 237–256.

Mott, J.J., Williams, J. Andrew, M.H. & Gillison, A.N. 1985. Australian savanna ecosystems. In: Tothill, J.C. & Mott, J.J. (eds) Ecology and Management of the World's Savannas. pp. 56–82. Australian Academy of Science, Canberra.

New, T.R. 1989. Associations Between Insects and Plants. New South Wales University Press, Kensington NSW. 113 pp.

Noske, R.A. 1990. Seed dispersal of black wattles (*Acacia auriculiformis*) by birds. North. Territory Natural. 12: 24–25.

O'Dowd, D.J. & Gill, A.M. 1984. Predator satiation and site alteration: mass reproduction of alpine ash (*Eucalyptus delegatensis*) following fire in southeastern Australia. Ecol. 65: 1052–1066.

O'Dowd, D.J. & Gill, A.M. 1986. Seed dispersal syndromes in Australian *Acacia*. In: Murray, D.R. (ed.) Seed Dispersal. pp. 87–121. Academic Press, Sydney.

Okello-Oloya, T., Spain, A.V. & John, R.D. 1985. Selected chemical characteristics of the mounds of two species of *Amitermes* (Isoptera, Termitinae) and their adjacent surface soils from northeastern Australia. Rev. Ecol. Biol. Soil 22: 291–311.

Orians, G. 1975. Diversity, stability and maturity in natural ecosystems. In: Van Dobben, W.H., Lowe-McConnell, W.H. & Lowe-McConnell, R.H. (eds) Unifying Concepts in Ecology. pp. 139–150. Dr. W. Junk, The Hague.

Partridge, L. 1978. Habitat selection. In: Krebs, J.R. & Davies, N.B. (eds) Behavioural Ecology: An Evolutionary Approach. pp. 351–376. Blackwell Scientific Publications, Oxford.

Peterson, H. & Luxton, M. 1982. A comparative analysis of soil fauna populations and their role in soil decomposition processes. Oikos 39: 287–388.

Redhead, T.D. 1979. On the demography of *Rattus sordidus colletti* in monsoonal Australia. Aust. J. Ecol. 4: 115–136.

Reichman, O.J. 1979. Desert granivore foraging and its impact on seed densities and distributions. Ecol. 60: 1085–1092.
Rice, B.L. & Westoby, M. 1981. Myrmecochory in sclerophyll vegetation of the West Head, NSW. Aust. J. Ecol.6: 291–298.
Ridpath, M.G. & Waithman, J. 1988. Controlling feral Asian water buffalo in Australia. Wildl. Soc. Bull. 16: 385–390.
Risch, S.J. & Carroll, C.R. 1986. Effects of seed predation by a tropical ant on competition among weeds. Ecol. 67: 1319–1327.
Rosenzweig, M.L. 1968. Net primary productivity of terrestrial communities: prediction from climatological data. Amer. Natural. 102: 67–74.
Sands, W.A. 1969. The association of termites and fungi. In: Krishna, K. & Weesner, F.M. (eds) Biology of Termites, Volume I. pp. 495–524. Academic Press, New York.
Schaefer, D.A. & Whitford, W.G. 1981. Nutrient cycling by the subterranean termites *Gnathamitermes tubiformans* in a Chihuahuan desert ecosystem. Oecologia 48: 277–283.
Scholes, R.J. 1990. The influence of soil fertility on the ecology of southern African dry savannas. J. Biogeog. 17: 415–419.
Seastedt, T.R. 1984. The role of microarthropods in decomposition and mineralisation processes. Ann. Rev. Entomol. 29: 25–46.
Spain, A.V., John, R.D. & Okello-Oloya, T. 1983. Some pedological effects of selected termite species at three locations in north-eastern Australia. In: Lebruhn, P., Andre, H.N., de Medts, A., Gregoire-Wibo, C. & Wauthy, G. (eds) New Trends in Soil Biology. pp. 143–149. Dieu-Brichart, Ottignes-Louvain-la-Neuve.
Spain, A.V. & McIvor, J.G. 1988. The nature of herbaceous vegetation associated with termitaria in north-eastern Australia. J. Ecol. 76: 181–191.
Spain, A.V. & Okello-Oloya, T. 1985. Variation in the growth of two tropical pasture plants on soils associated with the termitaria of *Amitermes laurensis* (Isoptera: Termitinae). Proceedings of the 4th Australasian Conference on Grassland Invertebrate Ecology. pp. 141–145. Caxton Press, Melbourne.
Stafford-Smith, D.M. & Morton, S.R. 1990. A framework for the ecology of arid Australia. J. Arid Environ. 18: 255–278.
Taylor, J.A. & Dunlop, C.R. 1985. Plant communities of the wet-dry tropics: the Alligator Rivers region, Northern Territory. Proceed. Ecolog. Soc. Aust. 13: 83–127.
Thompson, J.N. 1982. Interaction and Coevolution. Wiley-Interscience, New York. 179 pp.
Watkinson, A.R., Lonsdale, W.M. & Andrew, M.H. 1989. Modelling the population dynamics of an annual plant: *Sorghum intrans* in the wet-dry tropics. J. Ecol. 77: 162–181.
Wellington, A.B. & Noble, I.R. 1985. Seed dynamics and factors limiting recruitment of the malle *Eucalyptus incrassata* Labill, in semi-arid, south-eastern Australia. J. Ecol. 73: 657–666.
Whitford, W.G., Steinberger, Y. & Ettershank, G. 1982. Contributions of subterranean termites to the 'economy' of Chihuahuan desert ecosystems. Oecologia 55: 298–302.
Whittaker, R.H. 1975. Communities and Ecosystems (Second Edition). Macmillan, New York, 387 pp.
Wild, H. 1952. The vegetation of southern Rhodesian termitaria. Rhodesian Agric. J. 49: 280–292.
Williams, M.A.J. 1968. Termites and soil development near Brocks Creek, Northern Territory. Aust. J. Sci. 31: 153–154.
Woinarski, J.C.Z. 1990. Effects of fire on bird communities of tropical woodlands and open forests in northern Australia. Aust. J. Ecol. 15: 1–22.
Woinarski, J.C.Z. & Braithwaite, R.W. 1990. The terrestrial vertebrate fauna and vegetation of the Kakadu Conservation Zone. Report for Resource Assessment Commission, Canberra. 196 pp.
Wood, T.G. 1976. The role of termites (Isoptera) in decomposition processes. In: Anderson, J.M. & MacFayden, A. (eds) The Role of Terrestrial and Aquatic Organisms in Decomposition Processes. pp. 145–168. Blackwell Scientific Publications, Oxford.
Zimmerman, P.R., Greenberg, J.P., Wandiga, S.O. & Crutzen, P.J. 1982. Termites: a potentially large source of atmospheric methane, carbon dioxide, and molecular hydrogen. Science 218: 563–565.
Zimmerman, P.R. & Greenberg, J.P. 1983. Termites and methane. Nat. 302: 354–355.

CHAPTER 8

Impact of feral water buffalo

ANDREW J. SKEAT, T. JON EAST and LAURIE K. CORBETT

Abstract. Water buffalo were introduced into northern Australia from South-east Asia between 1826 and 1866. They became feral when early British settlements were abandoned. From these introductions they spread and occupied all major habitat types in the Northern Territory north of latitude 16 °S. There are no reliable historical estimates of population size. However, between the 1880s and 1956 buffalo were used intensively for hides and the large number of animals killed during that time almost certainly had a negative effect on population growth. Buffalo populations probably peaked around the middle of this century following the collapse of the hide industry and the ensuing reduction in harvesting. Since the 1980s buffalo numbers have declined dramatically as the result of an intensive effort to eradicate bovine tuberculosis of which feral buffalo and cattle are hosts. In Kakadu National Park buffalo populations declined from 5.6 to 1.2 animals km^{-2} between 1983 and 1988. Present population density is estimated at < 0.1 animal km^{-2}.

Buffalo have had major impacts on the biophysical environment of the Kakadu region. Overgrazing and trampling have resulted in a reduction of vegetation biomass, in some areas the complete removal of vegetation, and changes in species composition. Trampling, and the formation of pads and wallows have caused soil compaction. Vegetation loss and damage to the soil structure have contributed to accelerated soil erosion, particularly sheet erosion, rilling, piping and gullying. Buffalo induced erosion of levees flanking the freshwater reaches of rivers and those separating the tidal reaches from freshwater flood basins have resulted in changes to surface hydrology, including reduced retention of freshwater in flood basins and intrusion of salt water into freshwater swamps. The water quality of water bodies has suffered due to trampling, wallowing, grazing and the contamination by buffalo faeces and urine. The extent and intensity of impacts differed greatly between different landforms of the region. The uplands, which include the sandstone plateau of the Arnhem Land Escarpment were the least disturbed landform, while the most spectacular and intensive degradation of both land and vegetation was on the flood plains, particularly those in the lowland reaches of the South Alligator River. Flood plains and levees here are formed in silty and sandy alluvium which have areas of erodible duplex soils. The East Alligator and West Alligator River flood plains in contrast, are formed in sandy and clayey-sand soils and were much less severely degraded. The effects of buffalo on vegetation, soils and surface hydrology have also affected the native fauna. Changes in vegetation communities have caused magpie geese to change their nesting sites on the flood plains. Populations of dusky rats and other small vertebrates on the flood plains were also adversely affected by high densities of buffalo. The main reason probably related to the removal of vegetation cover which provided food and protection from predators.

While some studies point out that changes in the environment attributed to buffalo could have other reasons, such as fire, climate change or natural geomorphological processes, a large body of anecdotal evidence and photographic records seems to support research findings that buffalo at least contributed significantly to these changes.

1. Introduction

> 'It may not be out of place here, to call attention to the fact that these animals [buffalo] are increasing very rapidly, and if some preventative measures are not taken it is only a matter of time when they will so increase to become a serious evil.' (Carrington 1885)

Bubalus bubalis, the feral water buffalo (Fig. 1) has had a dominant place in the recent history of the Kakadu region. In a little over 150 years these animals increased from a few strays to vast herds (Fig. 2) and became the much loved and unofficial symbol of the northern part of the Northern Territory (Letts *et al.* 1979). They became an integral part of aboriginal diet and culture, and hunting for hides was a principal reason for European settlement in the region. The buffalo industry has made, at times, an important contribution to the Northern Territory economy. At the same time, overgrazing devastated vegetation communities and caused soil erosion, and buffalo became a reservoir for diseases which threaten Australia's livestock industry (Hein & Tomasovic 1981). When an eradication program for buffalo commenced in the 1980s it provoked a storm of protest

C.M. Finlayson and I. von Oertzen (eds), Landscape and Vegetation Ecology of the Kakadu Region, Northern Australia, 155-177.

Fig. 1. Feral buffalo, Kakadu National Park (photo A.N.C.A.).

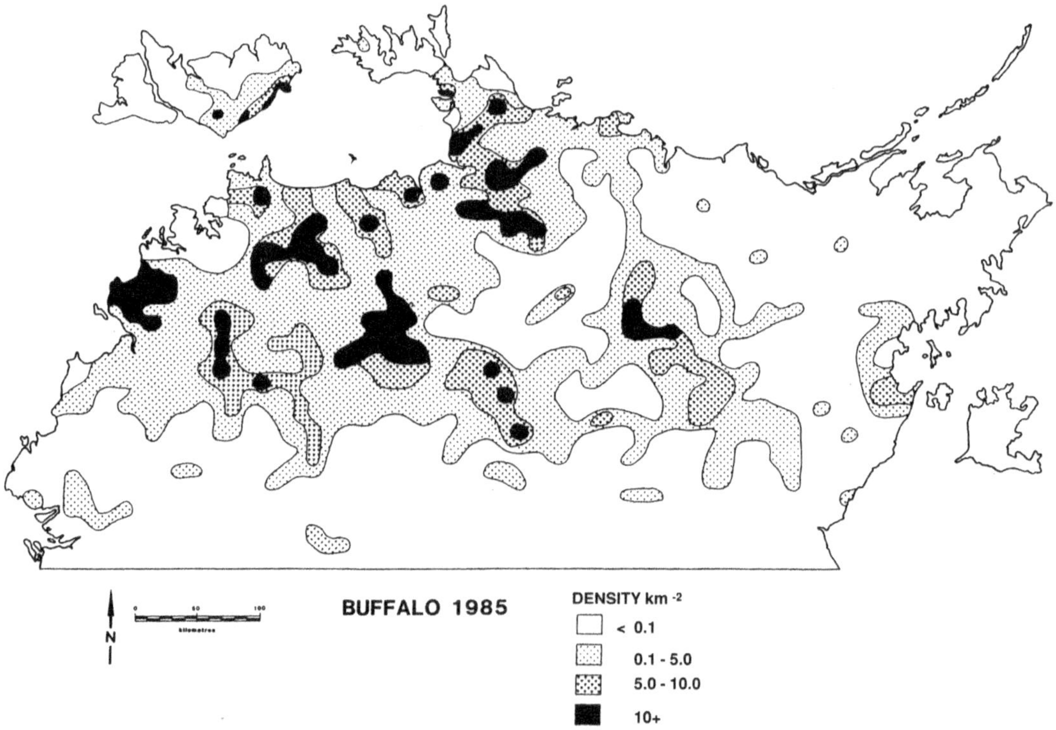

Fig. 2. Distribution of feral Asian buffalo, northern tropical Australia (from Bayliss & Yeomans 1989).

which has continued to this day. The ecological impact of the species is also still debated (Letts *et al.* 1979, Braithwaite *et al.* 1984).

Despite the interest in buffalo, their impact on the environment of northern Australia is not well documented. Aspects of the biology of buffalo and its impact on the environment are discussed in Tulloch (1970), Letts *et al.* (1979), Fogarty (1982), Braithwaite *et al.* (1984), Graetz (1989), East (1990) and Skeat (1990). In addition, the more dramatic effects on the landscape have been recorded photographically, and there is a large body of

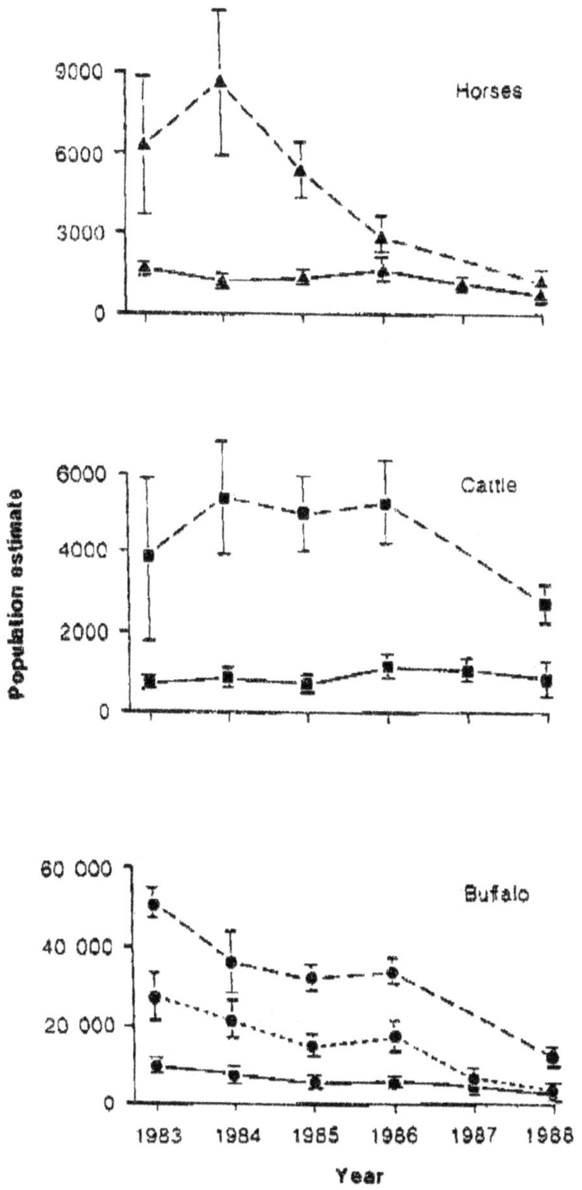

Fig. 3. Estimated size of feral buffalo, cattle and horse populations between 1983 and 1988 for Kakadu National Park Stages 1 (——), 2 (. . .) & 3 (- - -) (from Skeat 1990).

anecdotal evidence. This chapter reviews the history of the introduction and spread of buffalo in the region and examines the major impacts that the species has had on the vegetation, fauna and physical landscape.

The Kakadu region contains other feral animals, including feral cattle (*Bos taurus*), pigs (*Sus scrofa*) and horses (*Equus caballus*). Their numbers have generally been considerably lower than buffalo populations (Fig. 3) and their impact has been correspondingly smaller (Letts *et al.* 1979).

2. History of buffalo populations

Water buffalo were introduced into northern Australia from South-east Asia between 1826 and 1866 (Letts *et al.* 1979). When early settlements were abandoned, remaining animals rapidly became feral. Feral buffalo were first recorded in the Kakadu region by Ludwig Leichhardt, an early European explorer who, in 1845 noted the presence of animals on the eastern margin of the region, near the border of the present Kakadu National Park (Leichhardt 1847). Buffalo spread to occupy all major habitat types in the Northern Territory north of latitude 16° S. Population size in this area in 1985 was estimated to be more than 340 000 (Bayliss & Yeomans 1989). Buffalo numbers have since declined dramatically as a result of a control program commenced because feral buffalo and cattle are hosts of bovine tuberculosis (Hein & Tomasovic 1981). The aim of the program is to eradicate the disease by reducing infected populations of feral livestock to very low levels. This campaign is the largest animal disease eradication program undertaken in Australia, with an estimated final cost of A$705.8 million (Northern Territory Department of Primary Industry and Fisheries 1990).

Large buffalo populations probably did not become established in the Kakadu region until the late 1800s. John McKinley, an explorer who traversed the region in 1866, wrote:

> 'It is very astonishing that, although this is just the country for buffalo, and within miles of where Leichhardt [sic] saw first traces of them years ago, not a trace of one is visible.' (McKinley 1866)

Carrington, however, during an inspection for the Government in 1885 noted the following scene along the margins of the East Alligator River:

> 'Here large herds of buffalo numbering several hundreds were seen daily,...' (Carrington 1885).

Carrington also observed buffalo on the West Alligator River, but not on the South Alligator River (Carrington 1885). Similarly, another explorer and naturalist, Dahl in 1897, made no mention of them on the South Alligator River (Dahl 1897). However, by 1919 buffalo were well established throughout the region (Warburton 1944).

Following their establishment, it seems likely that buffalo population size was controlled to a large degree by hunting. Buffalo were utilised for hides between the 1880s and 1956, when the industry failed due to poor quality control and competition from synthetic fabrics. A meat industry has operated since 1959. In 1920, the Acting Administrator for the Territory wrote in his annual report in reference to buffalo:

> 'The above areas [range of buffalo at the time] have now been practically all shot out except on Melville Island and a small strip of coastal country on the mainland lying between the South Alligator and Adelaide Rivers.
>
> To indicate the havoc wrought by the buffalo shooters it is only necessary to record the accelerated rate of destruction. Over 10 000 hides were exported during the 25 years ended 20 December 1911.....' (quoted from Granger & Peisley 1945).

While the precise effects of harvesting on population size are not clear, the large numbers of buffalo killed almost certainly slowed the growth of populations. In the absence of reliable historical estimates of population size, it is not possible to determine when populations peaked in the region, but it is speculated this did not occur until as recently as the middle of the twentieth century, following a reduction in harvesting when the hide industry collapsed. This view is supported by sequential aerial photography showing changes in the extent of buffalo induced damage. A relatively recent date for high densities of buffalo would also help explain the dramatic recovery of degraded areas of the landscape which has occurred following reductions in population size. If the densities and level of environmental damage observed in the 1970s (e.g. Letts *et al.* 1979) had been present over a long period, it is unlikely that the regenerative capacity of the vegetation communities would have been maintained to such a high degree.

Reliable estimates of population size in the region, or elsewhere in northern Australia, were not available until the 1980s. Aerial surveys of Kakadu National Park were conducted annually between 1983 and 1988 (Fig. 3). Over this period buffalo population density for the Park declined from 5.6–1.2 animals km^{-2} as a result of harvesting (Skeat 1991). Populations have now been reduced to the point where animals are rarely encountered; present densities are possibly less than 0.1 animals km^{-2}. The surveys did not coincide with peak populations, as intensive control in the region commenced with sporadic, but intense shooting on the South Alligator River flood plain in the mid 1970s and was followed by more systematic eradication in 1979. Such overall estimates of density do not reflect the potential impact of buffalo on the environment, as buffalo strongly favour some habitats. Ridpath *et al.* (1983), for example, estimated the density of buffalo in forest vegetation along the margins of flood plains to be 34 animals km^{-2}, compared with a mean density of approximately 15 animals km^{-2} for the entire study area of *Eucalyptus* spp. dominant woodland.

An understanding of buffalo population dynamics provides insight into the effects of buffalo on vegetation and the physical landscape. Skeat (1990), using aerial surveys, found that rates of increase in large buffalo populations were correlated with the amount of dry season rainfall in the year of survey. When dry season rainfall was low and surface water scarce, the rates of increase slowed and, in some years, were negative. Freeland & Boulton (1990), using data gathered from nine social groups in the 1960s (D.G. Tulloch unpublished data), determined a similar relationship between dry season rainfall and rate of population increase. The periodic rapid decline in buffalo populations following low dry season rainfall has probably helped to reduce the impact of buffalo on the landscape.

In summary, the history of buffalo populations in the Kakadu region can be described as follows. While buffalo have been present in northern Australia for more than 150 years, they have spread relatively slowly into the region and populations were probably controlled by hunting. Peak densities which led to environmental changes (described below) may have been a recent phenomenon. Commencing in the late 1970s populations have been reduced to very low levels.

3. Impacts on vegetation, soils and hydrology

Buffalo have had major impacts on the bio-physical environment of the Kakadu region (East 1990). The range of impacts and interactions with processes in the ecosystem are summarised in Fig. 4. Primary effects due to buffalo activity, i.e. grazing, trampling, and the formation of pads and wallows include:

i) a reduction in vegetation biomass, changes in species composition (including the dispersal of weeds) and, locally, complete removal of vegetation;
ii) compaction of soils and soil erosion;
iii) changes to surface hydrology, including reduced retention of fresh water in flood basins and intrusion of salt water into freshwater swamps; and
iv) increased turbidity in water bodies resulting from trampling, wallowing and grazing, as well as their contamination by buffalo faeces and urine;

Effects on one part of the ecosystem will often compound effects on others, e.g. loss of vegetation and soil compaction both contribute to accelerated soil erosion, while soil compaction and changes in surface hydrology will affect the composition of the vegetation. The modification of habitats, in turn, has an impact on native fauna.

3.1. General impacts

The extent and intensity of impacts differed greatly between landform types (uplands, lowlands, flood plains and flood basins) of the region (East 1990). Each of these landforms is characterised by a distinct combination of soils, topography, drainage and vegetation that all influence the extent and intensity of impacts. East (1990) analysed the published information on the impacts in the different landform types.

Uplands include the sandstone plateau, outliers and escarpment of Arnhem Land in the east and south of the region, and isolated hills and ridges of resistant rock on the lowlands. Because of the largely inaccessible nature of this steep terrain, relatively little is known of the impacts of buffalo on the soils, flora and fauna of upland areas. It is likely, however, that the impact of buffalo on hill slopes and the plateau surface has been minimal because of the low buffalo populations which are associated with the steep, rocky and dissected terrain, the sparseness of vegetation and the scarcity of permanent water. Moreover, the potential for accelerated erosion on the plateau surface is low, because of stable low-gradient slopes with

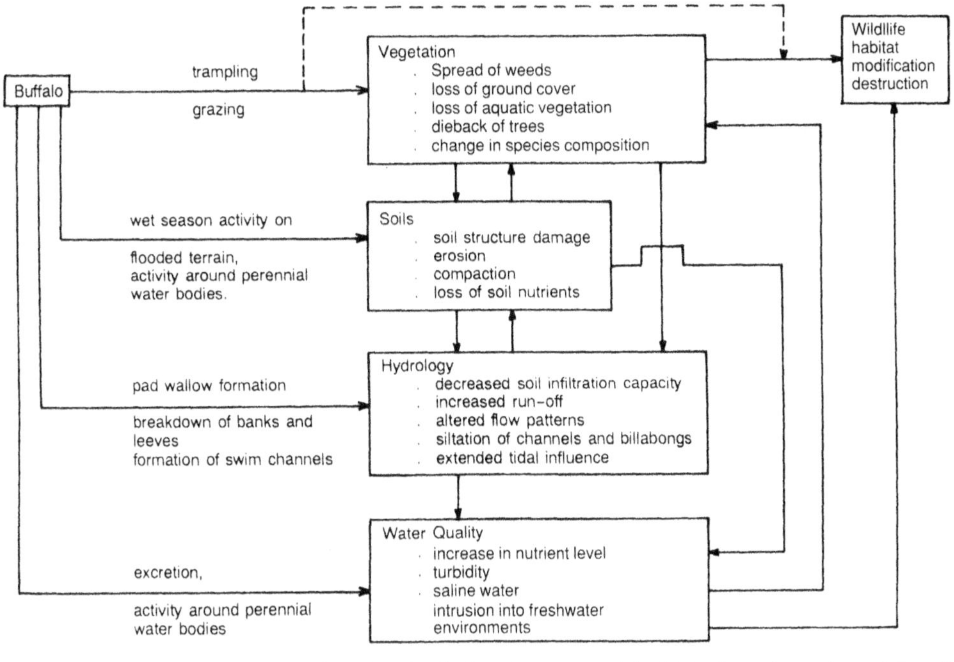

Fig. 4. Effects of buffalo grazing on processes of the biophysical environment of the Kakadu region (adapted from Fogarty 1982).

generally shallow soils, and large areas of bare rock.

Buffalo have caused localised erosion and destruction of vegetation in gorges and valleys in the plateau margins and surface. Heavy grazing and trampling of vegetation and soils have occurred in some of the wetter gorges and valleys, resulting in a deterioration in water quality. A survey of monsoon vine forests in the uplands by Russell-Smith (1984) showed that, of the five habitat types examined, only the 'topographically exposed perennial spring/streamline habitat' was significantly affected by feral animals. The soils in these areas were waterlogged, organic loams and sands, which were subject to erosion and depletion of nutrients as a consequence of buffalo activity. Destruction of the vegetation permitted the penetration of fire.

The hills and strike ridges on the lowlands surface have steep (up to 40°) rocky slopes with concave profiles, which have a high degree of erosional stability (Cull & East 1987, East 1990). Buffalo disturbance on these landforms is minor, and has been confined to the formation of pads aligned across-slope on the lower parts of hill slopes.

The lowlands are the most extensive of the landform types; they cover about 60% of the area of the region, forming the broad expanse of gentle (1–4%) slopes between the plateau and the flood plains of the rivers. The lowland vegetation is dominated by open woodland, which has a *Eucalyptus* spp. overstorey and an annual grass (*Sorghum* spp.) ground layer, and by small areas of denser riparian monsoon forest (Story 1976). Buffalo ranged throughout the lowlands, but were concentrated in monsoon forest along drainage lines and water holes, and on the margins of flood plains and flood basins. Buffalo disturbance included wallows and well-defined pads near waterholes, and the destruction of riparian vegetation which has facilitated the spread of weedy species. In places, buffalo formed wallows sequentially as the wet season waters receded. The result was a succession of wallows in a natural drainage depression linking to initiate a gully (Applegate *et al.* 1986). During the wet season, gully systems expand by gully head growth and the development of tributary gullies.

Flood plains of silty and sandy alluvium occur in the middle and upper reaches of the major regional rivers, and are subject to intermittent over bank flooding during the wet season. The vegetation is dominated by annual grasses, such as *Sorghum intrans*, and scattered trees including *Melaleuca* spp. and *Cathormion umbellatum*. The flood plains currently exhibit some spectacular and intensive feral animal land and vegetation degradation. This is because these habitats have been the last to have buffalo populations reduced to low densities (see Fig. 3, Stage 3), the flood plain soils are typically highly erodible, and have erosional features that do not rehabilitate quickly in these soil types (East 1990). The flood plains in the valley and lowland reaches of the South Alligator River between Coronation Hill and the Cooinda flood basin are characterised by the most severe erosion (Graetz 1989, East 1990). Flood plains and levees here are formed in silty and sandy alluvium which have areas of erodible duplex soils. The sandy and clayey-sand soils of the East Alligator and West Alligator River flood plains are much less severely degraded.

The impact of buffalo is more dramatic and widespread in the seasonally flooded black clay plains and the associated perennial billabongs that occur in the lowland freshwater reaches of the rivers than elsewhere in the region (Graetz 1989, East 1990). Major flood basin-billabong systems that have experienced intensive degradation include the large, approximately 100 km^2 flood basin in the vicinity of Cooinda in the South Alligator River catchment, and the 200 km^2 basin in the catchment of Magela Creek downstream of the settlement of Mudginberri. In addition, tidal channels have extended into low-lying freshwater backwater areas of the estuarine flood plain (Fogarty 1982). There are a number of scientific studies as well as a considerable body of historical record and anecdotal evidence which document the effects of buffalo on the wetlands (e.g. Stocker 1971, Letts *et al.* 1979, Fogarty 1982, O'Neill & Matthews 1982, Friend & Taylor 1984, Graetz 1989).

Until the recent large scale removal of buffalo in the late 1980s, many flood basins and billabongs were characterised by intensive pugging of surface sediments, the formation of swim channels, highly turbid water and the almost total destruction of the vegetation cover (Fig. 5). Ex-

Fig. 5. Buffalo degradation of wetland vegetation and soils, Shovel Billabong, South Alligator River catchment, 1990 (photo T.J. East).

tension of tidal channels into freshwater flood basins and billabongs resulted in many freshwater systems changing to saline conditions, and the deposition of tidal sediments. It is perhaps ironic that, since the recent large reduction in buffalo numbers, these most degraded parts of the landscape have recovered to the extent that the effect of buffalo is no longer obvious to the casual observer.

3.2. Impacts on the vegetation

Whereas the role of feral animals in initiating erosion is well understood and documented (see below), the extent and nature of the impacts on flora and fauna are less well understood. Cowie & Werner (1987), in an extensive survey of weeds in Kakadu National Park, concluded that buffalo disturbance of the natural vegetation facilitated the spread of many of the 87 species of exotic plants found. On the flood plains, *Hyptis suaveolens* and *Sida* spp. were found in dense and widespread 1.5 m high thickets in areas disturbed by buffalo. Weedy species tend to persist in the vicinity of billabongs and on flood plains following the removal of buffalo (Cowie & Werner 1987).

Several studies have shown that buffalo have had a significant impact on the floristics and structure of the lowland woodland communities. Braithwaite *et al.* (1984) in a study of 30 monsoon forest sites in Kakadu National Park, measured the effect of buffalo on faunal abundance and vegetation. They concluded that buffalo had changed some monsoon forest sites extensively. Vegetation density in the (less than 3 m high) understorey was negatively related to buffalo density during the dry season, but not in the wet season when an increase in annual weedy species such as *Cassia spp.* and *Hyptis suaveolens* replaced the depleted perennial species. Buffalo significantly increased the foliage height diversity (a measure of the number of vegetation strata weighted for abundance of foliage) in both dry and wet seasons. This was attributed to the death of large trees, as a result of buffalo trampling and changed site hydrology, which opened the tree canopy and enabled accelerated growth of mid-level perennial and ground level annual species.

Table I. Understorey plant species whose percentage cover has changed significantly since buffalo removal. Species are grouped by growth-form and exotics are flagged (*) (from Minchin & Dunlop, unpublished results)

(a) Species with higher cover on sites protected from buffalo.

Growth-form	Species
Perennial shrubs and young trees	*Buchanania obovata* *Petalostigma quadriloculare* *Livistona humilis* *Terminalia ferdinandiana*
Perennial grasses	*Alloteropsis semialata* *Heteropogon contortus* *Imperata cylindrica* *Chrysopogon latifolius* *Heteropogon triticeus*
Perennial herb with annual aerial parts	*Eragrostis cumingii* *Thaumastochloa* spp.
Perennial vines with annual aerial parts	*Cayratia trifolia* *Galactia* sp.

(b) Species with higher cover on buffalo exposed sites.

Growth-form	Species
Annual shrubs and small-shrubs	*Hyptis suaveolens* * *Tephrosia remotiflora* *Sida cordifolia* * *Triumfetta rhomboidea* *
Annual grasses	*Eragrostis cumingii* *Thaumastochloa* spp.
Annual herbs	*Borreria spp.* *Desmodium muelleri* *Desmodium* sp. 3 *Ploygala* sp. 3 *Stylosanthes humilis* * *Calogyne holtzeana* *Desmodium* sp. 2 *Desmodium trichostachyum* *Poranthera coerulea*

P.R. Minchin & C.R. Dunlop (unpublished results) at the CSIRO Kapalga research station located between the South Alligator and West Alligator Rivers, compared vegetation cover on either side of a buffalo proof fence; the buffalo were excluded from one side of the fence as far as was possible. (For a more detailed description of the Kapalga studies see 4.1.) Forty-five pairs of plots were chosen along the fence, each plot having an equivalent on the other side, and percentage cover estimates were made for all vascular plant species in the understorey. Vegetation changes following buffalo removal were investigated by analysis of variance of the cover of each species. The exclusion of buffalo was shown to have a significant effect (at the 0.05 level) on the percentage cover for 27 plant species, with some species increasing in cover and others decreasing. In the plots from which buffalo had been excluded, all species that increased in cover were perennials, and all those that decreased were annuals (Table I). For a number of annuals and perennials, the change in cover represented a substantial proportion of total cover – the cover of the perennial species increasing from 4–20 %, and that of annual species declining from 26–10 %.

The increase in cover of perennial grasses was interpreted as the recovery of these species from buffalo grazing. A change in the balance between annuals and perennials has implications for the lowland ecosystems; a change toward annuals will increase the potential for soil erosion, increase the frequency and intensity of dry season fires, and affect the population dynamics of insects, birds and small mammals that utilise grass seed. In a similar study at the same site P.A. Werner & I.D. Cowie (personal communication) examined the effects of buffalo grazing on lowland *Eucalyptus* spp., and concluded that buffalo significantly suppressed the recruitment of trees into the canopy.

The flood basin vegetation was also extensively modified. Heavy grazing and trampling combined with the transition to saline conditions in some areas, resulted in the death of *Melaleuca* spp. and the removal of grasses. The grasses *Hymenachne acutigluma* and *Phragmites karka*, for example, appear to have once been dominant across large areas of the subcoastal wetlands of the Northern Territory. Leichhardt (1847) described a vista at Cannon Hill near the East Alligator River as comprising '...large lagoons, surrounded with mangrove myrtles (*Stravadium* sp.), with *Pandanus*, and a belt of reeds and Nelumbiums.' Lewis (1922) makes reference to very tall reeds, presumably *Phragmites* sp., while trying to cross the East Alligator in 1875. Warburton, while hunting buffalo near Cannon Hill in 1919, described the scene thus:

'As far as we could see in every direction, were those wonderful grassy plains dotted with waterholes; the quantity of game about convinced me that we were looking at some of the finest country in the world. Pushing back to camp, Whittaker rode in front, and so high was the beautiful green, water couch that at times I almost lost sight of him.' (Warburton 1944)

In October (dry season) 1979, the condition of the plains of the flood basin surrounding Cannon Hill was in stark contrast to these earlier accounts (A.J. Skeat personal observation). The plains were almost completely bare of vegetation and were pug marked by buffalo to the point that driving across them necessitated travelling at walking speed. A plume of black dust was thrown up by the vehicle. Billabongs at Cannon Hill were completely devoid of aquatic vegetation and fringing vegetation was reduced to mature trees, largely *Barringtonia acutangula*.

Aboriginal traditional owners in Kakadu National Park have related accounts of changes in vegetation and soils on the South Alligator River. D. Lindner (personal communication) and O'Neill & Matthews (1982) have recorded recollections of older Aboriginal residents catching geese in the Ginjala area using spears manufactured from *Phragmites karka* which they described at that time as being common. It is now virtually absent on the South Alligator River flood plain except for a few clumps, most of which were planted in the 1980s.

Story (1976), following a vegetation survey of the Kakadu region in 1972 and 1973, reported that, because of disturbance by buffalo it was not possible to present a comprehensive picture of vegetation communities on estuarine alluvium. He noted that the scarcity of grasses found on these soils was at variance with the dense growth and excellent grazing reported for coastal plains by Christian & Stewart (1953). Also, *Hymenachne acutigluma* could not be found at Cannon Hill at the time of publication despite the fact that it had been reported as being common in 1946.

3.3. Soil degradation

The destruction of vegetation, changed soil hydrological properties and the concentration of runoff by wallows and pads have resulted in greatly increased erosion rates (East 1990). The flood plains and levees of the South Alligator River, in particular, are characterised by localised and intensive rill and gully erosion which is

Fig. 6. Severe sheet and rill erosion, and incipient gullying on the flood plain of the South Alligator River (photo T.J. East).

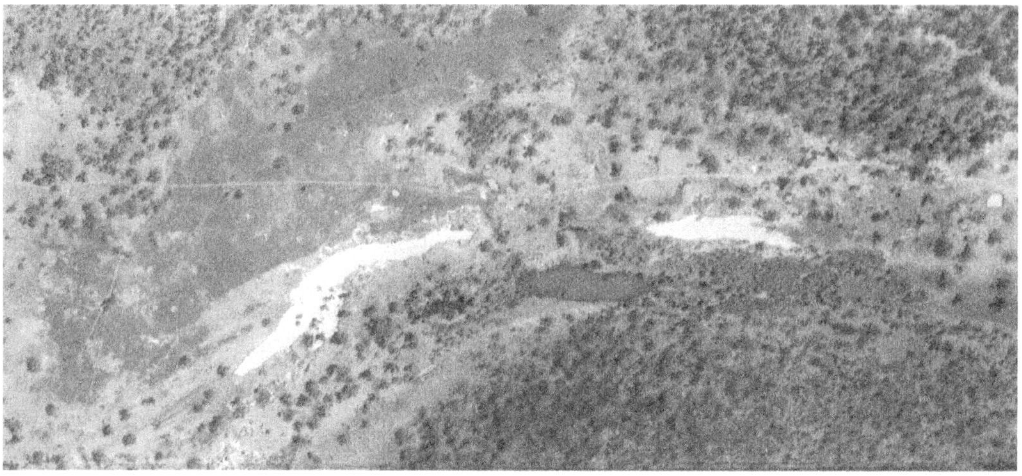

Fig. 7. Severe erosion and vegetation degradation associated with buffalo overgrazing, South Alligator River flood plain, July 1988; wallows and gullies are filled with water (photo AUSLIG).

associated with larger areas of sheet erosion and soil structure damage (Fig. 6). The most severely eroded areas are those closest to perennial or seasonal water supplies, such as riparian zones along flood plain margins, watercourses and billabongs. These are not only preferred buffalo habitats, but also tend to be characterised by erodible duplex and alluvial soils (Fig. 7). Erosional features are invariably associated with abundant evidence of a buffalo presence, including wallows, pads, pugged ground and damaged vegetation, providing compelling evidence of the causal link between the majority of degradation and buffalo.

Buffalo trampling along pads and other high-use areas has compacted the soil surface and damaged soil structure by pulverizing and reducing soil aggregate size. Such trampling causes compaction of soil surface horizons and changes soil hydrological properties. The resultant decreases in pore space and infiltration capacity give rise to increased runoff rates, increased erosion, and less soil water available for plant growth (Noble & Tongway 1986). Heavy grazing may reduce infiltration to about half that of the soil in an ungrazed condition and reduced recharge of groundwater may kill trees (Braithwaite *et al.* 1984). The fines produced through soil compaction have an increased susceptibility to removal by wind or water with ensuing loss of plant nutrients. Following rain, they form crusts which further reduces infiltration and increases runoff. The loamy-textured red and yellow earths, which occur throughout the lowlands of the region have a strong tendency to form surface seals or crusts.

Buffalo graze flood plains and other riparian areas all year round to a varying extent and, unlike other ungulates, tend to pull out the entire plant before biting off selected parts and discarding the remainder (Applegate *et al.* 1986). Ground denuded of vegetation by grazing and compacted by trampling is prone to sheet erosion and rilling by high intensity wet season rainfall (Fig. 6). The concentration of surface runoff in rills and along buffalo pads leads to gully erosion. Gullies were observed in alluvium of varying textures, including sandy, silty and clayey sand sediments. Duplex (texture contrast) soils formed in alluvium typically have a sandy and relatively coherent A horizon overlying a dispersive clayey B horizon. They are particularly prone to sheet and gully erosion as the buffalo pad or wallow disrupts the A horizon exposing the B horizon and leading to its eventual dispersion. This can lead to the formation of large dendritic gully systems with the soil eroded to a variable depth, exposing tree and grass roots, and forming soil pedestals (Fig. 8). Gullying was also associated with pipes, which were formed by the subsurface horizontal flow of water along lines of weakness in a dispersive B horizon. Gullies extend headwards following basal sapping and the collapse of pipes (Fig. 9), channelised flow in the gully subsequently removing the disturbed soil.

Fig. 8. Large buffalo-initiated gully system, South Alligator River flood plain; trees have locally stabilised eroding alluvium forming soil pedestals (photo T.J. East).

Fig. 9. Piping in dispersive B horizon of gullied duplex soil, flood plain of the South Alligator River (photo T.J. East).

Gully systems on flood plains are frequently of large and spectacular proportions. Flood plain gullies exceed 100 m in length, with an area of some hectares and an average depth of about 1.5 m. A typical gully in the flood plain of Fisher Creek occupied an area of over 2 ha with an average depth of about 1.5 m. It would have resulted in a loss of about 32 000 m^3 or 53 000 t of sediment to Fisher Creek. This quantity of sediment is comparable with an estimated annual sediment yield for the Fisher Creek catchment of 48 000 t (Dames and Moore 1989). Because of intensive erosion caused by buffalo in the Fisher Creek catchment, the annual sediment yield probably exceeds the pre-disturbance yield by a factor of 12 or more (East 1990). Some gullies remain filled with water at the end of the wet season, forming turbid perennial artificial billabongs (Fig. 9). In contrast to natural billabongs, they are largely devoid of aquatic flora.

3.4. Impact on catchment hydrology and water quality

3.4.1. Erosion rates, solutes and sediment loads

Catchments in the region exhibit varying degrees of buffalo accelerated erosion, depending on the size of buffalo populations and the extent of soil and vegetation degradation. Selected regional catchments are described here as either minimally, moderately or severely disturbed by buffalo, depending on the severity and extent of erosion as determined by field observation (see Table II).

Minimally disturbed catchments tend to have higher proportions of plateau and uplands which limit buffalo numbers and accelerate erosion. Erosion rates for these catchments are comparable with erosion rates for catchments in the seasonally wet tropics elsewhere in Australia and overseas (Table II).

Moderately and severely disturbed catchments have higher proportions of flood plains and other riparian areas, and hence, support larger buffalo populations (c.f. Graetz 1989). Erosion rates in the moderately disturbed catchments are greater than rates in morphologically comparable and minimally disturbed catchments by a factor of six, and for severely disturbed catchments by a factor of 35 or even higher (Table II). For example, the erosion rate for the South Alligator River catchment below its confluence with Fisher Creek (52.5 t km^2) was higher than that for the similarly plateau dominated upper Magela Creek catchment by about a factor of 12. Disturbed catch-

Fig. 10. Gully filled with wet season runoff forming a turbid perennial billabong; aquatic vegetation is absent (photo T.J. East).

Table II. Suspended sediment concentrations and erosion rates for East and South Alligator River catchments disturbed by feral animals.

Catchment characteristics	Mean suspended sediment conc. (mg L^{-1})	Erosion (t km^{-2} yr^{-1})	Upland areas (%)
Minimally disturbed			
East Alligator River			
Kawudjulah (7J) Creek	22[1]	15	40
Georgetown Creek 2	82	9	18
Magela Creek plateau catchment	13[2]	4.2	85
South Alligator River			
Koongarra Creek catchment	49[1]	38	24
Arnhem Land plateau		5[4]	100
Moderately disturbed			
East Alligator River			
Gulungul Creek	79	41	28
South Alligator River			
SAR above Gimbat	81[3]		91
SAR above Fisher Creek	42	> 25.2	90
SAR at Coronation Hill	25	91	
Koolpin Creek	24	> 13.5	93
Severely disturbed			
South Alligator River			
Fisher Creek	270[3]	>149.5	78
SAR below Fisher Creek	221.0	52.5	85
North Queensland catchments		5–15	
Amazon River catchments		4–83	

[1] Duggan (1988); K. Duggan (personal communication)
[2] Hart *et al.* (1986a)
[3] Summarised from Dames and Moore (1989)
[4] Roberts (1991)
[5] Douglas (1967)
[6] Gibbs (1967)

ments with high proportions of plateau may still have high erosion rates if the remaining area is dominated by flood plain rather than lowlands.

Accelerated erosion caused by feral animals has resulted in corresponding increases in solute and sediment loads and concentrations in some catchments (Table II & Fig. 11). The impact of feral animals is most pronounced in the South Alligator River catchment, whereas the East Alligator River catchment is presently relatively undisturbed (East 1990). Solute and suspended sediment concentrations in undisturbed catchments are low by world standards for the seasonally wet tropics (East 1990). Suspended sediment concentrations for severely disturbed catchments are substantially higher than those for moderately or minimally disturbed catchments (Table II). Fisher Creek (area 324 km^2), the catchment of which is severely degraded by buffalo, has a mean suspended sediment concentration (270 mg L^{-1}) about 11 times that (24 mg L^{-1}) of Koolpin Creek and about 20 times that (13 mg L^{-1}) of the upper Magela Creek, catchments which have comparable areas and proportions of sandstone plateau (Table II). The differences are attributed to differences in buffalo numbers and to the area of erodible texture contrast soils, both of which are greatest in the Fisher Creek catchment (East 1990).

The seasonal variation in sediment concentrations in catchments disturbed by feral animals

Fig. 11. Jim Jim Creek (clear water) flowing into the turbid waters of the South Alligator River, February 1988 (photo P. Wellings).

differs to that in the minimally disturbed catchments. Sediment concentrations in disturbed South Alligator River catchments are highest in the early wet season, because of the abundant supply of soil on ground that is disturbed and depleted of vegetation at the end of the dry season (Dames and Moore 1989). In undisturbed catchments, highest sediment concentrations and yields are associated with the greatest discharges, which generally occur late in the wet season (Hart *et al.* 1986b, 1987a,b, Duggan 1988).

The naturally high temporal variability of suspended sediment concentrations means that comparisons between catchments should be treated with caution. Mean suspended sediment concentrations are only broadly indicative of catchment condition, as additional factors such as catchment size, landforms and geology also affect sediment processes. However, the consistently large differences between disturbed and relatively undisturbed catchments do allow a measure of the extent of degradation caused by feral animals.

Mean annual solute concentrations for moderately disturbed catchments are generally higher than concentrations for minimally disturbed catchments, although increases are not of the same magnitude as for suspended sediment loads (Table III). The low value (6.5 mg L^{-1}) recorded for the undisturbed upper Magela Creek catchment reflects the high proportion (85%) of plateau sandstone in this catchment. Undisturbed lowland catchments have higher solute levels.

The South Alligator River catchments (dominated by plateau and upland) are more lithologically diverse than the Magela Creek catchment (dominated by Kombolgie Formation sandstone), and it is likely that the pre-disturbance South Alligator River solute levels lie between the two (plateau and lowlands) East Alligator River values. Solute concentrations in the South Alligator River, therefore, have increased by about as much as a factor of two as a result of feral animal induced erosion. Increased solutes are a result of the greater availability of solutes in eroding soils.

3.4.2. Channel morphology and saltwater intrusion

Buffalo in gaining access to water have formed numerous incised tracks on the sandy and silty levees that flank the freshwater reaches of rivers and their major tributaries. The tracks have been subsequently widened and deepened by runoff from the adjoining flood plain. The uncon-

Table III. Solute concentrations for East Alligator and South Alligator River catchments disturbed by feral animals. The solute concentrations are corrected for contribution from rainwater solutes, 11 mg L^{-1} (Noller 1983).

Catchment characterics	Solute concentration (mg L^{-1})
Minimally disturbed	
East Alligator River	
Kawudjulah (7J) Creek	28[1]
Georgetown Creek 2	24
Magela Creek (plateau upstream of (GS821009)	7[3]
South Alligator River	
Koongarra Creek	24[1]
Moderately disturbed	
East Alligator River	
Gulungul Creek	19[1]
Georgetown Creek 1	28
South Alligator River near Coronation Hill	57[2]
South Alligator River confluence with Camp Creek	54[4]

Sources:
[1] Recalculated from Duggan (1988)
[2] Recalculated from Dames and Moore (1988)
[3] Hart *et al.* (1986b)
[4] B. Gardiner (unpublished data)

solidated sands of the levees have facilitated the collapse of track sides, and the gullying and enlargement of the pad. Gullied tracks are typically 2–3 m deep, 5–8 m wide and may extend some tens of metres into the flood plain (Fig. 12). Where levees were not present, buffalo gained access to the channel by breaching the channel bank.

Levee banks separating the saline tidal reaches of rivers from the freshwater flood basins or swamps are normally vegetated by salt tolerant grasses such as the couch *Sporobolus virginicus* which is grazed by buffalo at high population densities. When denuded of grass and subject to pugging and wallowing, levees have been breached by strong outflows of freshwater at the end of the wet season. Breaching of the levees has had important consequences for the ecosystem:

i) Rapid headwards extension of the tidal channels has resulted in the incursion of salt water into flood basins killing vegetation species such as *Melaleuca* spp. The death of large areas of *Melaleuca* forest can be a visually striking feature in the landscape.
ii) Related to i), channels through breached levees allowed the 'premature' draining of freshwater flood basins at the end of the wet season.

Fig. 12. Channel bank of South Alligator River breached by buffalo pad and subsequently enlarged by gully erosion (photo T.J. East).

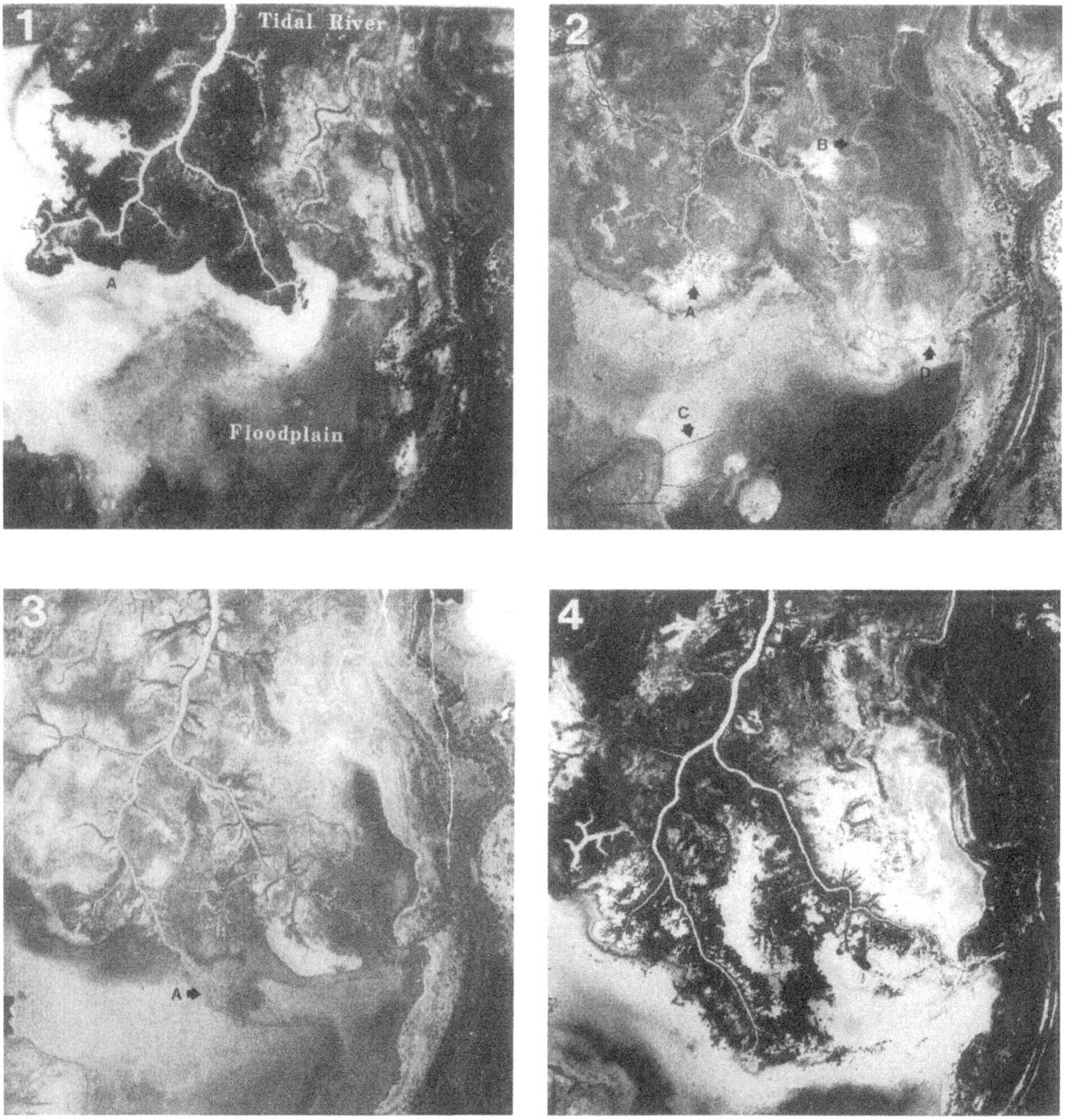

Fig. 13. Changes in tidal channels resulting from high buffalo populations (photos AUSLIG):
1: 1950 – Effects of buffalo on landscape are not readily apparent; tidal channels generally confined by vegetated levies, for example, point A.
2: 1964 – Landscape is being destabilised by buffalo activity; pads (A), wallows (B) and swim channels (C) indicate high buffalo density; channel heads are increasing in length, for example, A and D.
3: 1975 – Substantial landscape change is evident; tidal channels have greatly extended in length, for example, A; freshwater areas are subject to tidal inundation and premature draining at end of wet season.
4: 1985 – Recovery following buffalo removal has commenced; buffalo induced features are barely evident, there has been little further increase in channel length, channel heads are revegetating and retention of freshwater on flood plains is improving.

While less visually apparent than salt water intrusion, the biological effect of premature draining was probably greater. For example, fresh water retention in flood basins is essential for growth of a range of grass and sedge genera such as *Oryza* and *Eleocharis*, which in turn, are essential for the survival of magpie geese.

iii) Headward extension of tidal channels into billabongs has resulted in infilling of billabongs with tidal sediments. Where this has occurred, it has been a matter of concern for Aboriginal residents as the billabongs were a major food source, and to other residents and tourists as they had provided excellent fishing. Examples include Palms Billabong and Trap Billabong in the South Alligator River flood basin which have filled with saline estuarine sediments since the 1960s; both were previously favoured destinations for fishing parties.

These changes are illustrated in sequential aerial photographs of the tidal and freshwater reaches of the South Alligator River flood basin downstream of Cooinda (Fig. 13, interpretation after D. Lindner, personal comm.). In 1950, tidal channels were generally confined by vegetated levees which showed little evidence of impact from high buffalo densities. By 1964, pads, wallows and swim channels indicated increased buffalo numbers and impact; tidal channels at this time were bare of vegetation and increasing in length. The 1975 photograph shows great extensions in channel length and large areas of flood basin subject to premature drainage. By 1985, buffalo numbers had been greatly reduced and there was evidence of stabilisation and limited recovery, including only a minor further increase in channel length and the reforming and revegetation of levees.

Many stream channels, particularly in the freshwater reach of the South Alligator River are characterised by localised instability over the past two to three decades. Evidence of instability includes incision and lateral erosion and widening of the sand channel forming vertical banks and benches in more resistant strata, with trees and bamboo thickets stabilising localised sections of bank. Changes in channel geometry (width, depth, plan form) can be indicative of a change in catchment hydrology or in sediment supply and type (Leopold *et al.* 1964, Schumm 1977). In the South Alligator River catchment, the recent changes in channel morphology (East 1990) may reflect an increase in runoff rates due to either increased rates of runoff from sheet-eroded and compacted soils, or the enhanced drainage of flood plains due to gullied buffalo tracks on levees.

4. Impacts on fauna

4.1. Kapalga studies

The effects of buffalo on animal species in the Kakadu region have been little studied. Georges & Kennett (1988) investigated the effects of buffalo on the pig nosed turtle (*Carettochelys insculpta*) in pools and billabongs in the South Alligator River drainage system. Although the species was locally abundant, buffalo had a profound adverse effect on populations, directly through destruction of nests by trampling, and indirectly through habitat destruction. They concluded that, in 1988 recruitment to the population over large stretches of flood plain habitat was likely to be negligible as a result of buffalo activity. Research at the Kapalga Research Station during the 1980s concentrated on the effect of buffalo on flood plain and flood basin animal and plant communities. One study, conducted over nine years (1980–88), investigated the effect of buffalo on magpie goose (*Anseranas semipalmata*) breeding habitat and nesting success, and on populations of small flood plain vertebrates, especially the dusky rat (*Rattus colletti*). The results of these studies are summarised here.

At the commencement of the study in 1980 , buffalo numbers at Kapalga were high, with at least 9000 animals distributed across the 614 km^2 area, at a mean density of 15 km^{-2}. The greatest densities, of around 34 animals km^{-2} were around swamps in the interface zone between the flood plains and basins and the upland forests and woodlands (Ridpath *et al.* 1983). In 1982, a buffalo-proof fence was erected which divided Kapalga into halves. The buffalo population in the southern half (the study 'treatment' area) was reduced to less than 0.1% of its former size (Ridpath *et al.* 1983, Ridpath & Waithman 1988). Since then, helicopter and ground shooting has maintained populations at virtually zero. Buffalo numbers in the northern section were reduced by about 30% in 1982, by a further 30% in 1985 and by about 3000 head in 1987. At the end of the study in 1988, the remaining 1000 buffalo were mustered so that all of Kapalga became virtually free of buffalo.

Magpie geese were studied because they are a declining species. Prior to 1900 their breeding range included the entire Australian coastline (Frith & Davies 1961). It is presently restricted to the tropical north of Australia, mostly on the flood plains and basins of the coastal rivers of the Northern Territory. Surveys indicate that populations are still declining (Frith & Davies 1961, Tulloch & McKean 1983, Whitehead *et al.* 1987). They breed almost exclusively on the flood plains. Most nests are built in moderately dense stands of spike rush and wild rice (*Eleocharis* spp./*Oryza meridionalis (rufipogon)*) up to 1220 mm in height growing in water 330–900 mm deep (Frith & Davies 1961, Tulloch *et al.* 1988). The mean clutch is 7.3 eggs with hatching occurring after 24–25 days; goslings immediately leave the nest and are taken to feeding areas by parents. The primary food of downy goslings at Kapalga is wild rice seed. After goslings become independent towards the end of the dry season, geese also feed along rivers, lagoons and other permanent and semi-permanent waters.

Dusky rats also utilise the flood plains almost exclusively. During the dry season many find refugia in the deep cracks of the dry flood plains or in burrows along the margins of moist swampy areas. At the peak of rat population cycles ('plagues'), rats move into marginal habitats, and up creek lines to seepage zones and soaks within forest and woodland habitats. Many of the other small vertebrates (mammals, reptiles, amphibians) that were sampled in this study live largely in the wet marginal habitats and other moist areas, except for bandicoots which are widespread throughout the forest and woodland habitat.

4.2. Buffalo impact on magpie goose breeding habitat

The major vegetation types within a magpie goose breeding area on the South Alligator River flood plain at Kapalga were mapped from aerial photographs. The mapped area (1015 ha) encompassed both control and treatment areas. Six vegetation communities were identified: i) spike-rush *Eleocharis sphacelata;* ii) wild rice *Oryza meridionalis;* iii) associations of wild rice and other spike-rush species (*Eleocharis dulcis, Eleocharis brassii, Eleocharis spiralis*); iv) *Fimbristylis* spp.; v) *Hymenachne acutigluma*; and vi) other vegetation (*Sesbania* spp., *Germania* sp., *Melaleuca* sp. and other unidentified species). Each photographic frame was subdivided into 30 "quadrats" equivalent to a ground area of approximately 400 m^2 and the proportions of the six major vegetation types recorded.

Following the removal of buffalo from the treatment area in 1982, there were statistically significant increases in the overall abundances of *Eleocharis sphacelata* and *Hymenachne acutigluma* (Table IV). There were no significant changes in vegetation types for the control area between the

Table IV. The mean abundance of major vegetation types (mean % quadrats ± standard deviation) and comparisons of abundance in the control and treatment areas before and after buffalo were removed from the Kapalga Research Station in 1982.

Vegetation type	Control area[1]			Treatment area[2]		
	1980–82	1983–88	Difference	1980–82	1983–88	Difference
Eleocharis spp./*Oryza*	47.3 ± 9.4	40.7 ± 9.8	NS[3]	40.4 ± 20.3	41.5 ± 19.0	NS
Oryza meridionalis	19.6 ± 14.5	21.5 ± 10.5	NS	32.8 ± 18.8	22.3 ± 15.3	NS
Eleocharis sphacelata	23.5 ± 4.0	28.3 ± 6.6	NS	4.5 ± 0.7	13.1 ± 11.1	*
Fimbristylis spp.	4.0 ± 3.2	2.3 ± 2.0	NS	21.4 ± 17.3	11.6 ± 5.2	NS
Bare ground/open water	4.7 ± 2.8	6.1 ± 4.8	NS	1.0 ± 0.2	2.6 ± 2.9	NS
Hymenachne acutigluma	0	0.2 ± 0.3	*[4]	0	6.0 ± 4.8	*
Other vegetation	0.7 ± 1.0	0.9 ± 0.5	NS	0.03 ± 0.06	3.0 ± 3.1	*

[1] Buffalo-inhabited 1980–88
[2] Buffalo-free 1983–88
[3] NS = not significant, $P > 0.05$
[4] * = $P < 0.05$

two study phases, except for an increase in *Hymenachne acutigluma* which was first recorded in 1984, two years after buffalo numbers were reduced. This suggested *Hymenachne acutigluma* was suppressed only by very high densities of buffalo.

In 1984, when *Hymenachne acutigluma* was first recorded in both control and treatment areas, it was distributed throughout the flood plains, but in subsequent years it persisted and increased in abundance only in the deeper water areas along the edge of the flood plains. In doing so, it appeared to have displaced the deep water spikerush (*Eleocharis sphacelata*) which had, in turn, displaced the *Eleocharis* spp./*Oryza meridionalis* associations in that part of the flood plains. Those associations also increased in the same region, by replacing pure stands of *Oryza meridionalis*. The latter changes may be independent of buffalo effects since Tulloch *et al.* (1988) found that the *Eleocharis* spp./*Oryza meridionalis* association replaced areas of pure *Oryza meridionalis* in other regions of the flood plains depending on the annual variation in the quantity of rainfall.

4.3. Buffalo impact on magpie goose nesting

The number of magpie goose nests were counted in each of the 30 quadrats of the aerial photographic frames (described above). There was considerable annual variation in the numbers of nests in both the treatment and control areas (6–791 and 1–1211 respectively). However, trends in annual nest numbers over the 9 year study period were similar for both the control and treatment areas (Spearman's rank correlation coefficient $r_s = 0.73$, $P < 0.5$), and there were no statistically significant differences in nest density (nests per quadrat) between the two areas over the period of study. Comparisons of numbers of nests before and after buffalo removal (Mann-Whitney U Test, $P < 0.05$) were not significantly different, indicating that buffalo did not affect the numbers of goose nests.

However, buffalo did appear to affect the distribution of nests within the overall breeding area. Following the removal of buffalo there was a gradual shift over the next six years in the density of nests from being greatest in the middle of the flood plain to being greatest at the edge of the flood plain, adjacent to the marginal interface vegetation zone. This change in nest density was especially pronounced in the treatment area where geese nested in *Hymenachne acutigluma*, as well as in the preferred sedges. The change in nesting patterns appears to be related to buffalo grazing and trampling effects. Buffalo tend to rest in the marginal areas of flood plains and to graze adjacent areas of the flood plain most heavily (Tulloch 1970). The density of connecting trails between buffalo resting sites and feeding areas and wallows is also greatest in those areas of flood plain (Tulloch & Litchfield 1981). Grazing pressure and trampling effects declined when buffalo numbers were reduced. Consequently, the increased height of vegetation, particularly that of the preferred nesting vegetation *Eleocharis*/*Oryza meridionalis* association (Frith & Davies 1961, Tulloch *et al.* 1988), triggered more geese to nest in those areas than previously.

The effect of the annual rainfall pattern (distribution and amount) on goose breeding behaviour was also investigated and shown to be the major factor influencing the annual number of goose nests. Most nests (> 50% of the maximum number previously counted) were built when greater than 250 mm of rain fell each month within a continuous 3 month period between December and March of each year (Corbett 1988).

The overall conclusion is that buffalo adversely affected some vegetation communities, but did not affect the number of magpie goose nests built, only their location within the flood plains. However, it should be borne in mind that this study examined the effects of buffalo on nesting geese during the wet season when the flood plains are full of water and high concentrations of geese and buffalo are essentially physically separated. At this time, most buffalo movements were along defined swim channels. There may have been adverse buffalo effects on goose recruitment, and thus fluctuations in population numbers, at other times of the wet-dry cycle, especially when buffalo and geese were concentrated together in dry season refugia. Amongst other factors, Frith & Davies (1961) and Tulloch & McKean (1983) attributed the overall decline in goose populations between 1950–72 to alteration of dry season swamp refugia. At these sites, buffalo trample and compact the soil so that geese have difficulty in digging for their staple diet (*Eleocharis* corms) at that time of the year.

4.4. *Buffalo impact on dusky rats*

Dusky rats *(Rattus colletti)* were live-trapped in grids of 100 Elliot traps over 3 nights in margin and flood plain habitats in both control and treatment areas. Indices of abundance were recorded as numbers per 100 trap-nights. Populations of dusky rats fluctuated greatly over the period of the study (Table V). Trends in population peaks and troughs were similar for both the control and treatment areas (Spearman's rank correlation coefficient $r_s = 0.89$, $P < 0.001$), supporting Redhead's (1979) findings that population cycles were primarily determined by rainfall events.

In the period 1983–86, following removal of buffalo in the treatment area, numbers of rats increased significantly; mean indices of abundance for treatment and control areas were 13.4 and 7.9 respectively (Table V). Rats in the treatment area continued to breed in the transitional periods between the wet and dry seasons (April-June) ($P < 0.05$, Table V) and relatively more survived through the dry season (July-September). The increased numbers of rats resulted from the increased food supply and the protection from predators afforded by the re-establishment of a vegetative cover in the wet season, and its persistence into the dry season: the vegetation cover would have been formerly depleted by buffalo.

When buffalo numbers were reduced in the control area in 1987–88 there was a significant increase in rat numbers (mean indices of abundance: 7.9 vs. 22.4 and 13.4 vs. 31.8 for the periods 1983–86 vs 1987–88 respectively, Mann-Whitney Test, $P < 0.05$). There were no significant differences in rat numbers between the control and

Table V. Indices of abundance of dusky rats (mean no./100 trap-nights ± sd) and other small vertebrates (mean no./survey ± sd) in the control and treatment areas before and after buffalo were removed from the Kapalga Research Station in 1982.

Surveys	No. of Surveys	Dusky Rats			Other small vertebrates[1]		
		Control	Treatment	Difference[5]	Control	Treatment	Difference
1981–82[2]							
January-March[3]	2	0.5 ± 0.6	–[4]	–	0	–	–
April-June	5	0.2 ± 0.4	–	–	1.8 ± 1.3	–	–
July-September	4	3.2 ± 5.0	0	–	0.3 ± 0.5	0	–
October-December	3	7.7 ± 6.7	0	–	0	0	–
TOTAL 1981–82	14	2.7 ± 4.7	0	–	0.7 ± 1.1	0	–
1983–86							
January-March	6	16.8 ± 14.7	25.5 ± 15.2	NS	2.0 ± 1.8	3.0 ± 2.5	NS
April-June	5	0.8 ± 0.7	10.2 ± 8.9	*	0.8 ± 1.1	4.0 ± 7.2	**
July-September	6	0.7 ± 0.7	4.0 ± 7.2	NS	0.2 ± 0.4	2.2 ± 2.2	*
October-December	5	13.3 ± 17.9	13.5 ± 11.6	NS	0.6 ± 0.9	4.4 ± 4.2	*
TOTAL 1983–86	22	7.9 ± 13.0	13.4 ± 13.3	**	0.9 ± 1.3	3.3 ± 2.9	***
1987–88							
January-March	2	19.6 ± 20.6	30.2 ± 32.8	–	0	3.5 ± 0.7	–
April-June	2	10.1 ± 14.3	15.5 ± 21.9	–	1.5 ± 0.7	1.5 ± 0.7	–
July-September	1	20.9	37.2	–	9.0	17.0	–
October-December	1	53.9	62.4	–	3.0	0	–
TOTAL 1987–88	6	22.4 ± 19.7	31.8 ± 24.7	NS	2.5 ± 3.4	4.5 ± 6.3	NS

[1] Mammals, reptiles and amphibians (see text).

[2] 1981–82: high density of buffalo in both areas; 1983–1986: high density in control area, no buffalo in treatment area; 1987–88: low density in control area, no buffalo in treatment area.

[3] January-March = wet season months, April-June = wet-dry transition, July-September = dry season, October-December = dry-wet transition.

[4] No data or insufficient data for statistical analysis.

[5] NS = not significant, * = P<0.05, ** = P<0.01, *** = P<0.001

treatment areas in 1987–88 (Table V) indicating that high densities of buffalo are required to influence adversely rat populations.

4.5. *Buffalo impact on other small vertebrates*

Other small vertebrates – northern brown bandicoot *(Isoodon macrourus)*, grassland melomys *(Melomys burtoni)*, common planigale *(Planigale maculata)*, slaty-grey snake *(Stegonotus cucullatus)*, keelback snake *(Styporhynchus mairii)*, spotted tree monitor *(Varanus scalaris)*, the skink *(Sphenomorphus douglasi)*, and frogs *(Cyclorana australis, C. dahli, Litoria nasuta)* were live trapped in the same trap grids as dusky rats. Indices of abundance were recorded as the number (of all small vertebrates) per survey.

Following the removal of buffalo in the treatment area there were significant increases in numbers of small vertebrates throughout the year, with the exception of the wet season (Table V). For the control area in the late study years, numbers increased, showing a similar trend ($P < 0.05$) to that observed for the dusky rats. Numbers in the treatment area also increased (mean index of abundance: 3.3 vs. 4.5 for the periods 1983–86 and 1987–88 respectively), but was not significant. That may have been due to the high number of dusky rats in October-December 1988 which occupied most of the traps and thereby excluded other species. In contrast to dusky rats, the population fluctuations between the control and treatment areas over the study period were not significantly correlated (Spearman's rank correlation coefficient $r_s = 0.27$, $P > 0.05$).

These data suggest that high densities of buffalo were a major factor in suppressing populations of small vertebrates on the flood plains. The main reason probably related (as it did with dusky rats) to the removal of a vegetative cover which provides food and protection from predators.

5. Conclusions

While the impact on buffalo of the environment of northern Australia is not well documented, the information available, corroborated by observations from residents of the Kakadu region and aerial photography, suggests that buffalo have had a substantial impact on the ecosytems of the region. However, changes attributed to disturbance by buffalo, particularly changes to catchment hydrology, may have had other causes as well. Woodroffe *et al.* (1986), in a detailed study of the geomorphological history of the South Alligator River, point out that it is particularly dynamic and all the processes attributable to buffalo have occurred in the past. However, on the basis of a large body of anecdotal and photographic evidence the case for buffalo having greatly accelerated these processes seems well established.

Factors other than feral animals, such as fire and climate change, may have promoted channel instability in the region. The catchment of the South Alligator River is burnt by ANCA in the early dry season on a more or less annual basis (early dry season) with the aim of replicating the traditional burning patterns of Aboriginal inhabitants (ANPWS 1980, 1986). Fires affect soil properties and plant growth, which may in turn affect soil hydrological properties and erosion (Walker *et al.* 1986). Fires have also been shown to increase the water repellence (hydrophobicity) of red earths, while there are numerous accounts of increased rates of soil erosion following wildfire in Australian woodlands (e.g. Blong *et al.* 1982). If the frequency and intensity of the present fire regime differs significantly from prior burning practices, then it is possible that changes in soil hydrology have resulted in stream channel instability.

Stratigraphic evidence from the Magela Creek backwater plain (flood basin) has revealed regional climatic fluctuations on time scales of centuries during the past two thousand years (Wasson 1991). The fluctuations changed the species composition of aquatic vegetation on the plain, but did not alter sedimentation or sediment delivery rates. Consequently, it is considered unlikely that climate change has been responsible for the changed hydrological and sediment conditions in the region. The most likely mechanisms are then, either changes in fire practices, or the impact of buffalo over the past 150 years, or, as is most likely, a combination of both.

6. References

Applegate, R. J., Burgess, J.W., Duggan, K. & Tatzenko, S.P. 1986. Erosion assessment report on Stage I of Kakadu National Park, Conservation Commission of the Northern Territory Technical Report No. 29. 113 pp.

ANPWS 1980. Kakadu National Park Plan of Management. AGPS, Canberra. 410 pp.

ANPWS 1986. Kakadu National Park Plan of Management. Canberra Publishing & Printing Co., Canberra. 186 pp.

Bayliss, P. & Yeomans, K.M. 1989. Distribution and abundance of feral livestock in the Top End of the Northern Territory (1985–86), and their relation to population control. Aust. Wildl. Res. 16: 651–676.

Blong, R.J., Riley, S.J. & Crozier, P.J. 1982. Sediment yields from runoff plots following bushfire near Narrabeen Lagoon, New South Wales. Search 13: 36–38.

Braithwaite, R.W., Dudzinski, M.L., Ridpath, M.G. & Parker, B.S. 1984. The impact of water buffalo on the monsoon forest ecosystem in Kakadu National Park. Aust. J. Ecol. 9: 309–322.

Burch, G.J. 1986. Land clearing and vegetation disturbance. In: Russell, J.S & Isbell, R.F. (eds) Australian Soils: The Human Impact. pp. 159–184. University of Queensland Press, Brisbane.

Carrington, Capt. F. 1885. Half yearly report on Northern Territory to December 31 1885. In: South Australia, Adelaide Government Printer, Adelaide. 27 pp.

Christian, C.S. & Stewart, G.A. 1953. General report on survey of the Katherine-Darwin region, 1946. CSIRO Land Research Series No. 1. 156 pp.

Corbett, L.K. 1988. Fluctuations in the yearly numbers of magpie goose (*Anseranas semipalmata*) nests at Kapalga in Kakadu National Park with emphasis on the effects of feral buffalo and rainfall. Report to Australian National Parks and Wildlife Service, Canberra. 47 pp.

Cowie, I.D. & Werner, P.A. 1987. Weeds in Kakadu National Park – a survey of alien plants. Report to the Australian National Parks and Wildlife Service, Canberra. 127 pp.

Cull, R.F. & East, T.J. 1987. Geomorphic factors in the rehabilitation of earthen mine structures (with reference to uranium mining in the Northern Territory). Australian Mining Industry Council Environmental Workshop – 1987 Papers, pp. 63–76.

Dahl, K. 1897. Biological notes on North Australian mammalia. The Zoologist (Forth Series) 1: 189–216.

Dames and Moore 1988. Coronation Hill Gold, Platinum and Palladium Project, Draft Environmental Impact Statement. Prepared for Coronation Hill Joint Venture, December 1988. unpaginated.

Dames and Moore 1989. Coronation Hill Gold, Platinum and Palladium Project. Prepared for Coronation Hill Joint Venture, July 1989. unpaginated.

Douglas, I. 1967. Man, vegetation and the sediment yields of rivers. Nat. 215: 925-928.

Duggan, K. 1988. Mining and erosion in the Alligator Rivers Region of Northern Australia. Unpublished Ph.D. thesis, School of Earth Sciences, Macquarie University. 317 pp.

East, T.J. 1990. Erosion and sediment processes in the Kakadu Conservation Zone, South Alligator River catchment, Northern Territory. Resource Assessment Commission Kakadu Conservation Zone Inquiry Consultancy Series, AGPS, Canberra. 97 pp.

Fogarty, P. 1982. A preliminary survey of environmental damage associated with the activity of feral buffalo. Technical report to the Feral Animals Committee, Conservation Commission of the Northern Territory, Darwin. 88 pp.

Freeland, W.J. & Boulton, W.J. 1990. Feral water buffalo (*Bubalus bubalis*) in the Northern Territory, Australia: population growth, minimum viable population size and the Brucellosis and Tuberculosis Eradication Campaign. Aust. Wildl. Res. 17: 411–420.

Friend, G.R. & Taylor, J.A. 1984. Ground-surface features attributable to feral buffalo (*Bubalus bubalis*) II. Their relationship to the abundance of small animal species. Aust. Wildl. Res. 11: 311–323.

Frith, H.J. & Davies, S.J.F. 1961. Ecology of the magpie goose. Wildl. Res. 6: 91–141.

Georges, A. & Kennett, R. 1988. Dry-season distribution and ecology of the warradjan (*Carettochelys insculpta* Ramsay) in Kakadu National Park. Unpublished report submitted to Australian National Parks & Wildlife Service, Canberra. 62 pp.

Gibbs, R.J. 1967. The geochemistry of the Amazon River system I. The factors that control the salinity and the composition and concentrations of the suspended solids. Geol. Soc. Amer. Bull. 78: 1203-1232.

Graetz, R.D. 1989. Analysis of land degradation in Stage III of Kakadu National Park, including the Conservation Zone: Phase 1. Report to the Australian National Parks and Wildlife Service, Canberra. 126 pp.

Granger, W. & Peisley, H.R. 1945. Water buffaloes in north Australia, preliminary survey into procurement for the Directorate of Research and Civil Affairs. Australian Military Forces. 36 pp.

Hart, B.T., Ottaway, E.M. & Noller, B.N. 1986a. Nutrient and trace metal fluxes in the Magela Creek system, northern Australia. Ecol. Model. 31: 249–265.

Hart, B.T., Ottaway, E.M., Beckett, R. & Nollér, B.N. 1986b. Materials transport in the Magela Creek system: 1982/83 wet season. Water Studies Centre Technical Report No. 27, Chisholm Institute of Technology, Melbourne. 114 pp.

Hart, B.T., Ottaway, E.M. & Noller, B.N. 1987a. Magela Creek system, Northern Australia I: 1982–83 Wet season water quality. Aust. J. Mar. Freshw. Res. 38: 261–268.

Hart, B.T., Ottaway, E.M. & Noller, B.N. 1987b. Magela Creek system, Northern Australia II: Material budget for the floodplain. Aust. J. Mar. Freshw. Res. 38: 861–876.

Hein, W.R. & Tomasovic, A.A. 1981. An abattoir survey of tuberculosis in feral buffalo. Aust. Vet. J. 57: 543–547.

Leichhardt, L. 1847. Journal of an Overland Expedition in Australia from Moreton Bay to Port Essington. T & W Boone, London. 544 pp.

Leopold, L.B., Wolman, M.G. & Miller, J.P. 1964. Fluvial Processes in Geomorphology. W.H. Freeman and Company, San Francisco. 522 pp.

Letts, G.A., Bassingthwaite, A. & de Vos, W.L. 1979. Feral animals in the Northern Territory, Report of the board of inquiry 1979. Northern Territory Government Printer, Darwin. 234 pp.

Lewis, J. 1922. Fought and Won. W.K. Thomas Co., Adelaide. 243 pp.

McKinlay, J. 1866. J McKinlays Northern Territory explorations. S A P P, 2 October 1866. 21 pp.

Noble, J.C. & Tongway, D.J. 1986. Pastoral settlement in arid and semi arid rangelands. In: Russell, J.S. & Isbell, R.F. (eds) Australian Soils: The Human Impact. pp. 217–242. University of Queensland Press, Brisbane.

Noller, B.N. 1983. The physio-chemical nature of rainwater in the Magela and Nourlangie Creek ecosystems. In: Environmental Protection in the Alligator Rivers Region, Supervising Scientist for the Alligator Rivers Region Scientific Workshop Papers, 22. (unpaginated)

Northern Territory Department of Primary Industry and Fisheries 1990. Strategic plan for BTEC in the Northern Territory 1990–1992. Department of Primary Industry and Fisheries, Darwin. 55 pp.

O'Neill, G.C. & Matthews, D.V. 1982. An investigation of recent geochronological change on sections of the South Alligator River floodplain Kakadu National Park. Report to Australian National Parks and Wildlife Service, Canberra. 34 pp.

Redhead, T.D. 1979. On the demography of *Rattus sordidus colletti* in monsoonal Australia. Aust. J. Ecol. 4: 115–136.

Ridpath, M.G., Begg, R.J., Dudzinski, M.L., Forbes, M.A. & Graham, A. 1983. Counting the same populations of large tropical mammals from the ground and from the air. Aust. Wildl. Res. 10: 487–498.

Ridpath, M.G. & Waithman, J. 1988. Controlling feral Asian water buffalo in Australia. Wildl. Soc. Bull. 16: 385–390.

Roberts, R.G. 1991. Sediment budgets and Quaternary history of the Magela Creek catchment, tropical northern Australia. Unpublished Ph.D. thesis, University of Wollongong. 569 pp.

Russell-Smith, J. 1984. The Status and Condition of Monsoon Vine-forests in the Kakadu Region: A Management Report. Unpublished report to Australian National Parks & Wildlife Service, Canberra. 37 pp.

Schumm, S.A. 1977. The Fluvial System. John Wiley & Sons, New York. 338 pp.

Skeat, A.J. 1990. Feral buffalo in Kakadu National Park: survey methods, population dynamics and control. Unpublished M.Sc. Thesis, University of Canberra. 152 pp.

Skeat, A.J. 1991. Control of feral water buffalo in Kakadu National Park. Working papers of the 9th Australian Vertebrate Pest Conference, South Australian Animal and Plant control Commission, Adelaide, April 1991. unpaginated.

Stocker, G.C. 1971. The effects of water buffalo on paperbark forests in the Northern Territory. Aust. For. Res. 5: 29–34.

Story, R. 1976. Vegetation of the Alligator Rivers Area. In: Story, R., Galloway, R.W., McAlpine, J.R., Aldrick, J.M. & Williams, M.A.J. (eds) Lands of the Alligator Rivers Area, Northern Territory. pp. 89–111. CSIRO Land Research Series No. 38. CSIRO, Melbourne.

Tulloch, D.G. 1970. Seasonal movements and distribution of the sexes in the water buffalo, *Bubalus bubalis*, in the Northern Territory. Aust. J. Zool. 18: 399–414.

Tulloch, D.G. & Litchfield, R.T. 1981. Wallows for buffalo. Aust. Wildl. Res. 8: 555–565.

Tulloch, D.G. & McKean, J.L. 1983. Magpie goose populations on the coastal plains of the Northern Territory (1958–80). Corella 7: 32–36.

Tulloch, D.G., Cellier, K.M. & Hertog, A.L. 1988. The distribution of the nests of the magpie goose (*Anseranas semipalmata* Latham) at Kapalga NT: a four-year study. Aust. Wildl. Res. 15: 211–221.

Walker, J., Raison, R.J. & Khanna, P.K. 1986. Fire. In: Russell, J.S. & Isbell, R.F. (eds) Australian Soils: The Human Impact. pp. 185–216. University of Queensland Press, Brisbane.

Warburton, C. 1944. Buffaloes. Angus & Robertson Limited, Sydney. pp. 254.

Wasson, R.J. (ed.) 1991. Modern sedimentation and late quaternary evolution of the Magela Creek Plain. Supervising Scientist for the Alligator Rivers Region Research Report No. 6. 322 pp.

Whitehead, P.J., Bayliss, P. & Yoemans, K. 1987. Aerial survey of magpie goose populations and nesting in the Top End of the Northern Territory wet season 1987. Conservation Commission of the Northern Territory Technical Report No. 35: 1–10.

Woodroffe, C.D., Chappell, J.M.A., Thom, B.G. & Wallensky, E. 1986. Geomorphological Dynamics and Evolution of the South Alligator River and Plains, Northern Territory. Australian National University, North Australia Research Unit Mangrove Monograph No 3. 190 pp.

CHAPTER 9

Fire ecology and management

ALAN N. ANDERSEN

Abstract. Fire is a highly frequent and ecologically important phenomenon in the savanna landscape of the Kakadu region. The fire history of the region can be divided into three periods: i) pre-Aboriginal (before about 50 000 years ago) where ignitions were presumably restricted to lightning strikes immediately before and during the early part of the wet season; ii) Aboriginal, when fires were traditionally lit throughout the dry season for a variety of reasons; and iii) contemporary (since about 50 years ago), when there has been a perceived increase in the incidence of high intensity wildfires late in the dry season. The dominant fire management goal currently prevailing is to reduce the frequency and extent of high intensity wild fires through extensive burning early during the dry season, when fires tend to be low in intensity, and patchy. The ecological effects of fire have been poorly documented, but there are two major areas of concern: the effects of burning practices on monsoon rain forests and other fire-sensitive vegetation; and the effects of different fire regimes on the biota of the savanna.

The establishment of extensive research and monitoring programs is an obvious requirement for the resolution of conflict over fire management practices. However, there is also a need for development and articulation of fire management policies. To this end, clear strategic objectives, to which are directed tactical goals, and clarification of the relevance of traditional Aboriginal burning to contemporary fire management are required, along with continual fine-tuning of tactics in the face of great uncertainty over the ecological effects of different burning practices.

1. Introduction

Fire is a dominant force in tropical savanna landscapes throughout the world (Coutinho 1982, Gillon 1983, Stott 1986). Savannas have experienced frequent fires throughout their evolutionary history; indeed, frequent fire has played an integral role in their development from other vegetation types over geological time, and many savannas have been recently derived through repeated burning by people (Bourlière & Hadley 1983, Hopkins 1983). Savannas and fire are intimately linked in a positive feedback loop; the structure of savanna vegetation is maintained by fire, and it promotes further fire.

Most savannas outside Australia are intensively used and greatly modified by people, who light most of the fires (Gillon 1983). The fires are lit for a variety of reasons, including management of livestock production, land clearing, protection of property, hunting, and nature conservation. In many places, seasonal burning is an unquestioned tradition (Gillon 1983). Natural fires, caused predominantly by lightning, are still common in some places, but are limited in extent by burning earlier in the dry season.

Australian savannas have not experienced the intensive use and modification that characterises their overseas counterparts, but most fires are still lit by people. This has probably been the case for thousands of years, as fire has traditionally been an integral part of Aboriginal life and, in many cases, still continues to be (Stevenson 1985). Aborigines used fire for many reasons, some of which had obvious utilitarian value associated with walking through the bush, hunting, protecting edible plant species, and signalling (Jones 1969, Nicholson 1981, Haynes 1991). However, there were also more obscure reasons, such as a desire to 'clean up' and 'look after' the country (Jones 1980, Haynes 1985).

The Aboriginal population of north-western Australia (the northern parts of Western Australia and the Northern Territory) has declined markedly since European settlement early this century, and is now concentrated in permanent towns and outstations (Keen 1980). However, many fires are still lit by Aboriginal people, particularly around their settlements (Haynes 1985, Head *et al.* 1992). In the Kakadu region most fires are lit by

C.M. Finlayson and I. von Oertzen (eds), Landscape and Vegetation Ecology of the Kakadu Region, Northern Australia, 179-195.

Europeans and can be classified into: i) those that are prescribed and lit by management agencies, usually early during the dry season; and ii) those that are uncontrolled ('wildfires'), lit accidentally or otherwise, later in the dry season.

The ecology and management of fire in the Kakadu region is an extremely contentious issue (see Andersen & Braithwaite 1992). Most visitors to the region arrive with a generally negative attitude towards wildfire, due to their experience with the infrequent, high intensity fires characteristic of other fire-prone biomes (Gill 1977). Given that visitation is heavily concentrated in the dry (i.e. burning) season, most people are welcomed by charred landscapes and palls of smoke. Fire management practices are, therefore, constantly challenged by the public. The controversy also extends to the scientific arena, where the ecological effects of fire are hotly debated (Duff & Braithwaite 1990, Lonsdale & Braithwaite 1991, 1992, Andersen & Braithwaite 1992, Bowman 1992a).

2. Fire behaviour

As in other savanna landscapes, fire frequency in the Kakadu region is high, and often annual, because suitable fuel for fire is available every year. The profuse growth of grass and other herbage that is produced during the wet season becomes tinder dry over the long and severe dry season that follows. The resultant grassfires vary enormously in intensity; depending on fuel load, fuel moisture, wind speed and other factors, but fall in the low to moderate range for wild fires (Gillon 1983). Fires lit early during the dry season (May-June), when ground-layer vegetation is still moist, tend to be low in intensity, patchy, and limited in extent (Fig. 1; Haynes 1985, Braithwaite 1987). As the season progresses, and ground-layer vegetation cures, fire intensity tends to increase. Late season fires often completely incinerate ground-layer vegetation, cause substantial leaf scorch in the canopy, and cover large areas (Fig. 1;

Fig. 1. During and immediately after low intensity (a,b) and high intensity (c,d) fires at Kapalga Research Station in Kakadu National Park (Photos B. McKaige).

Braithwaite & Estbergs 1985, Day 1985a, Haynes 1985). Temperatures can exceed 500 °C (Gillon 1983, Stott 1986), but the spectacular canopy conflagrations known from fire-prone forested biomes rarely, if ever, occur.

The above relationship between timing and intensity does not always hold. Results from experimental fires at Kapalga Research Station in Kakadu National Park show that fire intensity can vary markedly at any given time of the year, and that fires early during the dry season can be just as intense as those occurring later the same year (R.J. Williams unpublished data). This occurs when prevailing weather conditions (temperature, humidity, wind speed) over-ride the effects of vegetation moisture.

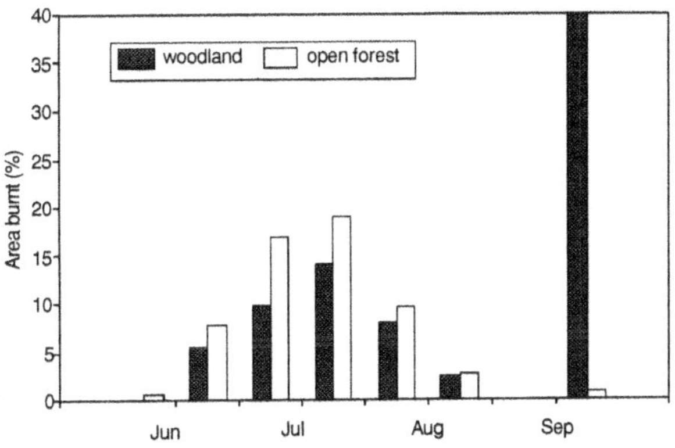

Fig. 2. Seasonal patterns of Aboriginal burning of savanna woodland and open forest near Maningrida, Arnhem Land, during 1976 (data from Haynes 1985).

3. Fire patterns

The fire history of the Kakadu region can be divided into three periods (Braithwaite & Estbergs 1985): i) pre-Aboriginal (before about 50,000 years ago); ii) Aboriginal; and iii) contemporary (since about 50 years ago). Even in pre-Aboriginal times, fire frequency was likely to be high because of the high incidence of lightning, but fire would have occurred almost exclusively during the build-up to, or early part of, the wet season. Thunderstorms occur, on average, one day in three from October to March, and during October and November lightning frequently occurs in the absence of substantial rain (Braithwaite & Estbergs 1985). Lightning-induced fires are still common (e.g. Bowman 1988), and were presumably far more extensive prior to dry season burning by people. However, the extent of such fires, and their frequency in different habitats, can only be speculated upon.

Despite the relatively recent displacement of Aboriginal people from north-western Australia, and the fact that many Aborigines still maintain close traditional ties with their land, traditional Aboriginal burning practices have been poorly documented. Indeed, there has been only one study in which such practices have been directly recorded; that of Haynes (1985) in the Maningrida area of Arnhem Land in the Northern Territory. Of the four major biomes in the region, flood plains and savanna woodlands were burnt more or less throughout the dry season, burning of open forests was restricted to the 'cool' months from June to August, and fire was actively excluded from monsoon rain forest. Haynes (1985) provided details of the spatial and temporal patterns of the burning of a 9000 ha block of savanna woodland and open forest during 1976. Burning commenced in June, peaked in July, and tailed off in August (Fig. 2). However, most of the remaining woodland was 'mistakenly' burnt by an extensive fire during September.

Important insights into traditional burning practices can be gained from the study of Haynes (1985), but it must be viewed in its proper context. As the study describes the situation for a small area in a single year, there is no indication as to what extent these findings are representative of the situation elsewhere in north-western Australia, nor during other years. Braithwaite (1991) has described seasonal patterns of Aboriginal burning throughout the Arnhem Land region by recording references to fire in the journals of nineteenth century European explorers. The results are consistent with those reported by Haynes (1985); burning peaked in July, tailed off in August, then showed another peak in October.

The available evidence, therefore, points to a clear temporal pattern of Aboriginal burning in the region: it occurred throughout the dry season, with a peak during the middle of the season (July), and another peak late in the season (September-October). There was also some burning during the early wet season (Haynes 1985, Braithwaite 1991). However, the areal extent of

Aboriginal burning, and regional variation in their burning practices, are virtually unknown.

Several studies have documented contemporary burning patterns in the Kakadu region, but even these are inadequately known. From an analysis of satellite imagery, Day (1985a) recorded burning patterns in Kakadu National Park for 1972 and 1980–82. The total area burnt varied markedly between landforms and years, ranging from 48–71% for lowland savanna woodlands and open forests, 38–48% for flood plains, and 8–68% for the sandstone country of the Arnhem Land escarpment and plateau (Table I). Seasonal burning patterns varied between landforms, with savannas and flood plains burnt throughout the dry season, and burning in the sandstone country occurring predominantly late in the dry season (Fig. 3; Day 1985a). They also varied within

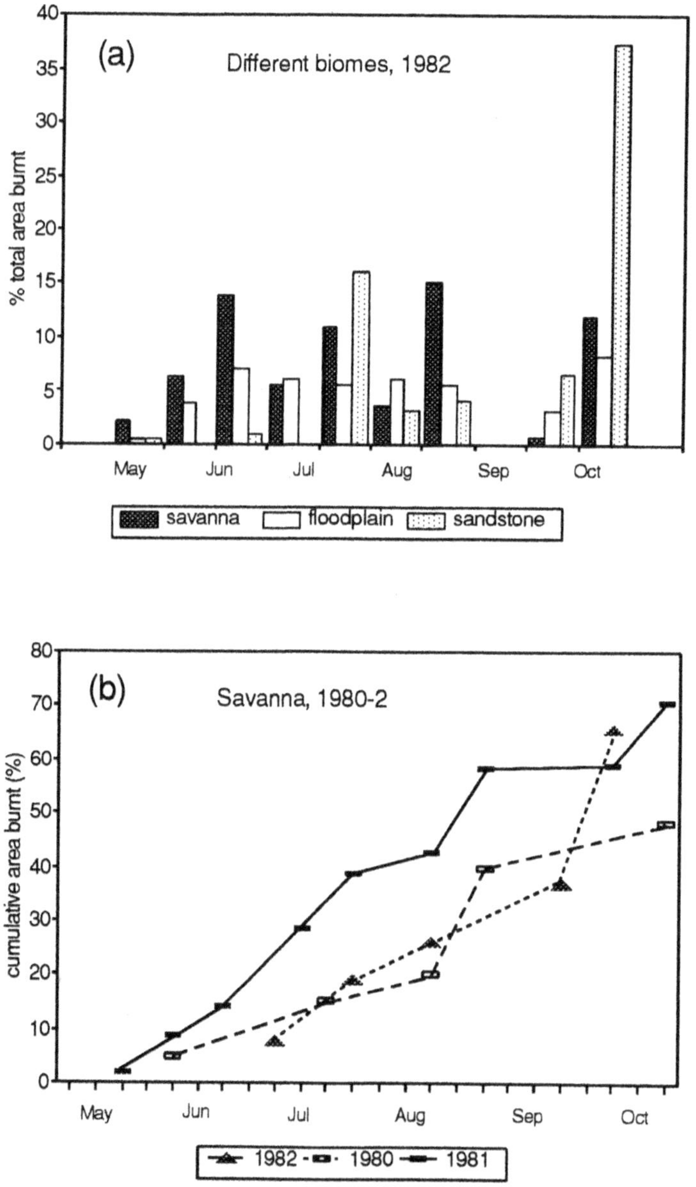

Fig. 3. Seasonal patterns of contemporary burning practices in Kakadu National Park, obtained from analysis of satellite imagery: a) 1982 data, comparing different biomes; b) lowland savanna during the years 1980–2. Data from Day (1985a).

Table 1. Percentage areas burnt in lowland savanna, flood plain, and sandstone escarpment and plateau regions of Kakadu National Park during the years 1972 and 1980–2 (data from Day 1985a)

Year	Savanna	Flood plain	Sandstone
1972	68	48	51
1980	66	38	8
1981	48	39	22
1982	71	46	68

landforms. For example, there tended to be more late season fires in savanna during years of less extensive early season burning (Fig. 3).

Braithwaite & Estbergs (1985) monitored fire patterns at 54 savanna woodland and open forest sites in Kakadu National Park during the years 1980–1982. In woodland, fires occurred between April and October, with the number peaking between June and August (Fig. 4). Fires tended to be later in open forest, occurring from May to November, and peaking in September (Fig. 4). Most woodland sites were burnt annually during the three year period, whereas the most common fire frequency in open forest was two out of three years (Braithwaite & Estbergs 1985). It should be noted, of course, that the number of fires occurring at any particular time of the year is not the same as areal extent, given that fires later in the season tend to burn greater areas.

Press (1988) used satellite imagery to compare fire patterns in Kakadu National Park with adjoining Aboriginal land to the east (Arnhem Land), and pastoral land to the west. Thirty lowland savanna sites were analysed in each area. Over 50% of all sites were burnt each year. Burning patterns in Kakadu National Park and Arnhem Land were similar to each other, but the pastoral lands tended to have more extensive early season burning, and consequently less extensive late season fires. In general, there was a strong inverse relationship between the extent of early and late burning in any one year.

There has been considerable speculation over the extent to which the contemporary fire regime differs from that prevailing prior to European settlement, with the common perception being that the frequency of extensive, late season fires has increased. This perception, however, is not supported by published evidence. If, as appears to be the case, Aboriginal burning in savanna woodland and open forest occurred throughout the dry season, with peaks during the middle and end of the season, then this seasonal pattern does not appear to be markedly different from that occurring today, especially in woodlands (Braithwaite & Estbergs 1985). Press (1988) found that fire patterns in savanna lands still under Aboriginal management were no different from those in Kakadu National Park (although the fact that the size and structure

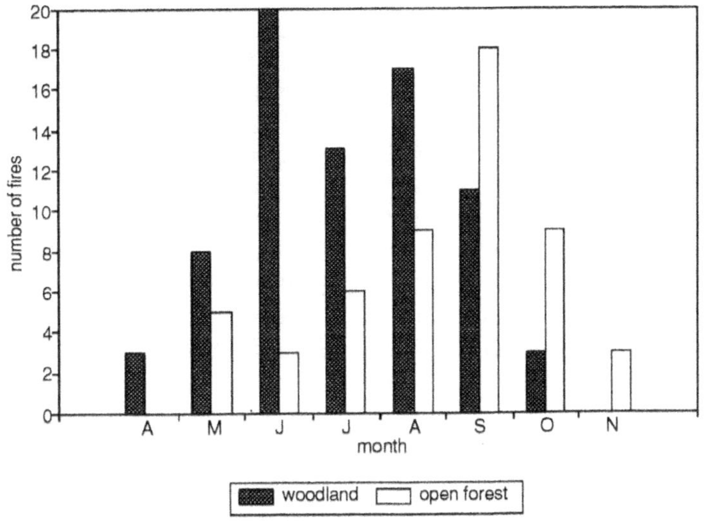

Fig. 4. Seasonal patterns of contemporary burning practices in savanna woodland and open forest in Kakadu National Park for the years 1980–2, obtained from direct observation of 54 plots (data from Braithwaite & Estbergs 1985).

of Aboriginal populations have changed markedly since European settlement must be noted).

No comparative information is available at all on Aboriginal versus contemporary burning practices on flood plains and in the sandstone country in the region. This is particularly unfortunate in the latter case, as most burning in the sandstone country in recent times has occurred late in the season (Fig. 3; Day 1985a), and, as outlined in this paper, it is this biome where concern over a perceived increase in the frequency of late season wildfires is greatest. A study of Aboriginal burning practices of flood plains in Kakadu National Park is currently underway (S. Roberts personal communication).

4. Ecological effects of fire

Fire has attracted considerable research attention in tropical savannas outside Australia, particularly in Africa, and its ecological effects have been reviewed extensively elsewhere (Coutinho 1982, Trollope 1982, Gillon 1983, Frost 1985). The extensive use of savannas for pastoralism is reflected in the concentration of research on the effects of fire on grass composition, biomass and productivity (West 1965, Gillon 1983, Pandey 1988, A Tchie & Gakahu 1989, Silva *et al.* 1991), or on underlying ecosystem processes such as nutrient cycling (Ram & Ramakrishnan 1992). This pastoral focus is also evident in Australia, where the best studied areas are the pastoral lands of Queensland and the Katherine region of the Northern Territory (Stocker & Mott 1981, Lacey *et al.* 1982, Gillison 1983, Mott *et al.* 1985).

Within the Kakadu region, the concern with fire is mainly directed towards its effects on nature conservation. Two themes have dominated research: i) boundaries between 'fire-resistant' eucalypt savannas and 'fire-sensitive' monsoon rain forests (Russell-Smith & Bowman 1991, Bowman 1992b); and ii) effects of fire on the savanna biota (Duff & Braithwaite 1990, Andersen & Braithwaite 1992).

4.1. Savanna-rain forest boundaries

The nature and dynamics of savanna-rain forest boundaries have long been of interest to tropical ecologists (Furley *et al.* 1992). Patches of monsoon rain forest occur throughout the vast expanses of savanna in north-western Australia. They are usually associated with permanent moisture or topographic positions affording protection from fire (Bowman *et al.* 1991, McKenzie *et al.* 1991, Russell-Smith 1991), but also occur within savanna vegetation without any discernible edaphic or moisture discontinuity (Stocker 1969, Bowman 1992b). Although the rain forest patches have received considerable attention because of their biogeographic significance (Kikawa *et al.* 1981, McKenzie *et al.* 1991, Menkhorst & Woinarski 1992, Gambold & Woinarski 1993, Woinarski 1993, Andersen & Reichel 1994), it is not clear to what extent they represent relictual fragments of a once extensive formation (Webb & Tracey 1981), or are more recent colonists of savanna vegetation (McKenzie *et al.* 1991).

Frequent fire enables eucalypt savanna to penetrate the boundaries of monsoon rain forest (Bowman & Dunlop 1986), and can eliminate monsoon rain forest entirely from otherwise suitable sites (Bowman & Wightman 1985). Conversely, in the absence of fire it has been suggested that monsoon rain forest will displace eucalypt savanna (Langkamp *et al.* 1981, Clayton-Greene & Beard 1985). However, although rain forest taxa can rapidly invade savanna on fertile soils elsewhere in the world (Rose-Innes 1972, Hopkins 1983, Kellman 1984), this appears to be a very slow process in monsoonal Australia. There was little evidence of incursions of rain forest taxa after 12 years of fire exclusion at Munmarlary in Kakadu National Park (Bowman *et al.* 1988a), and it was only slight at Solar Village near Darwin after 10 years (Fensham 1990). Large distances from source patches of rain forest possibly contributed to the slow rates in these two cases, but studies of a rain forest patch within a fire-protected savanna in Queensland indicate that expansion is slow even at rain forest boundaries (Bowman & Fensham 1991).

Fire, low soil fertility and seasonal drought obviously all play important roles in restricting the incursion of rain forest species into savanna, but there appear to be other contributing factors, such as grass competition, lack of suitable mycorrhizae, and possibly allelopathic inhibition (Stocker 1969, Bowman 1993, Bowman & Panton 1993a). The slow spread of rain forest taxa into savanna contrasts with the rather rapid recolonization following clearing of rain forest patches associated with permanent water (Panton 1993), and the sometimes rapid expansion of rain forest in the humid tropics of Queensland (Unwin *et al.* 1988).

The native cypress pine *Callitris intratropica* is a fire-sensitive species that has also attracted considerable research interest (Bowman & Panton 1993c). It differs from monsoon rain forest species in two important aspects (Bowman *et al.* 1988b): i) it is extremely drought tolerant; and ii) it expands far more rapidly into eucalypt savanna in the absence of fire. *Callitris* is typically restricted to sandstone escarpments offering protection from fire (Bowman & Wightman 1985, Bowman *et al.* 1990), but is also patchily distributed within eucalypt savanna, particularly on the least fertile (sandy) soils (Bowman *et al.* 1988b). The myrtaceous tree *Allosyncarpia ternata*, which forms monospecific stands on and around the Arnhem Land escarpment is another drought-tolerant (Fordyce 1992), but fire-sensitive species. At the bases of gorges it occurs as a floristic transition between wet monsoon forests and eucalypt savanna, with its savanna boundary apparently determined by fire (Bowman 1991a).

Persistence of neither monsoon rain forests, *Callitris* or *Allosyncarpia* requires total protection from fire. Indeed, *Allosyncarpia ternata* and most rain forest species have well-developed powers of vegetative recovery following fire (Bowman 1991a,b), and the fuel characteristics of stands of *Callitris intratropica* are such that fire intensity is generally low and therefore not fatal (Bowman & Wilson 1988). However, these vegetation types are sensitive to repeated burning, especially by high intensity fires late in the dry season (Bowman 1986, 1991a).

4.2. Savanna structure and dynamics

Resilience to fire is a striking feature of the savanna flora, with most woody plants having well-developed powers of vegetative recovery (Lacey 1974, Lacey & Whelan 1976, Lacey *et al.* 1982), and the dominant grasses regenerating vigorously, either vegetatively or by seed (Mott 1978, Lacey *et al.* 1981). It is not clear, however, whether these features evolved in response to fire or to seasonal drought (Dunlop & Webb 1991). Moreover, few if any savanna species have fire-dependent reproduction, in contrast to the situation in fire-prone vegetation types of southern Australia (Dunlop & Webb 1991). In southern tall open-forests, for example, eucalypts possess fire-promoting characteristics, such as decorticating bark and highly flammable leaves, and fire is required for the release of seeds from woody capsules (Ashton 1981). None of these characteristics occur in most savanna eucalypts. Indeed, serotiny (storage of seeds inside woody fruit) is a rare condition throughout the savanna flora.

There has been much speculation on the impact of fire on savanna vegetation based on contemporary floristic and structural patterns (Stocker 1966, Langkamp *et al.* 1981, Bowman & Dunlop 1986, Wilson & Bowman 1987, Bowman

Fig. 5. Effect of fire exclusion on vegetation structure at Munmarlary in Kakadu National Park: a) burnt annually; and b) unburnt for 14 years (Photos A. Andersen).

& Panton 1993b). Although it is generally accepted that fire is a modifier of broad patterns determined primarily by edaphic factors (Bowman & Minchin 1987), the extent of this modification is hotly debated. It is widely recognized that long-term fire-exclusion has a marked effect on the structure of savanna vegetation (Fig. 5; Stocker & Mott 1981, Bowman *et al.* 1988a) and associated faunal communities (Woinarski 1990, Andersen 1991), but this has limited relevance, given that fire frequency would be high even in the absence of humans (Bowman *et al.* 1988a, Fensham 1990). The controversy is over the effects of fires of different timing, intensity and frequency.

One view is that eucalypt savannas are so fire-adapted that it does not really matter how frequently, or how intensely, they are burnt (Bowman 1988). This was the conclusion from an analysis of a fire experiment at Munmarlary in Kakadu National Park, where it was claimed that neither fire frequency nor timing had a substantial effect on vegetation structure or floristics (Bowman *et al.* 1988a). This analysis has, however, been seriously questioned (Lonsdale & Braithwaite 1991, 1992). The opposing view is that the frequency, timing and intensity of fire are all potentially important factors influencing savanna structure and dynamics (Andersen & Braithwaite 1992).

4.2.1. Fire frequency

Fire frequency potentially has an important effect on the structure of savanna vegetation. A dominant structural feature of the savanna woodlands and open forests of the Kakadu region is the prominence of 'woody sprouts', representing the majority of woody species, in the ground layer (Fig. 5; Bowman 1986, Fensham & Bowman 1992, Bowman & Panton 1993b). They are mostly produced vegetatively (Lacey 1974), but can be derived directly from seedlings (Fensham 1992), and are maintained in a suppressed condition by frequent fire (Lacey & Whelan 1976).

There is a strong inverse relationship between shrub and litter cover on the one hand, and grass cover and the abundance of woody sprouts on the other (Bowman 1986). Fire is strongly implicated in this pattern, but the variable responses of different sites to changes in fire frequency (Bowman *et al.* 1988a, Fensham 1990) indicate that a complex fire/site interaction is involved. One model of such an interaction emphasises edaphic control (Bowman 1986), with sites of higher moisture availability producing a shrubby understorey, which then presents a barrier to fire. An alternative model emphasises the effect of fire: if a grassy site is protected from fire, grass cover will decrease and shrub and litter cover increase (Hoare *et al.* 1980), which will then reduce the susceptibility of that site to fire, such that the shrubby understorey is maintained. The rate at which the shrub layer develops, and therefore the fire-free period that is required, would be controlled by edaphic factors. These models share the notion that sites with more favourable edaphic conditions are more likely to support a less fire-prone ground layer, as has been hypothesised for neotropical savannas (Kellman 1984).

The continued suppression of woody sprouts by frequent fire has important demographic implications. It is common for savanna trees to have a bimodal size structure, comprising canopy trees on the one hand, and woody sprouts on the other (Braithwaite & Estbergs 1985, Werner 1986, Fensham & Bowman 1992), with virtually no saplings (Bowman & Panton 1993b). In some cases, some tree species are represented only by woody sprouts (Fensham & Bowman 1992). If woody sprouts are prevented from becoming saplings, then this would result in a gradual thinning of the overstorey, given that canopy trees suffer an annual mortality rate of at least 1% (Braithwaite *et al.* 1985) and sometimes considerably more (Lonsdale & Braithwaite 1991).

Frequent fire, therefore, has the potential to cause long-term structural degradation of woodlands and open forests (Hoare *et al.* 1980), much in the same way as introduced herbivores in parts of southern Australia have converted woodlands to grasslands by suppressing woody recruitment (Lange & Graham 1983, Auld 1990, Parsons 1990). It has been suggested that fire-free intervals of 3–5 years are required to prevent such degradation (Hoare *et al.* 1980). Fire has been implicated as a factor contributing to the major differences in forest structure between Elcho Island in Arnhem Land, where seedlings and saplings are common and canopy cover is 30%, and Gunn Point near Darwin, where seedlings are absent, saplings are rare, and canopy cover is only 15% (Bowman & Panton 1993b).

4.2.2. Fire intensity

Fires of different intensity obviously vary in their immediate and visible effects (Fig. 1), and this has often led to high intensity fires being described as 'destructive' compared to relatively 'benign' low intensity fires (Australian National Parks & Wildlife Service 1991, Salmon 1992, Anonymous 1992). The little evidence that is available, however, suggests that these descriptions are more superficial than real. For example, a particularly intense wildfire at the Kapalga Research Station in Kakadu National Park caused considerable tree and shrub mortality, but substantial numbers of all species survived, and the fire actually increased spatial diversity (Lonsdale & Braithwaite 1991). Studies of both invertebrate and vertebrate fauna indicate that, for many species, the short-term effect of high intensity fires is no more deleterious than that of lower intensity fires, and in fact, in some cases less so (A.N. Andersen & R.W. Braithwaite unpublished data).

It is the longer-term effects of high intensity fires on ecological processes, such as nutrient cycling, that may be of greater concern than their immediate impact. During biomass combustion, nutrients can be volatilised or otherwise transferred to the atmosphere as particulates (Cook 1992). The particulates are likely to be deposited on or near the site of fire, but not so elements lost in gaseous form. Moreover, nutrients remaining on the ground in ash and other residue after fire, along with deposited particulates, are highly susceptible to erosion through water runoff (Gillon 1983, Kellman *et al.* 1985). A preliminary analysis of the effects of fire on nutrient dynamics in the Kakadu region (Cook 1992), indicates that annual fires result in net losses of nitrogen, and possibly also potassium and magnesium, from the ecosystem. The rate of such losses is likely to increase with increasing fire intensity (as well as frequency). Even if nutrients are not lost entirely, their redistribution can have important effects. For example, fire can cause a transfer of nutrients from trees and shrubs on the one hand, to grasses on the other, due to different root depths (Coutinho 1982).

4.2.3. Fire timing

The effects of fire timing and intensity often confound each other because intensity tends to increase as the dry season proceeds. However, it is clear that the timing of fires can be ecologically important, independent of any confounding effects of intensity (Howe 1994). Studies elsewhere have shown that savanna fires have the greatest impact on plants during periods of active growth. This is because carbohydrate and nutrient reserves have been depleted, and new leaves are particularly susceptible to heat damage (West 1965). Thus, deciduous species are particularly sensitive to fires occurring just after leaf flush (Kennan 1971). The season when fires occur has also been shown to have a marked effect on the timing and intensity of grass seed production (Trollope 1982), and flowering phenology (Coutinho 1982). The effects of fire timing on the fauna have been poorly studied, but similar arguments are likely to apply, especially to species whose breeding or other activity is restricted to the ground at a specific time during the dry season.

4.3. Fire experiments

The fire debate in the Kakadu region has been criticised for being side-tracked by a preoccupation with the short-term and superficial effects of particular fires, based primarily on anecdotal and *ad hoc* observations (Andersen & Braithwaite 1992). As highlighted by Catling (1994) when discussing fuel-reduction burning in temperate forests, the real issue is the long-term ecological effects of different fire regimes, not the short-term impact of individual fires.

A proper understanding of the longer-term ecological effects of different fire regimes requires a rigorous experimental approach. The first attempt at this in the Kakadu region was at Munmarlary, where a series of 1 ha plots was established in 1972 jointly by CSIRO and the Conservation Commission of the Northern Territory (Hoare *et al.* 1980), and still operates (at the time of writing). There are four fire treatments (annual early-season burning, annual late-season burning, biennial early-season burning, and unburnt) in three replicate blocks at both a savanna woodland and open forest site. The experiment has provided useful information on the effects of different fire regimes on vegetation structure and floristics (Hoare *et al.* 1980, Bowman *et al.* 1988a), epiphytic orchids (Cook 1991), and communities of birds (Woinarski 1990) and ants (Andersen 1991). Its continued usefulness, however, is seriously limited by its small scale (Lonsdale & Braithwaite 1992). Important ecological processes which operate on larger scales, such as surface hydrology, nutrient dynamics, gene flow and faunal movements, cannot be studied (Andersen & Braithwaite 1992), and the behaviour of the fires themselves are seriously affected by small plot size (Lonsdale & Braithwaite 1991).

The CSIRO embarked upon an ambitious, landscape-scale fire experiment at Kapalga Research Station. The experimental units are 10–20 km^2 catchments based on seasonal creeklines, and capture the full topographic and moisture gradients occurring in the local landscape. There are four fire treatments (early season burning, late season burning, burning progressively through the dry season, and no burning), each replicated at least three times. The experimental fires commenced in 1990, and are planned to continue at least until 1995. The experiment has several strengths apart from its large scale, including a multi-disciplinary approach, the collection of extensive baseline data prior to the imposition of fire treatments, and intensive studies of the behaviour of the fires themselves (Andersen & Braithwaite 1992).

5. Fire management

Fire is used as a management tool in fire-prone ecosystems throughout the world (Habeck & Mutch 1973, Naveh 1975, Hobbs & Gimmingham 1987, Moore 1987), but it plays a particularly important role in land management in Australia (Gill 1977, Griffin 1992, Morton & Andrew 1987). The importance of fire management for nature conservation in the Kakadu region has been strongly emphasised (Walker *et al.* 1985, Press 1987).

Whatever the effects of extensive long-term fire exclusion might be, total fire protection over extensive areas is widely recognised as being an untenable management goal (Fensham 1990, Andersen & Braithwaite 1992) due to frequent ignitions by people (accidental or otherwise) and lightning. The 'burning question' is 'when' – at what times of the year and how frequently, and not 'if', the savannas should be burnt.

Two themes have dominated fire management strategies in the Kakadu region. The first is the reimposition of traditional Aboriginal fire regimes. The rationale for this strategy is that as the regional biota has experienced traditional Aboriginal burning practices over tens of thousands of years, then this is what it is adapted to, and such practices are the best way of ensuring their conservation. The second is the prevention of wildfires by extensive early burning. This strategy aims 'to create a mosaic of burnt and unburnt country, thereby maintaining habitat diversity and reducing the risk of widespread, destructive late dry season fires' (Salmon 1992). The latter is the primary strategy adopted by the Northern Territory Bush Fires Council (Anonymous 1993), which is responsible for fire management in natural areas in the region outside Kakadu National Park, whereas a combination of both strategies is adopted within the Park by the Australian Nature Conservation Authority (Australian National Parks and Wildlife Services 1991).

These two strategies are analysed below. A third management option, that of wet season burning, has had only limited use in the region, but is the subject of increasing interest (J. Day, K. Duggan, B. Bailey unpublished data, Russell-Smith personal communication). Wet season burning can eliminate the annual grass *Sorghum intrans*, which provides the bulk of the fuel for fire throughout much of the region. If annual *Sorghum* could be eliminated from strategic areas, then this would help minimise the extent of late dry season fires (Day 1985b). The rationale for this option is, therefore, similar to that for extensive early season burning.

5.1. Aboriginal fire management

The argument that contemporary fire management should be based on traditional Aboriginal burning practices is an intuitively attractive proposition, but does not appear to stand close analysis on ecological grounds. It is founded on a series of assumptions, all of which are open to question, as outlined below.

5.1.1. Influence of Aboriginal burning practices on the biota

The first assumption is that the regional biota has been strongly influenced by Aboriginal burning practices. The impact of Aboriginal burning on the Australian environment is in fact hotly disputed (Nicholson 1981). One view is that it has had a dramatic effect (Jones 1969), and was responsible, for example, for an extensive replacement of rain forests by sclerophyll vegetation over the past 40 000 years (Kershaw 1985). An opposing view is that the Australian biota was fire-adapted long before the arrival of Aborigines, such that Aboriginal burning did little more than cause slight modifications to an already fire-shaped biota (Cleland 1957, Horton 1982, Head 1989).

In reality, the impact of Aboriginal burning on the Kakadu landscape is poorly known and a largely intractable issue. It is certainly plausible that Aboriginal burning influenced the extent and distribution of fire-sensitive vegetation, but it is not clear which way: either reducing it because of increased fire frequency (Kershaw 1985), expanding it through limiting the extent of high intensity wildfires (Haynes 1985, Russell-Smith & Bowman 1991), or, perhaps, a combination of both spread across different areas. The extent to which Aboriginal burning shaped the savanna biota, which already had a long history of high fire frequency, is even more problematical. Whatever the effects were, it is unlikely that any rain forest or savanna elements are adapted to Aboriginal burning in any evolutionary sense (Andersen & Braithwaite 1992): any effects of Aboriginal burning are therefore on the distribution and abundance of existing biota.

5.1.2. Knowledge of traditional Aboriginal burning practices

The second assumption is that traditional Aboriginal burning practices are well-known. Our knowledge of such practices is, in fact, rather tenuous. As previously mentioned, a clear seasonal pattern has emerged, but it has been derived from a very small data base. Even less information is available on the spatial distribution of Aboriginal fires across the landscape, and, in particular, on their overall extent. It is difficult enough to document the patterns of contemporary Aboriginal burning practices (Head *et al.* 1992), let alone those occurring prior to European settlement (Gill 1977). Whatever is the case, the common perception that Aboriginal burning practices equate with extensive early season burning, is patently false (Braithwaite 1991, P. Wellings personal communication).

5.1.3. Appropriateness of traditional Aboriginal burning practices

The third assumption is that traditional Aboriginal burning practices are appropriate for nature conservation. The various reasons for Aboriginal burning, such as protection of food resources, improving hunting efficiency, facilitating travel, and signalling, are actually utilitarian ones. They were aimed at modifying the environment to support a subsistence life-style, not at nature conservation in the contemporary sense. A change in management goals, from utilitarianism to conservation, calls into question the continued appropriateness of traditional Aboriginal burning practices (Gallus 1994). Moreover, the environment that is to be managed has also changed (Braithwaite 1991), with both feral animals and weeds having a substantial impact on fire-vegetation relationships (Panton 1993, Cowie & Werner 1993).

5.1.4. Replicability of traditional Aboriginal burning practices

The fourth assumption is that, if details of Aboriginal burning practices are known, they can be replicated by current managers. Traditional Aboriginal fire 'management' was not a controlled operation by a single body with a single aim. Rather, it was an emergent property arising from numerous people in different parts of the landscape at different times of the year independently lighting fires for a variety of reasons (Braithwaite 1992). It is difficult to envisage such a system, even if it were known perfectly, being replicated by a centralised management agency with limited human resources. Moreover, the region now contains an extensive road network transporting hundreds of thousands of tourists each fire season, such that the inevitable fires lit by these people will seriously compromise any attempt to re-create traditional burning practices (Day 1985a, Andersen & Braithwaite 1992).

5.2. Wildfire prevention

There is considerable concern that the frequency of late season wildfires has increased since European settlement, and that this is causing serious degradation of the region's rain forest, *Callitris* and *Allosyncarpia* estate (Bowman 1988, Russell-Smith & Bowman 1991, Bowman & Panton 1993c). Moreover, the extent of fire damage to these vegetation types has been used as evidence that changed fire regimes since the demise of traditional Aboriginal burning practices have caused widespread ecological change, particularly on the Arnhem Land escarpment (Bowman *et al.* 1990, Russell-Smith & Bowman 1991, Bowman & Panton 1993c).

The information available to address this concern is equivocal. As previously outlined, there is no direct evidence that contemporary fire patterns differ markedly from those prevailing under Aboriginal management. However, this of course might just reflect the inadequacy of documented evidence, rather than the true situation. Circumstantial evidence suggests that at least some patches of monsoon rain forest have declined considerably in the relatively recent past (Stocker 1971, Russell-Smith 1985), but factors other than fire, such as climate change or feral animals, have been implicated. Fire has contributed to the marked contractions of rain forest patches near urban centres such as Darwin, but these patches have been subjected to atypically high fire frequencies, and have suffered severe disturbance by cyclones and weed invasion (Panton 1993, Bowman & Panton 1994).

Certainly, many rain forest patches outside urban areas show evidence of recent fire damage (Fig. 6; Russell-Smith 1991, Russell-Smith & Bowman 1991), but this is to be expected of any vegetation located in an extremely fire-prone environment. Without information on recovery after fire, and movement of rain forest/savanna boundaries, the implications of such fire damage for rain forest dynamics are not clear. For example, in a study of the rare palm *Ptychosperma bleeseri*, which is restricted to rain forest patches on the flood plain margins of the Howard and Adelaide rivers near Darwin, fire was claimed to have a 'severe detrimental effect' (Barrow *et al.* 1993). This was based on the observation that a late-season wildfire caused 22% adult mortality on one population. However, the author's own data show substantial seedling recruitment after the fire, plus a doubling in juvenile density over the following wet season. Without further information on recovery, the effects of the fire are not at all clear.

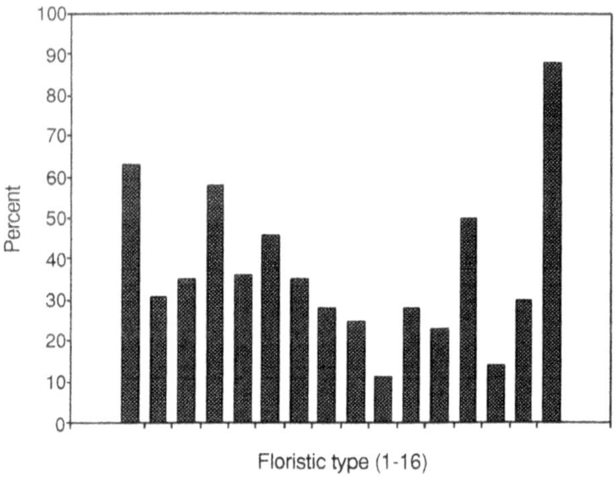

Fig. 6. Proportion of patches of monsoon rain forest in the Northern Territory, differentiated into floristic types, severely disturbed by fire (re-drawn from Russell-Smith & Bowman 1991).

Similarly, the severe fire damage evident in many *Callitris* populations does not necessarily indicate ecological change, as the extent of fire damage prior to European colonisation is unknown. Populations of *Callitris* in fire-prone situations would always show evidence of fire damage, no matter what regime prevailed: the question is, has this damage increased in the recent past? In the absence of information on the extent of past fire damage, this question needs to be addressed by developing demographic models for *Callitris* populations, and deducing the effects of observed fire damage on population size and persistence. Finally, the suggestion that the decline in certain rare and endangered animals, particularly granivorous birds, might be due to changed fire regime (Haynes 1985, Blakers *et al.* 1984) is purely speculation.

Despite the lack of documented evidence, the overwhelming personal experience of Kakadu National Park personnel is that damage to fire-sensitive vegetation associated with the sandstone country of the Arnhem Land escarpment is so apparent, that preventative action is urgently required (J. Russell-Smith personal communication). Through strategic early season burning, ANCA has successfully protected the sandstone country from extensive wildfires during the past few years (G. Spiers personal communication).

Throughout the extensive tracts of lowland savanna woodland and open forest, 'fire-sensitive' vegetation is generally absent. Here, as previously mentioned, the description of high intensity fires as 'destructive' appears to be based more on the visual appearance of burnt vegetation, than on known ecological effects of such fires. Moreover, the emphasis on intensity ignores the effects of timing and frequency. The real comparison is between annual early season fires of generally low intensity on the one hand, and less frequent, generally more intense fires later in the season on the other (Andersen & Braithwaite 1992). A key issue in this comparison is the effect of early burning practices on fire frequency; unfortunately, however, the longer-term incidence of late season fires in the absence of early season burning is unknown.

6. Conclusion

Fire is undoubtedly a potent force in the Kakadu landscape, and is an important tool for conservation management. However, there is a lack of adequate information, both on burning patterns, and the ecological effects of fire. The lack of information on burning patterns is particularly frustrating, especially in relation to the perceived increase in the incidence of late season wildfires. ANCA has now established an extensive fire monitoring program within Kakadu National Park (J. Russell-Smith personal communication), which will help redress this problem. In regard to ecological research, the focus needs to be on the large-scale, long-term effects of fire regimes, rather than on the immediate and possibly superficial effects of individual fires. In particular, there needs to be a shift towards mechanistic and process oriented research (Press 1987, Duff & Braithwaite 1990), such as detailed demographic studies of key species.

The controversy over fire management is also fuelled by a serious communication problem between fire managers and the general public. It is therefore fundamentally important that the strategic objectives of management be clearly established (Gill 1977, Braithwaite 1985), with the tactical goals of fire management explicitly directed toward them. This then needs to be communicated to the general public. Possible strategic objectives identified by Braithwaite (1985) include: i) re-creation of the pre-Aboriginal environment; ii) maintenance of the environment existing immediately prior to European settlement; and iii) maximisation of biotic diversity. It is important that tactical goals, such as early season burning, do not get confused with strategic objectives, and be seen as objectives themselves.

The relevance of traditional Aboriginal burning patterns to current management goals needs to be clearly articulated at both strategic and tactical levels; is the re-imposition of traditional burning practices a strategic objective (i.e. for socio-political reasons), or is a tactic used to achieve some biotic objective? If the latter, then its relevance requires close scrutiny. It has been suggested that the popular notion of harmonious environmental management through 'fire-stick' farming is little more than romanticism (Horton 1982, Redford 1991).

Land managers in the Kakadu region are placed in the awkward position of having to make decisions about fire in the absence of adequate ecological information. Whatever the merits of current burning practices, the bottom line is that their ecological effects are poorly known. In the short-term, managers have no alternative but to manage in the absence of substantial documented evidence. However, a lack of documentation is not credible in the longer-term. This is especially pertinent to fire management in savanna vegetation, where so-called 'benign' early season fires are used to prevent 'destructive' late season fires. It is also relevant to concerns over fire-sensitive vegetation in the sandstone country: the perception of an increased incidence of wildfires is not supported by published data, nor is there adequate documentation of a widespread contraction of fire-sensitive vegetation in recent times.

Once strategic objectives have been clarified, the setting of tactical goals such as early season burning begs further questions. What is an acceptable incidence of late season wildfires? How much preventative burning needs to be conducted to achieve this? What is the most effective pattern of preventative burning? Tactics need to be constantly fine-tuned so that tactical goals are achieved as effectively as possible. Similar tactical problems arise from fire management issues elsewhere in Australia, such as fuel-reduction burning in temperate forests (Gallus 1994).

It is clear that without human interference, fires would be restricted to the 'build-up' to, and early part of, the wet season, and that this was the regime prevailing during the evolution of the regional biota. However, details of this regime remain obscure. Whatever the 'natural' regime is, the high incidence of unauthorised burning during the dry season currently precludes it from being achieved. This is presumably why a management strategy based on the simulation of 'natural' lighting-induced fires (cf. Stander *et al.* 1993) is not considered an option by management authorities. The rationale of early season burning is not so much aimed at preventing lightning-induced fires, but to reduce the extent of unauthorised fires (although this needs to be clarified). Preventative tactics directed at the source of these fires may ultimately prove to be the most rewarding option.

7. Acknowledgements

This chapter benefited from stimulating discussions with Dave Bowman, Dick Braithwaite, Gordon Duff, Rod Fensham, Jeremy Russell-Smith and John Woinarski.

8. References

Andersen, A.N. 1991. Responses of ground-foraging ant communities to three experimental fire regimes in a savanna forest of tropical Australia. Biotropica 23: 575–585.

Andersen, A.N. & Braithwaite, R.W. 1992. Burning for conservation of the Top End's savannas. In: Moffatt, I. & Webb, A. (eds) Conservation and Development Issues in Northern Australia. pp. 117–122. North Australian Research Unit, Darwin.

Andersen, A.N. & Reichel, H. 1994. The ant fauna of Holmes Jungle, a rainforest patch in the seasonal tropics of Australia's Northern Territory. J. Aust. Entomol. Soc. 33: 153-158.

Anonymous 1992. Discussion paper to formalize a strategy for the management of wildfires in the Northern Territory. Bush Fires Council N.T. 25 pp.

Ashton, D.H. 1981. Tall open-forests. In: Groves, R.H. (ed.) Australian Vegetation. pp. 121–151. Cambridge University Press, Cambridge.

A Tchie, N. & Gakahu, G.C. 1989. Responses of important woody species of Kenya's rangeland to a prescribed burning. Afric. J. Ecol. 27: 119–128.

Auld, T.D. 1990. Regeneration in populations of the arid zone plants *Acacia carnei* and *A. oswaidii*. Proc. Ecol. Soc. Aust. 16: 267–272.

Australian National Parks and Wildlife Service, 1991. Kakadu National Park Plan of Management. Australian National Parks and Wildlife Service, Canberra. 139 pp.

Barrow, P., Duff, G. Liddle, D. & Rusell-Smith, J. 1993. Threats to monsoon rainforest habitat in northern Australia: the case of *Ptychosperma bleeseri*: Burrett (Arecaceae). Aust. J. Ecol. 18: 463–471.

Blakers, M., Davies, S.J.J.F. & Riley, P.N. 1984. The Atlas of Australian Birds. Melbourne University Press, Melbourne. 738 pp.

Boulière, F. & Hadley, M. 1983. Present-day savannas: an overview. In: Bourlière, F. (ed.) Ecosystems of the World 13: Tropical Savannas. pp. 1–17. Elsevier, Amsterdam.

Bowman, D.M.J.S. 1986. Stand characteristics, understorey associates and environmental correlates of *Eucalyptus tetrodonta* F. Muell. forests on Gunn Point, northern Australia. Vegetatio 65: 105–113.

Bowman, D.M.J.S. 1988. Stability amid turmoil?: towards an ecology of north Australian eucalypt forests. Proc. Ecol. Soc. Aust. 15: 149–158.

Bowman, D.M.J.S. 1991a. Environmental determinants of *Allosyncarpia ternata* forests that are endemic to western Arnhem Land, northern Australia. Aust. J. Bot. 39: 575–589.

Bowman, D.M.J.S. 1991b. Recovery of some northern Australian monsoon forest tree species following fire. Proc. Royal Soc. Qld 101: 21–25.

Bowman, D.M.J.S. 1992a. A burnt out case? Reply to Lonsdale and Braithwaite (1991). Aust. J. Ecol. 17: 103–106.

Bowman, D.M.J.S. 1992b. Monsoon forests in north-western Australia. II. Forest-savanna transitions. Aust. J. Bot. 40: 89–102.

Bowman, D.M.J.S. 1993. Establishment of two dry monsoon forest tree species on a fire-protected monsoon forest-savanna boundary, Cobourg Peninsula, northern Australia. Aust. J. Ecol. 18: 235–237.

Bowman, D.M.J.S. & Dunlop, C.R. 1986. Vegetation pattern and environmental correlates in coastal forests of the Australian monsoon tropics. Vegetatio 65: 99–104.

Bowman, D.M.J. & Fensham, R.J. 1991. Response of a monsoon forest-savanna boundary to fire protection, Weipa, northern Australia. Aust. J. Ecol. 16: 111–118.

Bowman, D.M.J.S. & Minchin, P.R. 1987. Environmental relationships of woody vegetation patterns in the Australian monsoon tropics. Aust. J. Bot. 35: 151–169.

Bowman, D.M.J.S. & Panton, W.J. 1993a. Factors that control monsoon rainforest seedling establishment and growth in north Australian *Eucalyptus* savanna. J. Ecol. 81: 297–304.

Bowman, D.M.J.S. & Panton, W.J. 1993b. Differences in the stand structure of *Eucalyptus tetrodonta* forests between Elcho Island and Gunn Point, northern Australia. Aust. J. Bot. 41: 211–215.

Bowman, D.M.J.S. & Panton, W.J. 1993c. Decline of *Callitris intratropica* K.T. Baker & H.G. Smith in the Northern Territory: implications for pre- and post-European colonization fire regimes. J. Biogeogr. 20: 373–381.

Bowman, D.M.J.S. & Panton, W.J. 1994. Fire and cyclone damage to woody vegetation on the north coast of the

Northern Territory, Australia. Aust. Geographer (in press).
Bowman, D.M.J.S. & Wightman, G.M. 1985. Small scale vegetation pattern associated with a deeply incised gully, Gunn Point, Northern Australia. Proc. Royal Soc. Qld 96: 63–73.
Bowman, D.M.J.S. & Wilson, B.A. 1988a. Fuel characteristics of coastal monsoon forests, Northern Territory, Australia. J. Biogeog. 15: 807–817.
Bowman, D.M.J.S., Wilson, B.A. & Davis, G.W. 1988b. Response of *Callitris intratropica* R.T. Baker & H.G. Smith to fire protection, Murgenella, Northern Australia. Aust. J. Ecol. 13: 147–159.
Bowman, D.M.J.S., Wilson, B.A. & Fensham, R.J. 1990. Sandstone vegetation pattern in the Jim Jim Falls region, Northern Territory, Australia. Aust. J. Ecol. 15: 163–174.
Bowman, D.M.J.S., Wilson, B.A. & Hooper, R.J. 1988a. Response of *Eucalyptus* forest and woodland to four fire regimes at Munmarlary, Northern Territory, Australia. J. Ecol. 76: 215–232.
Bowman, D.M.J.S., Wilson, B.A. & McDonough, L. 1991. Monsoon forests in northwestern Australia. I. Vegetation classification and the environmental control of tree species. J. Biogeog. 18: 679–686.
Braithwaite, R.W. 1985. Biological research for national park management. Proc. Ecol. Soc. Aust. 13: 323–333.
Braithwaite, R.W. 1987. Effects of fire regimes on lizards in the wet-dry tropics of Australia. J. Trop. Ecol. 3: 265–275.
Braithwaite, R.W. 1991. Aboriginal fire regimes of monsoonal Australia in the 19th Century. Search 22: 247–249.
Braithwaite, R.W. 1992. Black and green. J. Biogeog. 19: 113–116.
Braithwaite, R.W. & Estbergs, J.A. 1985. Fire patterns and woody vegetation trends in the Alligator Rivers Region of northern Australia. In: Tothill, J.C. & Mott, J.J. (eds) Ecology and Management of the World's Savannas. pp. 359–364. Australian Academy of Science, Canberra.
Braithwaite, R.W., Parker, B.S., Wood, J.T. & Estbergs, J.A. 1985. The distribution and abundance of termite-induced hollows in tropical forests of northern Australia. In: Braithwaite, R.W. (ed.) The Kakadu Fauna Survey: An Ecological Survey of Kakadu National Park. pp. 651–729. Report to the Australian National Parks and Wildlife Service, Canberra.
Catling, P. 1994. Bushfires and prescribed burning: protecting native fauna. Search 25: 37–40.
Clayton-Greene, K.A. & Beard, J.S. 1985. The fire factor in vine thicket and woodland vegetation of the Admiralty Gulf region, northwest Kimberley, Western Australia. Proc. Ecol. Soc. Aust. 13: 225–230.
Cleland, J.B. 1957. Our natives and the vegetation of southern Australia. Mankind 5: 149–162.
Cook, G.D. 1991. Effects of fire regimen on two species of epiphytic orchids in tropical savannas of the Northern Territory. Aust. J. Ecol. 16: 537–540.
Cook, G.D. 1992. The effects of fire on nutrient losses from Top End savannas. In: Moffatt, I. & Webb, A. (eds.) Conservation and Development Issues in Northern Australia. pp. 123–129. North Australia Research Unit, Darwin.
Coutinho, L.M. 1982. Ecological effects of fire in Brazilian cerrado. In: Huntley, B.J. & Walker, B.H. (eds.) Ecology of Tropical Savannas. pp. 273–291. Springer-Verlag, Berlin.
Cowie, I.D. & Werner, P.A. 1993. Alien plant species invasive in Kakadu National Park, tropical northern Australia. Biol. Conserv. 63: 127–135.
Day, J. 1985a. Application of satellite imagery for fire management in Kakadu National Park. In: Walker, J., Davis, J.R. & Gill, A.M. (eds.) Towards An Expert System for Fire Management at Kakadu National Park. pp. 122–137. CSIRO Technical Memorandum 85/2.
Day, J. 1985b. Wet season burning for fire management – some preliminary observations. In: Walker, J., Davis, J.R. & Gill, A.M. (eds.) Towards An Expert System for Fire Management at Kakadu National Park, pp. 92–95. CSIRO Technical Memorandum 85/2.
Duff, G.A. & Braithwaite, R.W. 1990. Fire and Top End forests – past, present and future research. In: Roberts, B.R. (ed.) Fire Research in Rural Queensland. pp. 84–97. University of Southern Queensland, Toowoomba.
Dunlop, C.R. & Webb, L.J. 1991. Flora and vegetation. In: Haynes, C.D., Ridpath, M.G. & Williams, M.A.J. (eds.) Monsoonal Australia: Landscape, Ecology and Man in the Northern Lowlands. pp. 41–60. A.A. Balkema, Rotterdam.
Fensham, R.J. 1990. Interactive effects of fire frequency and site factors in tropical *Eucalyptus* forests. Aust. J. Ecol. 15: 255–266.
Fensham, R.J. 1992. The establishment of eucalypt seedlings in tropical savanna forest. North. Territory Nat. 13: 30–36.
Fensham, R.J. & Bowman, D.M.J.S. 1992. Stand structure and the influence of overwood on regeneration in tropical eucalypt forest on Melville Island. Aust. J. Bot. 40: 335–352.
Fordyce, I.R. 1992. Some aspects of the ecophysiology and population biology of *Allosyncarpia ternata* S.T. Blake. Unpublished Graduate Diploma of Science Thesis, Northern Territory University. 50 pp.
Frost, P.G.H. 1985. The responses of savanna organisms to fire. In: Tothill, J.C. & Mott, J.J (eds.) Ecology and Management of the World's Savannas. pp. 232–237. Australian Academy of Science, Canberra.
Furley, P.A., Proctor, J. & Ratter, J.A. (eds.) 1992. Nature and Dynamics of Forest-Savanna Boundaries. Chapman & Hall, London. 616 pp.
Gallus, C. 1994. The New South Wales bushfires: developing a national park management strategy. Search 25: 34–36.
Gambold, N. & Woinarski, J.C.Z. 1993. Distributional patterns of herpetofauna in monsoon rainforests of the Northern Territory, Australia. Aust. J. Ecol. 18: 431–449.
Gill, A.M. 1977. Management of fire-prone vegetation for plant species conservation in Australia. Search 8: 20–26.
Gillison, A.N. 1983. Tropical savannas of Australia and the southwest Pacific. In: Bourliere, F. (ed.) Ecosystems of the World 13: Tropical Savannas. pp. 183–243. Elsevier, Amsterdam.
Gillon, D. 1983. The fire problem in tropical savannahs. In: Bourliere, F. (ed.) Ecosystems of the World 13: Tropical Savannas. pp. 617–641. Elsevier, Amsterdam.
Griffin, G.F. 1992. Will it burn – should it burn?: management of the spinifex grasslands of inland Australia. In: Chapman, G. (ed.) Desertified Grasslands: Their Biology and Management. pp. 63–76. The Linnean Society of London.

Habeck, J.R. & Mutch, R.W. 1973. Fire-dependent forests in the northern Rocky Mountains. J. Quatern. Res. 3: 408–424.
Haynes, C.D. 1985. The pattern and ecology of *munwag*: traditional Aboriginal fire regimes in nort central Arnhemland. Proc. Ecol. Soc. Aust. 13: 201–214.
Haynes, C.D. 1991. Use and impact of fire. In: Haynes, C.D. Ridpath, M.G. & Williams, M.A.J. (eds.) Monsoonal Australia: Landscape, Ecology and Man in the Northern Lowlands. pp. 61–71. A.A. Balkema, Rotterdam.
Head, L. 1989. Prehistoric Aboriginal impacts on Australian vegetation: an assessment of the evidence. Aust. Geog. 20: 37–46.
Head, L.M., O'Neill, A.L., Marthick, J.K. & Fullagar, R.L.K. 1992. A comparison of Aboriginal and pastoral fires in the north-west Northern Territory. In: Moffatt, I. & Webb, A. (eds.) Conservation and Development Issues in Northern Australia. pp. 130–144. North Australian Research Unit, Darwin.
Hoare, J.R.L., Hooper, R.J., Cheney, N.P. & Jacobsen, K.L.S. 1980. A report on the effect of fire in tall open forest and woodland with particular reference to fire management in Kakadu National Park in the Northern Territory. Australian National Parks and Wildlife Service, Canberra. 106 pp.
Hobbs, R.J. & Gimingham, C.H. 1987. Vegetation, fire and herbivore interactions in heathland. Adv. Ecol. Res. 16: 87–173.
Hopkins, B. 1983. Successional processes. In: Bourlière, F. (ed.) Ecosystems of the World 13: Tropical Savannas. pp. 605–616. Elsevier, Amsterdam.
Horton, D.R. 1982. The burning question: Aborigines, fire and Australian ecosystems. Mankind 13: 237–251.
Howe, H.F. 1994. Response of early- and late-flowering plants to fire season in experimental praires. Ecol. Applic. 4: 121–133.
Jones, R. 1969. Fire-stick farming. Aust. Nat. Hist. 16: 224–228.
Jones, R. 1980. Hunters in the Australian coastal savanna. In: Harris, D.R. (ed.) Human Ecology in Savanna Environments. pp. 107–146. Academic Press, New York.
Keen, I. 1980. The Alligator Rivers Stage II Land Claim. Northern Land Council, Darwin. 226 pp.
Kellman, M. 1984. Synergistic relationships between fire and low soil fertility in neotropical savannas: a hypothesis. Biotropica 16: 158–160.
Kellman, M., Miyanishi, K. & Hieburt, P. 1985. Nutrient retention by savanna ecosystems: 2. Retention after fire. J. Ecol. 73: 953–962.
Kennan, T.C.D. 1971. The effects of fire on two vegetation types of Matopos. Proc. Ann. Tall Timbers Fire Ecol. Conf. 11: 53–98.
Kershaw, A.P. 1985. An extended late Quaternary vegetation record from north-eastern Queensland and its implications for the seasonal tropics of Australia. Proc. Ecol. Soc. Aust. 13: 179–189.
Kikkawa, J., Webb, L.J., Dale, M.B., Monteith, G.B., Tracey, J.G. & Williams, W.T. 1981. Gradients and boundaries of monsoon forests in Australia. Proc. Ecol. Soc. Aust. 11: 39–52.
Lacey, C.J. 1974. Rhizomes in tropical eucalypts and their role in recovery from fire damage. Aust. J. Bot. 22: 29–38.
Lacey, C.J. & Whelan, P.I. 1976. Observations on the ecological significance of vegetative reproduction in the Katherine-Darwin region of the Northern Territory. Aust. For. 39: 131–139.
Lacey, C.I., Walker, J. & Noble, I.R. 1982. Fire in Australian tropical savannas. In: Huntley, B.J. & Walker, B.H. (eds.) Ecology of Tropical Savannas. pp. 246–272. Springer-Verlag, Berlin.
Lange, R. & Graham, C.R. 1983. Rabbits and the failure of regeneration in Australian arid zone *Acacia*. Aust. J. Ecol. 8: 377–382.
Langkamp, P.J., Ashton, D.H. & Dalling, M.J. 1981. Ecological gradients in forest communities on Groote Eylandt, Northern Territory, Australia. Vegetatio 48: 27–46.
Lonsdale, W.M. & Braithwaite, R.W. 1991. Assessing the effects of fire on vegetation in tropical savannas. Aust. J. Ecol. 16: 363–374.
Lonsdale, W.M. & Braithwaite, R.W. 1992. Fiddling while Rome burns: Nero much misunderstood. Reply to Bowman (1992). Aust. J. Ecol. 17: 106–108.
McKenzie, N.L., Johnson, R.B & Kendrick, P.J. (eds.) 1991. Kimberley Rainforests of Australia. Surrey Beattie & Sons, Sydney. 490 pp.
Menkhorst, K.A. & Woinarski, J.C.Z. 1992. Distribution of mammals in monsoon rainforests of the Northern Territory. Wildl. Res. 19: 295–316.
Moore, P.D. 1987. Burning issues in fire control. Nat. 25: 486.
Morton, S.R. & Andrew, M.J. 1987. Ecological impact and management of fire in northern Australia. Search 18: 77–82.
Mott, J.J. 1978. Dormancy and germination in five native grass species from savannah woodland communities of the Northern Territory. Aust. J. Bot. 26: 621–31.
Mott, J.J., Williams, J., Andrew, M.J. & Gillison, A.N. 1985. Australian savanna ecosystems. In: Tothill, J.C. & Mott, J.J. (eds.) Ecology and Management of the World's Savannas. pp. 56–82. Australian Academy of Science, Canberra.
Naveh, Z. 1975. The evolutionary significance of fire in the Mediterranean region. Vegetatio 29: 199–208.
Nicholoson, P.H. 1981. Fire and the Australian Aborigine – an enigma. In: Gill, A.M., Groves, R.H. & Noble, I.R. (eds.) Fire and the Australian Biota. pp. 55–76. Australian Academy of Science, Canberra.
Pandey, A.N. 1988. Short-term study of recovery of a tropical grassland following seasonal burning. Trop. Ecol. 29: 159–170.
Panton, W.J. 1993. Post World War II changes in the distribution and status of monsoon rainforest patches in the Darwin area. Aust. Geogr. 24: 50–59.
Parsons, R.F. 1990. Plant conservation problems at species level in Mallee lands. In: Noble, J.C., Joss, P.J. & Jones, G.K. (eds.) The Mallee Lands: A Conservation Perspective. pp. 25–28. CSIRO, Melbourne.
Press, A.J. 1987. Fire management in Kakadu National Park: the ecological basis for the active use of fire. Search 18: 244–248.
Press, A.J. 1988. Comparisons of the extent of fire in different land management systems in the Top End of the Northern Territory. Proc. Ecol. Soc. Aust. 15: 167–175.
Ram, S.C. & Ramakrishnan, P.S. 1992. Fire and nutrient

cycling in seral grasslands of Cherrapunji in north-eastern India. Inter. J. Wildl. Fire 2: 131–138.

Redford, K.H. 1991. The ecologically noble savage. Cultural Survival Quart. 15: 46–48.

Rose-Innes, R. 1972. Fire in West African vegetation. Proc. Tall Timbers Fire Ecol. Conf. 11: 147–173.

Russell-Smith, J. 1985. A record of change: studies of Holocene vegetation history in the South Alligator River region, Northern Territory. Proc. Ecol. Soc. Aust. 13: 191–202.

Russell-Smith, J. 1991. Classification, species richness, and environmental relations of monsoon rainforest vegetation in the Northern Territory, Australia. J. Veg. Sci. 2: 259–278.

Russell-Smith, J. & Bowman, D.M.J.S. 1991. Conservation of monsoon rainforest isolates in the Northern Territory, Australia. Biol. Conserv. 59: 51–63.

Salmon, A. (ed.) 1992. Australian National Parks and Wildlife Service Annual Report 1991–92. 216 pp.

Silva, J.F., Caswell, H. & Trevisan, M.C. 1991. Population responses to fire in a tropical savanna grass, *Andropogon semiberbis*: a matrix model approach. J. Ecol. 79: 345–356.

Stander, P.E., Nott, T.B. & Mentis, M.T. 1993. Proposed burning strategy for a semi-arid African savanna. Afric. J. Ecol. 31: 282–289.

Stevenson, P.M. 1985. Traditional Aboriginal resource management in the wet-dry tropics: Tiwi case study. Proc. Ecol. Soc. Aust. 13: 309–315.

Stocker, G.C. 1966. Effects of fire on vegetation in the Northern Territory. Aust. For. 30: 223–230.

Stocker, G.C. 1969. Fertility differences between the surface soils of monsoon and eucalypt forests in the Northern Territory. Aust. For. Res. 4: 31–38.

Stocker, G.C. 1971. The age of charcoal from old jungle fowl nests and vegetation change on Melville Island. Search 2: 28–30.

Stocker, G.C. & Mott, J.J. 1981. Fire in the tropical forests and woodlands of northern Australia. In: Gill, A.M., Groves, R.H. & Noble, I.R. (eds.) Fire and the Australian Biota. pp. 425–439. Australian Academy of Science, Canberra.

Stott, P. 1986. The spatial pattern of dry season fires in the savanna forests of Thailand. J. Biogeog. 14: 345–358.

Trollope, W.S.W. 1982. Ecological effects of fire in South African savannas. In: Huntley, B.J. & Walker, B.H. (eds.) Ecology of Tropical Savannas. pp. 292–306. Springer-Verlag, Berlin.

Unwin, G.L., Stocker, G.C. and Sanderson, K.D. 1988. Forest successions following European settlement and mining in the Rossville area, north Queensland. Proc. Ecol. Soc. Aust. 15: 303–305.

Walker, J., Davis, J.R. & Gill, A.M. (eds.) 1985. Towards an expert system for fire management at Kakadu National Park. CSIRO Land Resources Technical Memorandum 85/2, Canberra. 169 pp.

Webb, L.J. & Tracey, J.G. 1981. Australian rainforests: patterns and change. In: Keast, A. (ed.) Ecological Biogeography of Australia. pp. 607–694. W. Junk, The Hague.

West, O. 1965. Fire in vegetation and its use in pasture management with special reference to tropical and subtropical Africa. Commonwealth Bureau of Pastures and Field Crops, Hurley, Berkshire. 53 pp.

Wilson, B.A. & Bowman, D.M.J.S. 1987. Fire, storm, flood and drought: the vegetation ecology of Howards Peninsula, Northern Territory, Australia. Aust. J. Ecol. 12: 165–174.

Werner, P.A. 1986. Population dynamics and productivity of selected forest trees in Kakadu National Park. Unpublished report to the Australian National Parks and Wildlife Service, Canberra. 63 pp.

Woinarski, J.C.Z. 1990. Effects of fire on the bird communities of tropical woodlands and open forests in northern Australia. Aust. J. Ecol. 5: 1–22.

Woinarski, J.C.Z. 1993. A cut and paste community: birds of monsoon rainforests at Kakadu National Park, Northern Territory. Emu 93: 100–120.

List of contributors

A.N. Andersen
CSIRO Division of Wildlife and Ecology, Tropical Ecosystems Research Centre, PMB 44, Winnellie, NT 0821, Australia.

J.T. Arthur
Bureau of Meteorology, GPO Box 735, Darwin, NT 0801, Australia.

R.W. Braithwaite
CSIRO Division of Wildlife and Ecology, Tropical Ecosystems Research Centre, PMB 44, Winnellie, NT 0821, Australia.

I.J. Butterworth
Bureau of Meteorology, GPO Box 735, Darwin, NT 0801, Australia.

L.K. Corbett
CSIRO Division of Wildlife and Ecology, Tropical Ecosystems Research Centre, PMB 44, Winnellie, NT 0820, Australia.

I.D. Cowie
CSIRO Division of Wildlife and Ecology, Tropical Ecosystems Research Centre, PMB 44, Winnellie, NT 0820, Australia.
present address: Parks and Wildlife Commission of the Northern Territory, PO Box 496, Palmerston, NT 0831, Australia.

T.J. East
Bureau of Resource Sciences, Department of Primary Industries and Energy, PO Box E11, Canberra, ACT 2600, Australia.

C.M. Finlayson
Environmental Research Institute of the Supervising Scientist, Locked Bag 2, Jabiru, NT 0886, Australia.

C.V. McQuade
Office of the Supervising Scientist, GPO Box 462, Darwin, NT 0801, Australia.

J. Russell-Smith
Australian Nature Conservation Agency, PO Box 71, Jabiru, NT 0886, Australia.

A. Skeat
Australian Nature Conservation Agency, PO Box 71, Jabiru, NT 0886, Australia.
present address: Queensland Department of Environment and Heritage, PO Box 1579, Townsville, Qld 4810, Australia.

I. von Oertzen
C/- PO Box 406, Jabiru, NT 0886, Australia .

R. Williams
CSIRO Division of Wildlife and Ecology, Tropical Ecosystems Research Centre, PMB 44, Winnellie, NT 0821, Australia.

B.A. Wilson,
Parks and Wildlife Commission of the Northern Territory, PO Box 496, Palmerston, NT 0831, Australia.
present address: Department of Environment and Heritage, PO Box 7054, Mail Centre Toowoomba, Qld 4352, Australia.

C.D. Woodroffe
Department of Geography, University of Wollongong, Wollongong, NSW 2500, Australia.

C.M. Finlayson and I. von Oertzen (eds), Landscape and Vegetation Ecology of the Kakadu Region, Northern Australia, 197.

Index

C.M. Finlayson and I. von Oertzen (eds.), Landscape and Vegetation Ecology of the Kakadu Region, Northern Australia, 197.

Geobotany

1. J.B. Hall and M.D. Swaine (eds.): *Distribution and Ecology of Vascular Plants in a Tropical Rain Forest.* Forest Vegetation in Ghana. 1981 ISBN 90-6193-681-0
2. W. Holzner and M. Numata (eds.): *Biology and Ecology of Weeds.* 1982 ISBN 90-6193-682-9
3. N.J.M. Gremmen: *The Vegetation of the Subantarctic Islands Marion and Prince Edward.* 1982 ISBN 90-6193-683-7
4. R.C. Buckley (ed.): *Ant-Plant Interactions in Australia.* 1982 ISBN 90-6193-684-5
5. W. Holzner, M.J.A. Werger and I. Ikusima (eds.): *Man's Impact on Vegetation.* 1983 ISBN 90-6193-685-3
6. P. Denny (ed.): *The Ecology and Management of African Wetland Vegetation.* 1985 ISBN 90-6193-509-1
7. C. Gómez-Campo (ed.): *Plant Conservation in the Mediterranean Area.* 1985 ISBN 90-6193-523-7
8. J.B. Faliński: *Ecological Studies in Białowieża Forest.* 1986 ISBN 90-6193-534-2
9. G.A. Ellenbroek: *Ecology and Productivity of an African Wetland System.* The Kafue Flats, Zambia. 1987 ISBN 90-6193-638-1
10. J. van Andel, J.P. Bakker and R.W. Snaydon (eds.): *Disturbance in Grasslands.* Causes, Effects and Processes. 1987 ISBN 90-6193-640-3
11. A.H.L. Huiskes, C.W.P.M. Blom and J. Rozema (eds.): *Vegetation Between Land and Sea.* Structure and Processes. 1987 ISBN 90-6193-649-7
12. G. Orshan (ed.): *Plant Pheno-morphological Studies in Mediterranean Type Ecosystems.* 1988 ISBN 90-6193-656-X
13. B. Dell, J.J. Havel and N. Malajczuk (eds.): *The Jarrah Forest.* A Complex Mediterranean Ecosystem. 1988 ISBN 90-6193-658-6
14. J.P. Bakker: *Nature Management by Grazing and Cutting.* 1989 ISBN 0-7923-0068-8
15. J. Osbornová, M. Kovářová, J. Lepš and K. Prach (eds.): *Succession in Abandoned Fields.* Studies in Central Bohemia, Czechoslovakia. 1990 ISBN 0-7923-0401-2
16. B. Gopal (ed.): *Ecology and Management of Aquatic Vegetation in the Indian Subcontinent.* 1990 ISBN 0-7923-0666-X
17. B.A. Roberts and J. Proctor (eds.): *The Ecology of Areas with Serpentinized Rocks.* A World View. 1991. ISBN 0-7923-0922-7
18. J.T.A. Verhoeven (ed.): *Fens and Bogs in the Netherlands.* Vegetation, History, Nutrient Dynamics and Conservation. 1992 ISBN 0-7923-1387-9
19. Woo-seok Kong and D. Watts: *The Plant Geography of Korea.* With an Emphasis on the Alpine Zones. 1993 ISBN 0-7923-2068-9
20. R. Aerts and G.W. Heil (eds.): *Heathlands.* Patterns and Processes in a Changing Environment. 1993 ISBN 0-7923-2094-8
21. W. van Vierssen, M. Hootsmans and J. Vermaat (eds.): *Lake Veluwe, a Macrophyte-dominated System under Eutrophication Stress.* 1994 ISBN 0-7923-2320-3
22. Y. Laumonier: *The Vegetation and Physiography of Sumatra.* 1996 ISBN 0-7923-3761-1
23. C.M. Finlayson and I. von Oertzen (eds.): *Landscape and Vegetation Ecology of the Kakadu Region, Northern Australia.* 1996 ISBN 0-7923-3770-0

KLUWER ACADEMIC PUBLISHERS – DORDRECHT / BOSTON / LONDON

GPSR Compliance

The European Union's (EU) General Product Safety Regulation (GPSR) is a set of rules that requires consumer products to be safe and our obligations to ensure this.

If you have any concerns about our products, you can contact us on ProductSafety@springernature.com

In case Publisher is established outside the EU, the EU authorized representative is:

Springer Nature Customer Service Center GmbH
Europaplatz 3
69115 Heidelberg, Germany

Batch number: 10762189

Printed by Printforce, the Netherlands